Introduction to Statistics in Metrology

Stephen Crowder • Collin Delker • Eric Forrest
Nevin Martin

Introduction to Statistics in Metrology

 Springer

Stephen Crowder
Sandia National Laboratories
Albuquerque, NM, USA

Collin Delker
Sandia National Laboratories
Albuquerque, NM, USA

Eric Forrest
Sandia National Laboratories
Albuquerque, NM, USA

Nevin Martin
Sandia National Laboratories
Albuquerque, NM, USA

ISBN 978-3-030-53328-1   ISBN 978-3-030-53329-8   (eBook)
https://doi.org/10.1007/978-3-030-53329-8

© Springer Nature Switzerland AG 2020

This work is subject to copyright. All rights are reserved by the Publisher, whether the whole or part of the material is concerned, specifically the rights of translation, reprinting, reuse of illustrations, recitation, broadcasting, reproduction on microfilms or in any other physical way, and transmission or information storage and retrieval, electronic adaptation, computer software, or by similar or dissimilar methodology now known or hereafter developed.

The use of general descriptive names, registered names, trademarks, service marks, etc. in this publication does not imply, even in the absence of a specific statement, that such names are exempt from the relevant protective laws and regulations and therefore free for general use.

The publisher, the authors, and the editors are safe to assume that the advice and information in this book are believed to be true and accurate at the date of publication. Neither the publisher nor the authors or the editors give a warranty, expressed or implied, with respect to the material contained herein or for any errors or omissions that may have been made. The publisher remains neutral with regard to jurisdictional claims in published maps and institutional affiliations.

This Springer imprint is published by the registered company Springer Nature Switzerland AG
The registered company address is: Gewerbestrasse 11, 6330 Cham, Switzerland

To Lee, Leah, Stephen, Colleen, and Anna—S.V.C.
To Kim, David, and Shawn—C.J.D.
To Lisa, Cari, and Lyla—E.C.F.
To Zane, Andy, and Josie—N.S.M.

Preface

This book is the result of many years of collaboration between the Primary Standards Laboratory and the Statistical Sciences Department at Sandia National Laboratories. Project work together, publications, and many discussions regarding how to best use statistics in metrology have culminated in this manuscript. With this book, we wish to present statistical best practices to both students and practitioners of metrology. The book brings together in one place many of the basic statistical methods that have been applied to problems in metrology, plus much more. It not only includes methods presented in the JCGM 100 "*Guide to the Expression of Uncertainty in Measurement*" (aka, the GUM), but also presents topics in metrology seldom covered elsewhere. These topics include the design of experiments and statistical process control in metrology, uncertainties in curve fitting, assessment of binary measurement systems, and sample size determination in metrology studies. The book is not intended as a replacement for the GUM or other guiding documents from metrology bodies. Rather, it is intended as a companion resource for the student, technologist, engineer, or scientist involved in measurement studies. The chapters were chosen to provide a blend of topics that will both inform and challenge students and practitioners of metrology.

As a textbook, it is intended for junior or senior level college students studying engineering, statistics, or metrology within a specific discipline. It can also be used at the graduate level for students in instrumentation and measurement classes who are learning the basics of metrology and the statistical methods behind uncertainty analyses. As a prerequisite, readers should have a basic knowledge of calculus and probability and statistics. Related readings that go beyond the scope of the book are included in each chapter. We have also included exercises at the end of each chapter to further illustrate and emphasize material in the body of the book.

Statistical techniques are emphasized throughout, with appropriate engineering and physics background provided as needed. Most of the methods covered in the book are illustrated with case studies from our work in the Nuclear Security Enterprise. The case studies should provide the reader with a solid foundation for applying the techniques to a wide variety of metrology problems. Many end-of-chapter exercises also rely on these case studies.

The statistical topics in metrology are presented by first introducing the basic theory and models necessary to complete an uncertainty analysis. These topics are then followed by case studies illustrating the approach.

Noteworthy highlights of the book include:

- Measurement uncertainty as a part of everyday life.
- Basic measurement terminology and types of measurement.
- Role of measurement uncertainty in decision-making.
- Direct and indirect measurement models.
- Analytical methods for the propagation of uncertainties.
- Design of experiments in metrology.
- Uncertainties in curve fitting.
- Statistical process control in metrology.
- Evaluation of binary measurement systems.
- Sample size determination and allocation in metrology experiments.
- R-Code and Python Uncertainty Calculator used in metrology studies.

Of course, we have not covered all possible topics involving statistics in metrology. For example, we have chosen not to cover interlaboratory comparisons or proficiency tests, as these topics are more relevant for calibration laboratories and are well-covered in other sources such as the NCSLI's RP-15. We have also chosen not to cover in detail the metrology of system-level measurements. A system-level approach would include a broader understanding of topics such as frequency responses, sampling rates, aliasing, sensor placement and mounting, cables, and connectors. These topics are well-covered in various books and short courses. Other fields such as healthcare and analytical chemistry will have specialized extensions of statistics in metrology that are beyond the scope of this book. The many intricacies of discipline-specific metrology practices such as these are learned only through years of training and hands-on experience.

Chapter 1 of the book includes a brief history of measurement and the development of measurement science and technology. In Chap. 2, we introduce measurement terminology, types of measurement, and sources of uncertainty. Chapter 3 covers the International System of Units (SI), traceability, and calibrations. The SI base units and derived units are presented, along with the notion of unit realization. Measurement standards and various aspects of calibration are also presented. These three chapters are included to establish the background and language of metrology used throughout the book.

An introduction to probability and statistics is given in Chap. 4. Topics include types of data, summary statistics, graphical displays of data, and an introduction to the probability distributions most often used in metrology. In Chap. 5, we provide an overview of measurement uncertainty in decision-making, including risk, error probabilities, test uncertainty ratios, and guardbanding.

Chapter 6 develops both direct and indirect measurement models and their roles in an uncertainty analysis. Type A and Type B uncertainty evaluations, standard uncertainties, combined standard uncertainties, and expanded uncertainties are introduced here. The GUM approach to quantifying uncertainty is presented, and the methods are illustrated with an uncertainty analysis of a neutron yield measurement. Chapter 7 presents the analytical methods used to propagate uncertainties through an indirect measurement model, including both first-order and higher order models, with both uncorrelated and correlated inputs. Measurement examples are given for each case.

Chapter 8 introduces the Monte Carlo method for uncertainty analysis, beginning with a discussion of random number generation followed by a discussion of the techniques found in the JCGM 101 (aka, the GUM Supplement 1). Measurement examples and a case study are used to illustrate this approach. Chapter 9 presents the basic experimental designs that can be used in the evaluation of uncertainty. Emphasis is on full factorial, fractional factorial, and ANOVA-based designs. A step-by-step approach to designing an experiment is given, along with case studies to illustrate the design and analysis techniques.

In Chap.10, we present the methods for determining uncertainties in fitted curves, including both linear and nonlinear least squares. The Monte Carlo method is also applied to curve fitting, and examples are given for each approach. Finally, in Chap. 11, we cover special topics in metrology that have been important in our work. These topics include statistical process control applied to a measurement process, evaluation of binary measurement systems, sample size determination and allocation in metrology experiments, and an introduction to Bayesian analysis in metrology.

Throughout this book, R-Code is provided alongside many of the examples to give the reader an important tool that can be used to perform uncertainty analyses. R is an open-source programming language whose popularity stems primarily from the number of packages that are available for a wide range of statistical methods, including Monte Carlo sampling, linear and nonlinear regression, ANOVA, and more. R can be downloaded from the Comprehensive R Archive Network (CRAN) at www.r-project.org and it is available for Windows, Unix-Like, and Mac operating systems.

The Sandia Uncertainty Calculator (SUNCAL) is also being made available as open-source software. It was developed by the Primary Standards Lab at Sandia to perform propagation of uncertainty analyses and other statistical techniques in metrology. It computes uncertainties using both the GUM and Monte Carlo methods. Partial derivatives are solved symbolically to provide the analytical formulas used in the calculations. The calculator can handle units conversion and unlimited input variables and uncertainty components. In addition to uncertainty propagation, SUNCAL provides calculations for curve fitting uncertainty, analysis of variance, and false accept/reject risk. SUNCAL was written in Python, a multipurpose language popular among engineers because of its ability to perform data analysis along with tasks such as communicating with measurement equipment, interfacing with databases, and accessing the internet. SUNCAL can be used through a graphical user interface available for Windows and Mac, or as an importable

Python package for programmers. It is released under the GNU General Public License, with source code and executables available at https://sandiapsl.github.io.

Appendix A covers common acronyms and abbreviations used in metrology, followed in Appendix B by guidelines for valid measurements. Appendix C includes a traceability chain and uncertainty budget case study, presented in more detail than those in the body of the book. Appendix D includes a quick reference for the GUM propagation of uncertainty technique and a table of references for common topics in metrology. Finally, Appendix E provides information regarding the installation of R software and existing R packages used in metrology.

The acknowledgments are given to those from the Primary Standards Laboratory and the Statistical Sciences Department who have contributed their expertise and case studies in this collaborative effort. Our former and present colleagues in this work include Stuart Kupferman, Tom Wunsch, Bud Burns, Lisa Bunting Baca, Greg Guidarelli, Andrew Mackrory, David Sanchez, Edward O' Brien, Jesse Whitehead, Mark Benner, Meghan Shilling, Donavon Gerty, Stefan Cular, Harold Parks, Elizabeth Auden, Ricky Sandoval, Otis Solomon, Roger Burton, Hy Tran, Raegan Johnson, Allie Wichhart, Andrew Wofford, Lauren Wilson, and Dan Campbell. Special thanks to David Walsh for providing the lead probe case study and to Elbara Ziade for providing the CMM case study. Finally, this book would not have been possible without the support of Justin Newcomer and Adele Doser, Managers of the Statistical Sciences Department, Meaghan Carpenter, Senior Manager of the Primary Standards Lab, and Marcey Hoover, Director of Quality Assurance at Sandia.

Albuquerque, NM, USA	Stephen Crowder
Albuquerque, NM, USA	Collin Delker
Albuquerque, NM, USA	Eric Forrest
Albuquerque, NM, USA	Nevin Martin

*A false balance is an abomination to the LORD, but a just weight is his delight
Proverbs 11:1*

Contents

1 **Introduction** 1
   1.1   Measurement Uncertainty: Why Do We Care? 1
   1.2   The History of Measurement 2
   1.3   Measurement Science and Technological Development 2
   1.4   Allegations of Deflated Footballs ("Deflategate") 3
   1.5   Fatality Rates During a Pandemic 8
   1.6   Summary 13
   1.7   Related Reading 14
   References 15

2 **Basic Measurement Concepts** 19
   2.1   Introduction 19
   2.2   Measurement Terminology 19
        2.2.1   General Measurement Terminology 20
        2.2.2   Error Approach Terminology 23
        2.2.3   Uncertainty Approach Terminology 24
        2.2.4   Terminology of Calibration 28
   2.3   Types of Measurements 29
        2.3.1   Physical Measurements 29
        2.3.2   Electrical Measurements 30
        2.3.3   Other Types of Measurements 30
   2.4   Sources of Uncertainty 31
        2.4.1   Evaluating Sources of Uncertainty 33
   2.5   Summary 34
   2.6   Related Reading 34
   2.7   Exercises 35
   References 38

3 **The International System of Units, Traceability, and Calibration** .. 41
   3.1   History of the SI and Base Units 41
        3.1.1   SI Constants 42
        3.1.2   Time: Second (s) 43

		3.1.3	Length: Meter (m)	43
		3.1.4	Mass: Kilogram (kg)	43
		3.1.5	Electric Current: Ampere (A)	44
		3.1.6	Temperature: Kelvin (K)	44
		3.1.7	Quantity of Substance: Mole (mol)	44
		3.1.8	Luminous Intensity: Candela (cd)	44
	3.2	Derived Units		45
	3.3	Unit Realizations		45
		3.3.1	Gauge Block Interferometer	46
		3.3.2	Josephson Volt	46
	3.4	Advancements in Unit Definitions		46
		3.4.1	Kibble (Watt) Balance	47
		3.4.2	Intrinsic Pressure Standard	48
	3.5	Metrological Traceability		48
	3.6	Measurement Standards		49
		3.6.1	Certified Reference Materials	49
		3.6.2	Check Standards	50
	3.7	Calibration		51
		3.7.1	The Calibration Cycle	51
		3.7.2	Legal Aspects of Calibration	52
		3.7.3	Technical Aspects of Calibration	52
		3.7.4	Calibration Policies and Requirements	53
	3.8	Summary		54
	3.9	Related Reading		55
	3.10	Exercises		56
	References			57
4	**Introduction to Statistics and Probability**			**59**
	4.1	Introduction		59
	4.2	Types of Data		60
	4.3	Exploratory Data Analysis		60
		4.3.1	Calculating Summary Statistics	61
		4.3.2	Graphical Displays of Data	63
	4.4	Probability Distributions		68
		4.4.1	Identification of Probability Distributions	69
		4.4.2	Estimating Distribution Parameters	75
		4.4.3	Assessing Distributional Fit	76
	4.5	Related Reading		77
	4.6	Exercises		78
	References			79
5	**Measurement Uncertainty in Decision Making**			**81**
	5.1	Introduction		81
	5.2	Measurement Uncertainty and Risk		81

		5.2.1	Measurement Uncertainty and Risk in Manufacturing	82
		5.2.2	Measurement Uncertainty and Risk in Calibration	93
	5.3	Summary		96
	5.4	Related Reading		96
	5.5	Exercises		97
	References			101
6	**The Measurement Model and Uncertainty**			103
	6.1	Introduction		103
	6.2	Uncertainty Analysis Framework		103
		6.2.1	Standard Uncertainty	104
		6.2.2	Type A Uncertainty Evaluation	104
		6.2.3	Type B Uncertainty Evaluation	104
		6.2.4	Combined Standard Uncertainty	105
		6.2.5	Confidence Level and Expanded Uncertainty	105
	6.3	Direct Measurements and the Basic Measurement Model		107
		6.3.1	Case Study: Voltage Measurement	109
		6.3.2	Discussion	114
	6.4	Indirect Measurements and the Indirect Measurement Model		114
		6.4.1	Case Study: Neutron Yield Measurement	117
		6.4.2	Discussion	123
	6.5	Related Reading		125
	6.6	Exercises		127
	References			128
7	**Analytical Methods for the Propagation of Uncertainties**			131
	7.1	Introduction		131
	7.2	Mathematical Basis		132
	7.3	The Simple Case: First-Order Terms with Uncorrelated Inputs		133
		7.3.1	Measurement Examples	135
	7.4	First-Order Terms with Correlated Inputs		137
		7.4.1	Covariance, Correlation, and Effect on Uncertainty	138
		7.4.2	Measurement Examples	139
	7.5	Higher-Order Terms with Uncorrelated Inputs		142
		7.5.1	Measurement Examples	144
	7.6	Multiple Output Quantities		145
	7.7	Limitations of the Analytical Approach		146
	7.8	Related Reading		147
	7.9	Exercises		147
	References			151

8 Monte Carlo Methods for the Propagation of Uncertainties ... 153
8.1 Introduction to Monte Carlo Methods ... 153
8.1.1 Random Sampling Techniques and Random Number Generation ... 154
8.1.2 Generation of Probability Density Functions Using Random Data ... 157
8.1.3 Computational Approaches ... 157
8.2 Standard Monte Carlo for Uncertainty Propagation ... 159
8.2.1 Monte Carlo Techniques ... 159
8.3 Comparison to the GUM ... 166
8.3.1 Quantitative GUM Validity Test ... 167
8.4 Monte Carlo Case Studies ... 169
8.4.1 Case Study: Neutron Yield Measurement ... 169
8.4.2 Case Study: RC Circuit ... 173
8.5 Summary ... 175
8.6 Related Reading ... 176
8.7 Exercises ... 176
References ... 179

9 Design of Experiments in Metrology ... 181
9.1 Introduction ... 181
9.2 Factorial Experiments in Metrology ... 181
9.2.1 Defining the Measurand and Objective of the Experiment ... 183
9.2.2 Selecting Factors to Incorporate in the Experiment ... 183
9.2.3 Selecting Factor Levels and Design Pattern ... 183
9.2.4 Analysis of CMM Errors via Design of Experiments (2^4 Full Factorial) ... 184
9.2.5 Finite Element Method (FEM) Uncertainty Analysis via Design of Experiments (2^{7-3} Fractional Factorial) ... 196
9.2.6 Summary of Factorial DOEx Method ... 205
9.3 ANOVA Models in Metrology ... 206
9.3.1 Random Effects Models ... 206
9.3.2 Mixed Effects Models ... 208
9.3.3 Underlying ANOVA Assumptions ... 209
9.3.4 Gauge R&R Study (Random Effects Model) ... 210
9.3.5 Voltage Standard Uncertainty Analysis (Mixed Effects Model) ... 215
9.3.6 Summary of ANOVA Method ... 220
9.4 Related Reading ... 220
9.5 Exercises ... 222
References ... 224

10	**Determining Uncertainties in Fitted Curves**		227
	10.1	The Purpose of Fitting Curves to Experimental Data	227
		10.1.1 Resistance vs. Temperature Data	228
		10.1.2 Considerations When Fitting Models to Data	229
	10.2	Methods for Fitting Curves to Experimental Data	230
		10.2.1 Linear Least Squares	231
		10.2.2 Uncertainty in Fitting Parameters	231
		10.2.3 Weighted Least Squares: Non-constant $u(y)$	233
		10.2.4 Weighted Least Squares: Uncertainty in Both x and y	233
	10.3	Uncertainty of a Regression Line	234
		10.3.1 Uncertainty of Fitting Parameters	235
		10.3.2 Confidence Bands	235
		10.3.3 Prediction Bands	235
	10.4	How Good Is the Model?	238
		10.4.1 Residual Analysis	238
		10.4.2 Slope Test	238
		10.4.3 Quantitative Residual Analysis	239
	10.5	Uncertainty in Nonlinear Regression	241
		10.5.1 Nonlinear Least Squares	241
		10.5.2 Orthogonal Distance Regression	243
		10.5.3 Confidence and Prediction Bands in Nonlinear Regression	244
	10.6	Using Monte Carlo for Evaluating Uncertainties in Curve Fitting	245
		10.6.1 Monte Carlo Approach	245
		10.6.2 Markov-Chain Monte Carlo Approach	246
	10.7	Case Study: Contact Resistance	247
	10.8	Drift and Predicting Future Values	249
		10.8.1 Uncertainty During Use	249
		10.8.2 Validating Drift Uncertainty	252
	10.9	Calibration Interval Analysis	258
	10.10	Summary	260
	10.11	Related Reading	261
	10.12	Exercises	261
	References		265
11	**Special Topics in Metrology**		267
	11.1	Introduction	267
	11.2	Statistical Process Control (SPC)	267
		11.2.1 Case Study: Battery Tester Uncertainty and Monitoring Via SPC	269
		11.2.2 Discussion	273
	11.3	Binary Measurement Systems (BMS)	274
		11.3.1 BMS Overview	274
		11.3.2 BMS Case Study Introduced	275

	11.3.3	Evaluation of a BMS	275
	11.3.4	Sample Sizes for a BMS Study	282
11.4		Measurement System Analysis with Destructive Testing	284
11.5		Sample Size and Allocation of Samples in Metrology Experiments	285
11.6		Summary of Sample Size Recommendations	291
11.7		Bayesian Analysis in Metrology	292
11.8		Related Reading	296
11.9		Exercises	299
References			301

Appendix A: Acronyms and Abbreviations 303

Appendix B: Guidelines for Valid Measurements 305
Related Reading: Electrical Measurements 305
Related Reading: Time and Frequency Measurements 305
Related Reading: Physical Measurements 306
Related Reading: Temperature Measurement 306
Related Reading: Radiation 306
Related Reading: General Measurement and Instrumentation Techniques 307

Appendix C: Uncertainty Budget Case Study: CMM Length Measurements 309
Coordinate Measuring Machine (CMM) Measurements 309
    Product Acceptance Uncertainty: Dimensional Part Inspection with a CMM 309
    Radius of Curvature of a Spherical Mirror 310
    The Measurement Model 310
    Measurement Considerations 311
    ROC Measurement 313
    Uncertainty Analysis 314
Related Reading 319

Appendix D: Uncertainty Quick Reference 321
GUM Method for Measurement Uncertainty 321
Percentage Points of the t Distribution 322
Guardbanding 323
    Symmetric Specification Limits 323
    Asymmetric Specification Limits 323
    One-Sided Specification Limits 323
Metrology Reference Table 324

Appendix E: R for Metrology 325
    Introduction 325
        Installation of R 325
        R Packages 326
    R for Metrology 326
    Summary 327

References 329

Index 341

About the Authors

Stephen Crowder is a Principal Member of Technical Staff in the Statistical Sciences department at Sandia National Laboratories with over 30 years of experience working in industrial statistics and metrology. He received his B.S. degree in Mathematics from Abilene Christian University and his M.S. and Ph.D. in Statistics from Iowa State University. He has previously done research and published in the fields of statistical process control, system reliability, and statistics in metrology.

Collin Delker is a Senior Member of Technical Staff at Sandia National Laboratories, working in the Primary Standards Laboratory. He received his B.S. degree in Electrical Engineering from Kansas State University and his Ph.D. in Electrical Engineering, with an emphasis on microelectronics and nanotechnology, from Purdue University. He specializes in developing techniques for the calibration of microwave frequency devices in addition to providing software solutions for metrology.

Eric Forrest is a Principal Member of Technical Staff at Sandia National Laboratories, working in the Primary Standards Laboratory where he leads the Radiation and Optics Project. He received his B.S., M.S., and Ph.D. in Nuclear Science and Engineering from MIT, where he was a National Nuclear Security Administration Fellow. His research focused on high-speed optical/infrared imaging and development of nanoengineered surfaces for enhanced heat transfer in nuclear thermal hydraulics applications. He specializes in uncertainty analyses for complex experimental measurements.

Nevin Martin is a Member of Technical Staff in the Statistical Sciences Department at Sandia National Laboratories. She received her B.S. degree in Finance from the University of Arizona and her M.S. degree in Statistics from the University of New Mexico. She collaborates on a wide range of projects that include work in statistical computing with R, data visualization and modeling, and uncertainty quantification. She teaches a short course on "Introduction to Statistical Computing in R" and develops R-Code for the application of statistics in metrology.

Chapter 1
Introduction

Statistics and metrology, that is, the science of measurement, permeate mod-ern engineering, society, and culture. This chapter provides a brief historical perspective on the importance of metrology, along with introducing the core concept of *uncertainty*: a key element that spans both statistics and metrology. The importance of uncertainty is demonstrated through two contemporary case studies. First, this chapter considers how uncertainty may have adversely affected the outcome of pressure measurements on footballs in a 2015 sports scandal often referred to as "Deflategate." Second, the importance of an accurate measurement model in obtaining reliable estimates of case fatality rate during a pandemic is discussed.

1.1 Measurement Uncertainty: Why Do We Care?

Measurements are an important part of everyday life. Measurements drive decision-making in nearly all aspects of modern society. Do you ever question the validity of a measurement result? You should, considering that every measurement has uncertainty. Knowledge of this associated uncertainty is *necessary* for making *informed* decisions. Yet, uncertainty analysis and propagation constitute a subject area that rarely receives proper attention in science and engineering curriculums or in public discourse.

Uncertainty analysis and propagation are central to metrology, the science of measurement. And metrology is fundamentally interlinked with statistics. This textbook helps bridge the gap between statistics, metrology, and uncertainty analysis, not just for the measurement practitioner, but also for those who utilize measurement data. In this chapter, a brief historical backdrop is provided, along with two introductory case studies based on contemporary issues. Case studies demonstrate the importance of a holistic approach to measurement uncertainty and highlight several concepts and methods that will be introduced in later chapters.

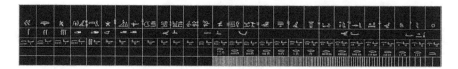

Fig. 1.1 A replica of the ancient Egyptian cubit rod. A cubit is a unit of length equivalent to the distance between the elbow and the tip of the middle finger of the ruling Pharaoh at the time

1.2 The History of Measurement

The importance of proper measurement and its associated measurement uncertainty have long been recognized. The ancient Egyptians and ancient Mesopotamians established the earliest known recorded system of measure while constructing the pyramids and other architectural feats in the 4th to 3rd millennia BCE (Clarke and Engelbach 1990). Figure 1.1 shows an example of a cubit rod used by the ancient Egyptians. The cubit rod was used as a unit of length and represented the length of a Pharaoh's forearm. In ancient Israel, a well-established system of measurement was used to ensure that trade of food items and other goods was carried out fairly and justly. Prior to the introduction of a standardized system, some would intentionally offset weights and measures to try to cheat their neighbors in the sale or trade of goods.

Prior to 500 BCE, the ancient Greeks developed a system of official weights and measures and employed a method of calibration using reference standards. The Roman Empire adapted the earlier Hellenic system to create a well-documented, sophisticated system of measurement that employed standards and calibration in commerce, trade, and engineering (Smith 1851). However, many of the official weights and measures varied from region to region. In the third century BCE, the Greek librarian and scientist Eratosthenes measured the circumference of the Earth simply by observing the lengths of shadows in two locations. He determined the circumference to be 250,000 stades. The stade was a unit based on the length of a typical Greek stadium, but there were several regional variations on the definition of a stade (Walkup 2005). Unfortunately, because of the different definitions in use at the time, it may never be known exactly how close Eratosthenes's measurement came to the modern accepted circumference of the Earth. Not until many centuries later did weights and measures become internationally standardized.

1.3 Measurement Science and Technological Development

Throughout history, advances in measurement science and standards have been a prerequisite for the practical implementation of scientific developments. For example, the concept of interchangeable parts, which revolutionized manufacturing and led to numerous other technological advances, could only be implemented successfully with improvements in measurement science. In the late 1700s and early 1800s, complex mechanical devices, such as firearms, required careful handfitment of parts by experienced gunsmiths. In 1801, when Eli Whitney presented his concept of interchangeable parts for a musket to the United States War Department (Hays 1959), a key framework, such as dimensional standards, was not yet available. Unbeknownst to the War Department, the muskets presented were specially prepared and individual parts were not standardized or interchangeable with other muskets of the same type. It took decades to successfully realize the vision of interchangeable parts for modern manufacturing. In fact, gauge blocks, a type of dimensional standard introduced in the 1890s by Swedish machinist C.E. Johannson (Althin 1948), were a key facilitator for modern production and fabrication methods, enabling the use of interchangeable parts and common tooling for assembly line manufacturing.

The importance of measurement standardization and measurement traceability for technological advancement and economic growth was realized early in the Technological Revolution. In 1875, 17 countries, including the USA, signed the Metre Convention, which established the International Bureau of Weights and Measures (BIPM) and defined international standards for mass and length.

Improvements in standards and measurement would continue to drive major revolutions in machining, fabrication, and production of industrial, commercial, and consumer goods throughout the 1900s. The advent of digital computing in the late 1940s, along with advances in air and space travel throughout the last century, necessitated improvements in all areas of measurement, with associated reductions in uncertainty. Despite the agreed-upon importance of a universal system of units and maintenance of consistent standards, treatment of measurement data and its associated uncertainties remained relatively ambiguous until the 1990s. To this day, measurement uncertainty is often misunderstood by engineers and scientists performing measurements.

The criticality of accurate measurements in the marketplace has never been greater. Measurement inaccuracies in food and fuel purchases alone place billions of consumer dollars at risk each year. Measurement uncertainties in manufacturing increase the risk of accepting bad product or rejecting good product, each resulting in lost productivity and profits. Measurement uncertainties in medical diagnostics can result in both missed or incorrect diagnoses, with major public health implications. Measurements are the basis for legal decisions and evidence in trials and form the basis for science-based policies across the globe. The accuracy of such measurements relies heavily on measurement standardization and measurement traceability and is the forefront of topics discussed in this handbook. We will demonstrate the importance of measurements, and their associated uncertainties, with two modern real-world case studies highlighting important measurement uncertainty concepts detailed in later chapters.

1.4 Allegations of Deflated Footballs ("Deflategate")

During the 2014 American Football Conference (AFC) Championship Game on January 18, 2015, between the Indianapolis Colts and the New England Patriots of the National Football League (NFL), allegations arose regarding the New England Patriots intentionally deflating their game balls to provide an unfair competitive advantage over the Colts. Measurements of football air pressure at halftime became the central part of a subsequent NFL investigation and disciplinary hearings. In addition to becoming a public spectacle, the outcome of the investigations resulted in major penalties for the New England Patriots and their quarterback, Tom Brady. The Patriots were ultimately fined $1 million and forced to forfeit a first-round draft pick in 2016 and a fourth-round draft pick in 2017, with Tom Brady (Fig. 1.2). being suspended for four games. However, a general lack of understanding of measurement uncertainty may have led to erroneous conclusions based on the football air pressure data.

The controversy centered on the following requirement in the NFL rulebook (Goodell 2014):

> The ball shall be made up of an inflated (12½ to 13½ pounds) urethane bladder enclosed in a pebble grained, leather case (natural tan color) without corrugations of any kind. It shall have the form of a prolate spheroid and the size and weight shall be: long axis, 11–11¼ inches; long circumference, 28–28½ inches; short circumference, 21 to 21¼ inches; weight, 14 to 15 ounces.

While the requirement, as written, does not specify proper units, it is interpreted to mean internal football air pressure shall be between 12.5 pounds per square inch gauge (psig) and 13.5 psig. Proper and consistent use of *units* is paramount when specifying a measurement requirement and when reporting measurement results (introduced in Chap. 3).

Following an interception by the Colts in the first half, suspicions arose of underinflated Patriots' game balls. At halftime, two NFL officials measured air

Fig. 1.2 Tom Brady in 2011 as quarterback of the New England Patriots. Untraceable measurements of football air pressure, with large uncertainties, were central to allegations that he directed the deflation of footballs prior to the 2014 AFC Championship Game. Photograph by Jeffrey Beall/CC-BY-SA-3.0

1.4 Allegations of Deflated Footballs ("Deflategate")

Fig. 1.3 Pressure gauges used by the championship game officials to measure internal football air pressure at halftime. The gauge on the left is referred to as the "non-logo gauge" and the gauge on the right was referred to as the "logo gauge" (Exponent 2015)

Fig. 1.4 Pressure gauges used by the championship game officials to measure internal football air pressure at halftime. Pressure gauges (model CJ-01) distributed by Wilson sporting goods were likely manufactured by Jiao Hsiung Industry Corp. (Exponent 2015)

pressure of eleven Patriots' game balls. Two pressure gauges, provided by another official, were used for the measurements. The pressure gauges were not *calibrated* and therefore lacked *traceability* (see Chap. 3). In addition, the pressure gauges were of unknown origin aside from the fact that one had a Wilson logo, whereas the other did not (Figs. 1.3 and 1.4).

The officials demonstrated an understanding of measurement variability, and more specifically, repeatability and reproducibility (see Chap. 2): two measurements were taken on each game ball, with a different gauge and operator used for each. However, applying a t-distribution and looking at the t-table (introduced in Chap. 4) show that taking only two independent measurements (one degree of freedom) are generally inadequate and greatly increase the expanded measurement uncertainty (see Chap. 6 and Table 1.1).

We can perform a Type A uncertainty evaluation (see Chaps. 2 and 6) of the game-day internal football pressure data that was recorded by officials. In general, 20–30 independent measurements are desirable to properly assess repeatability and reproducibility (see Chap. 11). However, this may not always be achievable. When the sample size is less than 30, the t-distribution is typically used. For a limited sample size ($n = 2$), the coverage factor for a 95% level of confidence (see Chaps. 2 and 6) becomes large (12.7), resulting in a much larger expanded uncertainty for the

Table 1.1 Game-day data for internal air pressure of eleven different Patriots' footballs taken at halftime by officials

Ball	Air pressure, non-logo gauge (psig)	Air pressure, logo gauge (psig)	Average air pressure (psig)	Type A std. uncertainty (psig)
1	11.50	11.80	11.65	0.15
2	10.85	11.20	11.03	0.18
3	11.15	11.50	11.33	0.18
4	10.70	11.00	10.85	0.15
5	11.10	11.45	11.28	0.18
6	11.60	11.95	11.78	0.18
7	11.85	12.30	12.08	0.23
8	11.10	11.55	11.33	0.23
9	10.95	11.35	11.15	0.20
10	10.50	10.90	10.70	0.20
11	10.90	11.35	11.13	0.23

The measurement repeatability is calculated using a Type A uncertainty analysis

measurement. While some might propose treating the eleven different footballs as the same sample, there is no expectation that the true value of the air pressure in different footballs will be the same. Variability between footballs would provide insight into variability in the fill process, but not necessarily the measurement uncertainty.

Repeatability and reproducibility are only one aspect of measurement uncertainty. Type B evaluations (see Chaps. 2 and 6) must also be applied to capture elements of the pressure measurement uncertainty such as pressure gauge resolution, inherent pressure gauge uncertainty, and environmental factors. Use of uncalibrated measuring and test equipment is not recommended for quality-affecting measurements and precludes the ability to properly determine total uncertainty. Nonetheless, uncertainty estimates can be made from manufacturer specification sheets, although these cannot always be trusted. The pressure gauges used by the officials did not have associated specification sheets. They were likely both produced by Hsiung Industry Corp. for Wilson, which does not have a stated accuracy for these gauges. While the display of the digital gauges read to ± 0.05 psig, resolution is rarely indicative of total uncertainty, although it is a contributor. The manufacturer's specified uncertainty for similar handheld pressure gauges is $\pm 1\%$ of full scale (20 psig), or no better than ± 0.20 psig. Without a traceable calibration, this specification is difficult to prove, and based on performance between gauges measuring the same football, it is likely worse.

Ultimately, we must combine the uncertainties from the Type A and Type B evaluations for this direct measurement (see Chap. 6). Without going into detail and assuming the Type A and Type B uncertainties are uncorrelated, we have for ball #1:

$$u_c = \sqrt{u_A^2 + u_B^2} = \sqrt{0.15^2 + \left(0.20/\sqrt{3}\right)^2} = 0.19 \text{ psig} \quad (1.1)$$

This combination of terms will be discussed in detail in later chapters. We must still determine the expanded uncertainty at a desired level of confidence. This is done by

1.4 Allegations of Deflated Footballs ("Deflategate")

multiplying our uncertainty in Eq. (1.1) by an appropriate coverage factor. The coverage factor is determined by calculating effective degrees of freedom (see Chap. 6). The degrees of freedom for the Type A uncertainty is relatively straightforward: the number of measurements minus (n-1). Since the gauges were not calibrated, and the specification sheet for the specific gauges used was not available, our estimate of the Type B uncertainty could have large variability (say up to 50%). Therefore, the Type B degrees of freedom will be low, as determined from Eq. (11.22) (see Chap. 11). Assuming 50% relative uncertainty gives us two degrees of freedom. Ultimately, we can compute an expanded measurement uncertainty at a 95% level of confidence for ball #1:

$$U = t_{95}(\nu_{\text{eff}}) \times u_c = 4.3 \times 0.19 \text{ psig} = 0.81 \text{ psig} \quad (1.2)$$

Therefore, the *complete* measurement result for the internal pressure of ball #1 is 11.7 psig ± 0.81 psig at a level of confidence of 95% ($k = 4.3$).

The test uncertainty ratio (TUR, introduced in Chap. 5) provides a means of determining suitability of the measurement when compared to a given requirement. We can calculate a TUR for the measurement on ball #1 by comparing the total measurement uncertainty for football air pressure against the requirement in the NFL rulebook (13.0 psig ± 0.5 psig):

$$\text{TUR} = \frac{\text{Specification Limit}}{\text{Total measurement uncertainty}} = \frac{\pm 0.5 \text{ psig}}{\pm 0.81 \text{ psig}} = 0.62 \quad (1.3)$$

Typically, a TUR of 4 or greater is required to mitigate false accept and false reject risk (concepts introduced in Chap. 5). A TUR of 0.62 indicates the measurement equipment and process is not sufficiently accurate to determine whether or not the requirement was met. The use of uncalibrated gauges and this uncertainty analysis tells us that the football pressure data alone was not adequate to conclusively determine whether the true value fell within or outside the requirement.

Ideally the referees would have performed a well-designed Gauge R&R study (see Chap. 9) to separate out the individual contributors to measurement uncertainty. Such a study would have resulted in an analysis of variance (ANOVA) to model uncertainties due to operators and gauges and a Type A evaluation of uncertainty with many more degrees of freedom.

The saga of football pressure measurement does not end there, however. In the subsequent investigation (Wells Jr. et al. 2015), a firm was hired to characterize the pressure gauges used for the game-day measurements. The firm procured a "master" pressure gauge, shown in Fig. 1.5, with "NIST traceable calibration" from an *unaccredited* vendor in an attempt to calibrate the game-day gauges after the fact. Any vendor, laboratory, or individual can claim traceability to the National Institute of Standards and Technology (NIST). However, laboratory accreditation to a standard such as ISO/IEC 17025, through a reputable accrediting body, is necessary to demonstrate competence in calibration (concept discussed in Chap. 3).

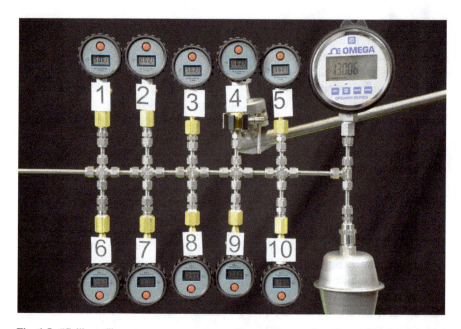

Fig. 1.5 "Calibrated" master gauge experiment. Traceability was based on calibration provided from an unaccredited laboratory (Exponent 2015)

In addition, the so-called calibration of the handheld game-day pressure gauges after the fact is not a valid practice (see Chap. 3) and only constitutes a characterization. There is no way to guarantee that the gauges performed the same on game-day due to drift and other factors. The uncertainty of the "master" gauge, and uncertainties in general, was not considered or incorporated. An uncertainty or tolerance must be assigned to a unit under test (UUT) during a valid calibration.

While other evidence, such as interviews with players, officials, and equipment personnel, along with text message conversations ultimately weighed on the outcome of the investigation and sanctions by the NFL, the centerpiece of the case was untraceable measurements of internal football air pressure using equipment with an unacceptably low test uncertainty ratio. As seen in this example, concepts of measurement uncertainty, uncertainty propagation, calibration, and traceability have important implications in sports and legal investigations but are unfortunately not always applied properly. Decisions made on measurement data are only as good as the uncertainties that come with it.

1.5 Fatality Rates During a Pandemic

An infectious disease is spreading around the globe, with dire predictions of lethality. Shortly after the World Health Organization (WHO) announces a Phase 6 pandemic alert and the USA declares a Public Health Emergency, fatality

1.5 Fatality Rates During a Pandemic

rate estimates are as high as 5.1%. The U.S. Centers for Disease Control and Prevention (CDC) is releasing supplies from the Strategic National Stockpile. School closures and community level social distancing are being implemented in certain areas of the USA. The CDC is recommending that colleges suspend classes through the Fall. Certain countries have instituted travel restrictions and quarantine requirements. Panic buying of food items and consumer goods is rampant.

This is not 2020. This is 2009, and the Swine Flu pandemic, caused by a novel strain of the H1N1 influenza virus (H1N1/09), is underway. Despite initial reports of fatality rates up to 5.1%, with an estimate of 0.6% across all countries considered (Vaillant et al. 2009), the final estimated fatality rate for the 2009 pandemic was 0.02% (Simonsen et al. 2013; Baldo et al. 2016). The difference in preliminary and final estimated fatality rate represents a *30-fold decrease*. Given the extraordinary importance of predicted fatality rate in determining appropriate response to a spreading pandemic at national, regional, and local levels, *how could the initial estimates have been so far off?* The answer is because of measurement uncertainty and sampling bias.

Underestimation of fatality rate in the initial stages of a pandemic may prevent government leaders and policymakers from implementing appropriate mitigation and quarantine strategies, leading to millions of excess deaths. Overestimation of fatality rate can lead to panic, unnecessary quarantines at national, regional, and local levels, along with irreversible damage to the economy and the livelihoods of millions of people. Proper estimation of fatality rate during a pandemic, along with calculation and communication of associated uncertainties and measurement limitations, is critical for proper decision-making. Yet we see limited attention given to these important aspects of the problem.

Determination of fatality rate due to a disease represents an *indirect* measurement (introduced in Chap. 6). Even with the most accurate measurements of input parameters, uncertainty in the *measurement model* itself frequently can lead to grossly inaccurate estimates of a measurand (see Chap. 2). Here we will begin by formulating a simple measurement model ("Model 1") for fatality rate that was used in initial estimates for H1N1:

$$\text{CFR} = \frac{N_{\text{deaths}}}{N_{\text{cases}}} \times 100. \tag{1.4}$$

Here the *CFR* is the "case fatality rate" in percent. *CFR* is crucial for predicting clinical outcomes in patients infected with a disease and estimating disease burden on society. The term is somewhat of a misnomer, as it does not constitute a rate, although the numerator and denominator are usually derived over some time period. Per the U.S. CDC, the *CFR* is (Dicker et al. 2012):

> The proportion of persons with a particular condition (e.g., patients) who die from that condition. The denominator is the number of persons with the condition; the numerator is the number of cause-specific deaths among those persons.

N_{deaths} and N_{cases} represent the number of deaths from disease X and the number of cases of disease X, respectively. Simple enough? This represents the measurement model and is effectively the model used by Vaillant et al. (2009) for initial *CFR* estimates of H1N1/09 infections during the Swine Flu pandemic.

The standard combined uncertainty (introduced in Chap. 6) for the *CFR* based on this model will be

$$u_{CFR} = \sqrt{\left(\frac{100}{N_{cases}}\right)^2 u_{N_{deaths}}^2 + \left(-\frac{100 \times N_{deaths}}{N_{cases}}\right)^2 u_{N_{cases}}^2}. \quad (1.5)$$

Taking the data for the USA, Vaillant computed a *CFR* of 0.6% based on $N_{deaths} = 211$ and $N_{cases} = 37{,}246$, where the number of deaths and cases were taken from data reported in CDC bulletins up to July 16, 2009. While no uncertainties are provided, we can look at the effects of relative uncertainties of inputs to determine if this could lead to the gross error in the initial estimate. An uncertainty of $\pm 25\%$ in each input parameter yields a standard uncertainty in *CFR* of $\pm 0.20\%$ ($\sim 0.40\%$ at $k = 2$). Does the true value for fatality rate fall in this interval? Probably not, based on later revised estimates. What are we missing?

The error could be in the model itself. While for developed countries with adequate testing capability, the term in the numerator should be somewhat representative of the reality, the term in the denominator may not be. N_{cases} is typically derived from the confirmed positive case count (via a positive result from diagnostic testing for disease *X*). Can you think of any problems with this approach? For a rapidly spreading disease, if *CFR* is calculated using aggregate numbers at a single point in time, estimates can be misleading due to the non-negligible number of infected patients whose outcome (death or survival) is unknown (Ghani et al. 2005). In addition, for a disease such as influenza, where a number of cases are mild, or even asymptomatic, the measurement model in Eq. (1.4) is not adequate in determining true fatality rate. The true number of cases will likely be significantly higher, even orders of magnitude higher, skewing the fatality rate. This effectively represents a selection bias (or sampling bias), whereby only the sickest patients are tested for disease *X*, thereby artificially increasing the fatality rate.

Even with adjustment and more complex models, *CFR* estimates can be misleading, especially early in an epidemic or pandemic. Using the number of individuals infected in Mexico by late April, WHO estimated a *CFR* of 0.4% (range: 0.3–1.8%) for H1N1/09 (Fraser et al. 2009). These estimates are still an order-of-magnitude higher than final estimates due to a *significant* undercount of the true number of infections.

Later estimates by the U.S. CDC (Reed et al. 2009; Shrestha et al. 2011) and in other countries (Kamigaki and Oshitani 2009; Baldo et al. 2016; Simonsen et al. 2013) incorporated a measurement model more akin to the following:

$$\text{CFR} = \frac{N_{deaths}}{a \times N_{cases}} \times 100 \quad (1.6)$$

1.5 Fatality Rates During a Pandemic

where a is a multiplier that adjusts the reported number of cases to more adequately estimate the actual number of cases. This model (which we will call "Model 2") is illustrated in Fig. 1.6.

Specifically, for determining number of deaths and number of H1N1/09 cases in the USA, Reed et al. (2009) and Shrestha et al. (2011) used a more complex model, relying on the number of hospitalizations from select hospitals (across 60 counties) participating in the CDC Emerging Infections Program (EIP) surveillance as a more reliable estimator:

$$\text{CFR} = \frac{c_1 \times c_2 \times \left(\sum_{i=1}^{60} n_{\text{deaths},i} \right)}{c_3 \times c_4 \times c_5 \times \left(\sum_{i=1}^{60} n_{\text{hospitalizations},i} \right)} \quad (1.7)$$

In this expression, $\sum_{i=1}^{60} n_{\text{deaths},i}$ = the number of reported fatalities from H1N1/09 over the 60 counties sampled, including deaths outside of hospitals, $\sum_{i=1}^{60} n_{\text{hospitalizations},i}$ = the number of reported hospitalizations from H1N1/09 over the 60 counties sampled, c_1 = a factor to correct for underestimate of deaths, c_2 = a factor to extrapolate the number of sampled deaths to a national estimate, c_3 = a factor to correct for underestimate of hospitalizations, c_4 = a factor to extrapolate the number of sampled hospitalizations to a

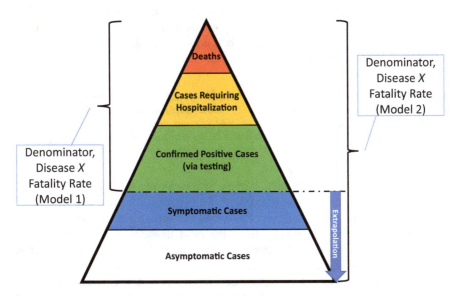

Fig. 1.6 Illustration depicting different measurement model inputs for fatality rate of disease X. Failure to account for unreported cases can lead to significant overestimates of fatality rate. However, extrapolation is required to estimate unreported case count and leads to large uncertainties. Adapted from Reed et al. (2009), Shrestha et al. (2011), and Verity et al. (2020)

national estimate, and c_5 = a factor to extrapolate the actual number of hospitalizations to the actual number of H1N1/09 infections.

In the CDC measurement model, the sampled number of deaths was derived from the Aggregate Hospitalizations and Deaths Reporting Activity (AHDRA) surveillance system. Reed et al. (2009) and Shrestha et al. (2011) assumed deaths were underreported to the same extent and used a value of 2.74 for both c_1 and c_3, as derived from a probabilistic model they developed. For c_4, a value of 12.7 and 14.6 was used by Reed et al. (2009) and Shrestha et al. (2011), respectively. For c_5, Reed et al. (2009) utilized a value of 222 to extrapolate from corrected hospitalization count to total the number of cases. Ultimately, Reed et al. (2009) showed that every reported case of H1N1/09 likely represented 79 actual cases, with a 90% probability range for this multiplier being 47–148.

Failure of early models to properly consider unreported symptomatic and asymptomatic cases from H1N1/09 led to excessively high estimates of the case fatality rate. Revised estimates, formulated with more reliable data and a better picture of impact after the pandemic had passed, still have incredibly large uncertainties. Ultimately, the U.S. CDC estimated 12,469 deaths (range: 8868–18,306) and 60,837,748 cases (range: 43,267,929–89,318,757) from H1N1/09 from April 12, 2009 through April 10, 2010 (Shrestha et al. 2011) in the USA. This results in a case fatality rate of 0.02% in the USA from H1N1, with an uncertainty range of 0.01–0.04%. While the measurement model of Eqs. (1.6) and 91.70 more correctly represents reality compared to that in Eq. (1.5), the uncertainties are still large in the final estimate. What is causing this?

A downside of these "corrected" measurement models is that they require *extrapolation* (discussed in Chap. 10). Extrapolation is rarely recommended but may be necessary in situations where measurement data is incomplete or unavailable. Extrapolation should only be performed if all measurement options are exhausted and with the understanding that uncertainties will generally be large. To quote Mandel (1984):

> With regard to extrapolation, i.e., use of a formula beyond the range of values for which it was established, extreme caution is indicated. It should be emphasized in this connection that statistical measures of uncertainty, such as standard errors, confidence regions, confidence bands, etc., apply only to the region covered by the experimental data; the use of such measures for the purpose of providing statistical sanctioning for extrapolation processes is nothing but delusion.

In epidemiology, the determination of the total caseload typically requires extrapolation due to the general unavailability of data (statistical censoring). But as we saw, the correct measurement model prevents over-prediction of fatality rate by more than an order-of-magnitude. Even the most thorough knowledge of input parameters and uncertainties is utterly meaningless if the measurement model is wrong, as in initial fatality rate estimates.

As we write this textbook in 2020, a new infectious disease is spreading around the globe. The WHO estimated a fatality rate of 3.4% in March of 2020 (World Health Organization 2020), one week before announcing a global health pandemic. Schools have been shuttered in the USA for the remainder of the academic year. Panic buying has been widespread. Retail businesses, restaurants, and other establishments have been ordered closed. In a 4-week period, an unprecedented 22 million people have filed for unemployment. Extreme measures have been implemented in nearly every state, with fines or imprisonment possible for those violating stay-at-home orders. Such measures have resulted, at least in part, from the WHO's initial estimated fatality rate and similar estimates that predicted 2.2 million deaths in the USA if no further actions were taken (Ferguson et al. 2020). Economists at the International Monetary Fund have predicted these measures will result in the worst economic downturn since the Great Depression (Gopinath 2020).

The infectious disease is coronavirus disease 2019 (COVID-19), caused by the novel coronavirus severe acute respiratory syndrome coronavirus 2 (SARS-CoV-2). The WHO's initial estimate came from applying measurement "Model 1" in Eq. (1.4) to data coming from China. Some, including the director of the U.S. National Institute of Allergy and Infectious Diseases (NIAID), recognized early on that such estimates could significantly overestimate the actual case fatality rate (Fauci et al. 2020, italicization added):

> If one assumes that the number of asymptomatic or minimally symptomatic cases is several times as high as the number of reported cases, *the case fatality rate may be considerably less than 1%*. This suggests that the overall clinical consequences of Covid-19 may ultimately be more akin to those of a severe seasonal influenza (which has a case fatality rate of approximately 0.1%) or a pandemic influenza (similar to those in 1957 and 1968) rather than a disease similar to SARS or MERS, which have had case fatality rates of 9 to 10% and 36%, respectively.

Indeed, revised models and studies report much lower fatality rates of 0.657% (range: 0.389–1.33%) when undiagnosed cases are estimated and then included in the estimate of fatality rate (Verity et al. 2020). Note that Verity et al. (2020) refer to this fatality rate as the "infection fatality ratio" to differentiate the denominator from positive confirmed cases, which, has become accepted terminology but is essentially equivalent to the CFR as defined by Dicker et al. (2012) and Fauci et al. (2020). Preliminary, non-peer reviewed serological studies, looking for the presence of antibodies due to immune response to SARS-CoV-2 in previously undiagnosed cases (the bottom two regions in Fig. 1.6), have already suggested substantially lower case fatality rates. A serological survey of an entire municipality in Germany estimated a case fatality rate of 0.37% (Regalado 2020), and a serological study in Santa Clara County, California estimated a fatality rate of 0.12–0.2% (Bendavid et al. 2020; Mallapaty 2020). More reliable data and improved measurement models will ultimately provide more realistic estimates of the true COVID-19 fatality rate.

1.6 Summary

Ancient societies realized measurement improvement and standardization were necessary for stability and development. To this day, measurement advances are closely tied to new technology implementation. And yet, measurements and their associated uncertainties are poorly understood by the general public and receive only limited consideration in college and university programs.

An understanding of measurement uncertainty is critically important in all fields of science, engineering, statistics, healthcare, economics, business, and any discipline that relies on measurement or measurement data for decision-making. Measurement uncertainty also affects our everyday lives. Failure to properly account for measurement uncertainty has led to major engineering disasters, loss of human life, and trillions of dollars in economic losses.

As we have seen in two contemporary case studies, measurement uncertainty has important implications in sports, legal settings, and the study of diseases. These case studies have presented, at a high level, a number of concepts and methods, blending statistics and metrology, that are the focus of this book.

1.7 Related Reading

The history of metrology and its role in modernizing the world is an interesting and ongoing area of study that cannot be given fitting coverage in this book. *The Science of Measurement: A Historical Survey* by Klein (1988), first published in 1974, provides a detailed look into the history of unit standardization. With a focus on historical developments in the International System of Units (SI), Klein's writing style and storytelling through the eyes of historical figures provide for a compelling read. Understandably, there are gaps in more recent developments in metrology given it was first written in 1974. Klein also offers historical perspectives on measurement developments in the areas of thermal and ionizing radiation, which are not found in other texts.

Lugli (2019) describes how revolutions in measurement processes in medieval Europe brought about revolutions in other areas of science and society. With a focus on medieval Italy up through the Renaissance, Lugli paints an extraordinary picture of the impacts of measurement standardization on modernization from a sociological perspective. Lugli provides numerous rich primary sources that help illustrate the tumultuous development of measurement standardization in medieval Italy.

For a reference text on historical and modern units of measurement, Gyllenbok (2018) offers perhaps the most comprehensive collection to date of measurement systems and units, by country, from ancient to modern times. Consisting of three volumes, the encyclopedia is well-organized and would serve as an excellent desk reference for the modern metrologist or historian. Gyllenbok also provides an immense collection of references and related reading.

Fluke Calibration also offers a short (~40 min) free webinar, titled "A Brief History of Metrology and Advancements in the Science of Measurement," available at https://us.flukecal.com/history-of-Metrology.

Another recommended reading that provides a brief historical perspective on metrology with a detailed bibliography and serves as a useful companion to this entire textbook is *The Handbook of Metrology* by Bucher (2012). While the primary audience is metrologists and calibration laboratory personnel, Bucher's coverage of quality systems and metrology concepts is useful for all practitioners of metrology.

As seen in the second case study of this chapter, metrology, statistics, and uncertainty are critically important in the field of epidemiology. While this textbook provides the fundamental tools for performing uncertainty analyses using any measurement model, epidemiological models and studies are not a focus of this book. For further reading in this area, *Modern Epidemiology* by Rothman et al. (2008) provides thorough coverage of statistical concepts for application in epidemiology, along with discussions on means of improving the accuracy of studies in this field. Stewart (2016) provides an excellent introduction to statistics concepts for those with no prior experience and basic concepts for applying statistics to epidemiological studies.

References

Althin, T.K.W.: C.E. Johansson 1864–1943. The Master of Measurement. Nordsk Rotogravyr, Stockholm (1948)

Baldo, V., Bertoncello, C., Cocchio, S., Fonzo, M., Pillon, P., Buja, A., Baldovin, T.: The new pandemic influenza A/(H1N1)pdm09 virus: Is it really "new"? J. Prev. Med. Hyg. **57**(1), E19–E22 (2016)

Bendavid, E., Mulaney, B., Sood, N., Shah, S., Ling, E., Bromley-Dulfano, R., Lai, C., Weissberg, Z., Saavedra-Walker, R., Tedrow, J., Tversky, D., Bogan, A., Kupiec, T., Eichner, D., Gupta, R., Ioannidis, J.P.A., Bhattacharya, J.: COVID-19 Antibody Seroprevalence in Santa Clara County, California (2020). https://doi.org/10.1101/2020.04.14.20062463. Accessed 29 Apr 2020

Bucher, J.L.: The Metrology Handbook, 2nd edn. ASQ Quality Press, Milwaukee (2012)

Clarke, S., Engelbach, R.: Ancient Egyptian Construction and Architecture. Dover Publications, Inc., New York (1990)

Dicker, R.C., Coronado, F., Koo, D., Parrish, R.G.: Principles of Epidemiology in Public Health Practice: An Introduction to Applied Epidemiology and Biostatistics, 3rd edn. U.S. Center for Disease Control and Prevention (CDC), Atlanta (2012)

Exponent: The effect of various environmental and physical factors on the measured internal pressure of nfl footballs. Exponent Engineering (2015)

Fauci, A.S., Lane, H.C., Redfield, R.R.: Covid-19 — navigating the uncharted. N. Engl. J. Med. **382**(13), 1268–1269 (2020)

Ferguson, N., Laydon, D., Nedjati Gilani, G., Imai, N., Ainslie, K., Baguelin, M., Bhatia, S., Boonyasiri, A., Cucunubá, Z., Cuomo-Dannenburg, G., Dighe, A., Dorigatti, I., Fu, H., Gaythorpe, K., Green, W., Hamlet, A., Hinsley, W., Okell, L.C., van Elsland, S., Thompson, H., Verity, R., Volz, E., Wang, H., Wang, Y., Walker, P.G.T., Walters, C., Winskill, P., Whittaker, C., Donnelly, C.A., Riley, S., & Ghani, A.C.: Impact of non-pharmaceutical Interventions (NPIs) to reduce COVID-19 mortality and healthcare demand. (2020). https://doi.org/10.25561/77482

Fraser, C., Donnelly, C.A., Cauchemez, S., Hanage, W.P., Van Kerkhove, M.D., Hollingsworth, T. D., Griffin, J., Baggaley, R.F., Jenkins, H.E., Lyons, E.J., Jombart, T., Hinsley, W.R., Grassly, N.C., Balloux, F., Ghani, A.C., Ferguson, N.M., Rambaut, A., Pybus, O.G., Lopez-Gatell, H., Alpuche-Aranda, C.M., Chapela, L.B., Zavala, E.P., Guevara, D.M.E.,

Checchi, F., Garcia, E., Hugonnet, S., Roth, C.: Pandemic potential of a strain of influenza A (H1N1): Early findings. Science. **324**(5934), 1557–1561 (2009)

Ghani, A.C., Donnelly, C.A., Cox, D.R., Griffin, J.T., Fraser, C., Lam, T.H., Ho, L.M., Chan, W.S., Anderson, R.M., Hedley, A.J., Leung, G.M.: Methods for estimating the case fatality ratio for a novel, emerging infectious disease. Am. J. Epidemiol. **162**(5), 479–486 (2005)

Goodell, R.: 2014 official playing rules of the National Football League. National Football League. (2014)

Gopinath, G.: The great lockdown: Worst economic downturn since the great depression. International Monetary Fund (2020). https://blogs.imf.org/2020/04/14/the-great-lockdown-worst-economic-downturn-since-the-great-depression/. Accessed 29 Apr 2020

Gyllenbok, J.: Encyclopaedia of Historical Metrology, Weights, and Measures, vol. 1–3. Birkhäuser, Basel (2018)

Hays, W.: Eli Whitney and the Machine Age. F. Watts, New York (1959)

Kamigaki, T., Oshitani, H.: Epidemiological characteristics and low case fatality rate of pandemic (H1N1) 2009 in Japan. PLoS Curr. Influenza. **2009**, 1 (2009)

Klein, H.A.: The Science of Measurement: A Historical Survey. Dover Publications, Inc., New York (1988)

Lugli, E.: The Making of Measure and the Promise of Sameness. University of Chicago Press, Chicago (2019)

Mallapaty, S.: Antibody tests suggest that coronavirus infections vastly exceed official counts. Nature. **20**, 01095 (2020). https://doi.org/10.1038/d41586-020-01095-0

Mandel, J.: The Statistical Analysis of Experimental Data. Dover Publications, Inc., New York (1984)

Reed, C., Angulo, F.J., Swerdlow, D.L., Lipsitch, M., Meltzer, M.I., Jernigan, D., Finelli, L.: Estimates of the prevalence of pandemic (H1N1) 2009, United States, April–July 2009. Emerg. Infect. Dis. **15**(12), 2004–2007 (2009)

Regalado, A.: Blood tests show 14% of people are now immune to covid-19 in one town in Germany. MIT Technology Review (2020). https://www.technologyreview.com/2020/04/09/999015/blood-tests-show-15-of-people-are-now-immune-to-covid-19-in-one-town-in-germany/. Accessed 29 Apr 2020

Rothman, K.J., Greenland, S., Lash, T.L.: Modern Epidemiology, 3rd edn. Lippincott Williams & Wilkins, Philadelphia (2008)

Shrestha, S.S., Swerdlow, D.L., Borse, R.H., Prabhu, V.S., Finelli, L., Atkins, C.Y., Owusu-Edusei, K., Bell, B., Mead, P.S., Biggerstaff, M., Brammer, L., Davidson, H., Jernigan, D., Jhung, M.A., Kamimoto, L.A., Merlin, T.L., Nowell, M., Redd, S.C., Reed, C., Schuchat, A., Meltzer, M.I.: Estimating the Burden of 2009 Pandemic Influenza A (H1N1) in the United States (April 2009–April 2010). Clin. Infect. Dis. **52**(suppl_1), S75–S82 (2011)

Simonsen, L., Spreeuwenberg, P., Lustig, R., Taylor, R.J., Fleming, D.M., Kroneman, M., Van Kerkhove, M.D., Mounts, A.W., Paget, W.J.: Global mortality estimates for the 2009 influenza pandemic from the GLaMOR project: A modeling study. PLoS Med. **10**(11), 1–17 (2013)

Smith, W.: A New Classical Dictionary of Greek and Roman Biography, Mythology and Geography. Harper & Brothers, New York (1851)

Stewart, A.: Basic Statistics and Epidemiology: A Practical Guide, 4th edn. CRC Press, Boca Raton (2016)

Vaillant, L., La Ruche, G., Tarantola, A., Barboza, P.: Epidemiology of fatal cases associated with pandemic H1N1 influenza 2009. Eurosurveillance. **14**(33), 1–6 (2009)

Verity, R., Okell, L.C., Dorigatti, I., Winskill, P., Whittaker, C., Imai, N., Cuomo-Dannenburg, G., Thompson, H., Walker, P.G.T., Fu, H., Dighe, A., Griffin, J.T., Baguelin, M., Bhatia, S., Boonyasiri, A., Cori, A., Cucunubá, Z., FitzJohn, R., Gaythorpe, K., Green, W., Hamlet, A.,

References

Hinsley, W., Laydon, D., Nedjati-Gilani, G., Riley, S., van Elsland, S., Volz, E., Wang, H., Wang, Y., Xi, X., Donnelly, C.A., Ghani, A.C., Ferguson, N.M.: Estimates of the severity of coronavirus disease 2019: A model-based analysis. Lancet Infect. Dis. **20**, 30243 (2020)

Walkup, N.: Eratosthenes and the Mystery of the Stades. Mathematical Association of America (2005). https://www.maa.org/press/periodicals/convergence/eratosthenes-and-the-mystery-of-the-stades-how-long-is-a-stade. Accessed 29 Apr 2020

Wells, Jr., TV., Karp, B.S., Reisner, L.L.: Investigative Report Concerning Footballs Used During the AFC Championship Game on January 18, 2015. Paul, Weiss, Rifkind, Wharton & Garrison LLP (2015)

World Health Organization: WHO Director-General's opening remarks at the media briefing on COVID-19 - 3 March 2020 (2020). https://www.who.int/dg/speeches/detail/who-director-general-s-opening-remarks-at-the-media-briefing-on-covid-19--3-march-2020. Accessed 29 Apr 2020

Chapter 2
Basic Measurement Concepts

Consistent and understandable terminology is essential in metrology. Uniform terminology facilitates comparison of results and uncertainties between metrology organizations and measurement practitioners. Despite the importance of correct terminology, misinterpretation of common terminology still abounds in scientific, engineering, and academic communities. This chapter covers basic measurement terminology and concepts that are necessary to proper understanding of measurement data and associated uncertainties. The difference between accuracy, precision, resolution, repeatability, reproducibility, uncertainty, and other terms are discussed. The International Vocabulary of Metrology (referred to as "the VIM"), fundamentally a dictionary of metrology terms, serves as the basis for common terminology.

2.1 Introduction

In this chapter we cover basic measurement concepts, including common terminology, types of measurements, the importance of uncertainty in measurement, and typical sources of uncertainty. The terminology and concepts provided here are not comprehensive, but references are provided for further reading. The material here, as well as the definitions outlined in Chap. 3, are intended to help provide a common language for practitioners of metrology.

2.2 Measurement Terminology

To properly analyze measurements and uncertainty, consistent terminology must be used. Prior to the first issuance of the GUM in 1993 (ISO/IEC Guide 98 1993), approaches to uncertainty analysis and the associated terminology were largely ambiguous (Kline 1985). Maintained by the same organization as the GUM, the

International Vocabulary of Metrology—Basic and General Concepts and Terms (JCGM 200 2012), known as the VIM, provides detailed definitions of many terms relevant to metrology and uncertainty analysis.

The VIM uses what it calls the "Uncertainty Approach" to measurement terminology. The "Error Approach" was the historical norm prior to publication of the VIM and was used to estimate a value as close as possible to a single true value. The Uncertainty Approach does not attempt to determine a single true value, rather, it assumes that information from the measurement permits an interval of reasonable values. The uncertainty approach is the preferred terminology for use by metrology organizations. Therefore, it is only natural that this textbook references definitions provided in the VIM. While the Error Approach terminology is sometimes still encountered in references such as American National Standards Institute (ANSI) and American Society of Mechanical Engineers (ASME) standards (ANSI/ASME PTC 19.1 1983; Abernathy et al. 1985), caution must be exercised when mixing older terminology with that prescribed by the VIM.

Basic terminology is reviewed in Sect. 2.2.1. General measurement terminology, along with terms from the uncertainty approach and the older error approach are given, followed by terms specific to calibration. For definitions of other terms not included here, the reader should refer to the VIM. Terminology presented here is not necessarily in alphabetical order; rather, frequently used terms are listed first, followed by less frequently used terminology. Verbatim definitions from the VIM are italicized, with additional clarification and description provided. Some definitions from the VIM have been merged for conciseness and readability.

2.2.1 General Measurement Terminology

2.2.1.1 Measurand

Quantity intended to be measured (VIM 2.3, p. 17).

The "measurand" refers to the physical quantity or object which is being measured.

2.2.1.2 True Value (True Value of a Quantity)

Quantity value consistent with the definition of a property of a phenomenon, body, or substance, where the property has a magnitude that can be expressed as a number and a reference (VIM 1.1, p. 2; VIM 2.11, p. 20).

The true value is the unique actual value or set of unique values representing the parameter of interest. Unfortunately, in practice, the true value is not known and cannot be determined because all measurements have uncertainty.

2.2 Measurement Terminology

2.2.1.3 Measurement Accuracy

Closeness of agreement between a measured quantity value and a true quantity value of a measurand (VIM 2.13, p. 21).

"Accuracy" refers to the degree to which the result of a measurement conforms to the true value. Higher accuracy indicates a smaller difference between the measured value and the true value, whereas lower accuracy indicates a larger difference between the measured value and the true value. While the term "accuracy" is commonplace when speaking of measurements or equipment, it is only appropriate to use as a relative or qualitative term. Accuracy is never equivalent to resolution or precision. Because the true value of a measurand can never be known exactly, accuracy cannot be determined quantitatively.

2.2.1.4 Measurement Precision

Closeness of agreement between indications or measured quantity values obtained by replicate measurements on the same or similar objects under specified conditions (VIM 2.15, p. 22).

"Precision" refers to the degree of agreement between nominally identical, repeat measurements. Higher precision indicates repeat measurements are in better agreement (smaller spread between measurements), whereas lower precision indicates there is less agreement (larger spread between measurements). Precision says nothing about bias or offset but is a contributor to overall uncertainty. While use of the term is commonplace, it is only appropriate to use as a relative term.

Figure 2.1 illustrates the terms "accuracy" and "precision" using an analogy to target shooting. The bullseye represents the true value of the parameter of interest, and the points of impact represent the measured values. In the first frame, the shots

Fig. 2.1 Illustration of precision and accuracy

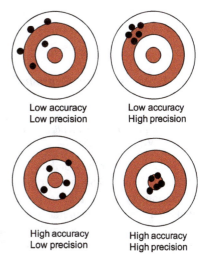

are found close together (more precise) but are far from the intended target (less accurate). In the second frame, the shots are found spread apart (less precise) but are centered around the intended target (more accurate). In the last frame, the shots are found close together (more precise) and are close to the intended target (more accurate). Note that the terms represent a qualitative, or relative, metric for describing the quality of a measurement.

2.2.1.5 Resolution

Smallest change in a quantity being measured that causes a perceptible change in the corresponding indication (VIM 4.14, p. 42).

The resolution of an instrument represents the smallest change at which the measured value can be read or recorded. For an instrument with an analog readout, display, or scale, this represents the finest increment at which the measurand can be determined and depends on the spacing between consecutive markings and width of the indicator. For a digital readout or display, resolution represents the least significant digit to which the value can be read or recorded. In nearly every measurement discipline with any instrument, resolution never indicates overall accuracy or total uncertainty, but can be a contributor.

2.2.1.6 Measurement Repeatability

Measurement precision under a set of conditions that includes the same measurement procedure, same operators, same measuring system, same operating conditions, and same location, and replicate measurements on the same or similar objects over a short period of time (VIM 2.20, p. 23; VIM 2.21, p. 24).

"Measurement repeatability" describes the degree of mutual agreement among a series of individual measurements, values, or results with the same operator, measuring system, and procedure over a short period of time (typically hours to days).

2.2.1.7 Measurement Reproducibility

Measurement precision under a set of conditions that includes different locations, operators, measuring systems, and replicate measurements on the same or similar objects (VIM 2.24, p. 24; VIM 2.25, p. 25).

"Measurement reproducibility" describes the degree of mutual agreement among a series of individual measurements, values, or results with different operators, measuring systems, and/or procedures. Measurement reproducibility is often evaluated using measurements made over longer time periods. Measurement reproducibility and repeatability can be quantified using statistical techniques such as Analysis of Variance (see Chap. 10).

2.2 Measurement Terminology

2.2.1.8 Independence of Measurements

The concept of "independence of measurements" is related to the concept of "statistical independence." To properly assess repeatability and reproducibility, a series of measurements (or observations) must not be related to one another. Typically this requires that repeat measurements are taken far enough apart in time that they are not related (or correlated in time). The goal is to allow only small random fluctuations on the measurement process. The conditions needed to achieve independence of repeat measurements are usually determined by subject matter experts in the measurement area. Section F.1.1 of the GUM (JCGM 100 2008) provides details regarding this concept.

2.2.2 Error Approach Terminology

The Error Approach to measurement uncertainty attempts to estimate the measurand as closely as possible to the single true value. Because the true value can never be known, the magnitude of the error can never be calculated exactly, and as such, use of the Error Approach terminology is discouraged. These definitions are included here because they are still frequently encountered in the literature, and they highlight the issue of inconsistent terminology. The terminology is summarized in Fig. 2.2.

2.2.2.1 Measurement Error

Measured quantity value minus a reference quantity value used as a basis for comparison (VIM 2.16, p. 22, VIM 5.18, p. 53).

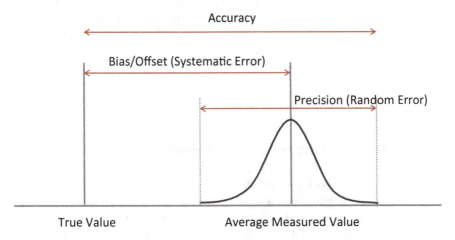

Fig. 2.2 Illustration summarizing terms for the older, discouraged "error approach" to measurement terminology

"Measurement error" is a term representing a quantitative value, indicating the magnitude difference between the measured value and the true value. However, since the true value can never be known, the exact error of a measurement can never be known. Therefore, use of the term "error," while commonplace, is not favored in the approach presented in the GUM or in this textbook.

2.2.2.2 Systematic Measurement Error

Component of measurement error that in replicate measurements remains constant or varies in a predictable manner (VIM 2.17, p. 22).

"Systematic error" is an error that, when fully quantified, is predictable. Systematic error can be a bias or offset in the measurement, or an error based on an independent parameter, such as a known temperature dependency. If the systematic error is known, a correction should be applied to the measurement result. Again, since the true value can never be known, use of the term "systematic error" is discouraged unless referring to a quantitative, known offset that can be corrected.

2.2.2.3 Random Measurement Error

Component of measurement error that in replicate measurements varies in an unpredictable manner (VIM 2.19, p. 23).

"Random error" is an error that, when fully quantified, is caused by factors that can vary from one measurement to another. Random error results from inherent variability in a measurement, process, or system, and cannot be corrected for since it derives from unpredictable variation in the measured value. Random error provides a quantitative measure of repeatability and reproducibility in a measurement and is typically determined stochastically. Use of the term "random error" is discouraged unless referring to a quantitative, known variability in a measurement system. Random error may result from noise or fundamental inherent variation from the measurement process, such as that observed when counting measurements of radioactive decay processes.

2.2.3 Uncertainty Approach Terminology

The Uncertainty Approach provides the preferred set of terminology for discussing measurement uncertainty. Instead of attempting to determine a "true" value, this approach assigns a range of likely values that could be attributed to the measurand based on the finite amount of knowledge of the measurement equipment and process.

2.2 Measurement Terminology

2.2.3.1 Measurement Uncertainty

Non-negative parameter characterizing the dispersion of the quantity values being attributed to a measurand, based on the information used (VIM 2.26, p. 25).

Uncertainty means "doubt," and in a broad sense, the term "measurement uncertainty" means doubt in the validity of a measurement result. Since the true value can never be known, all measurements will possess uncertainty. Therefore, a measurement reported without an associated uncertainty is incomplete. Measurement uncertainty includes components arising from the uncertainties in the equipment and standards used to make the measurement, as well as from the measurement model and process itself.

Measurement uncertainty always includes Type A and Type B evaluations of measurement uncertainty. These concepts will be formally defined in Chap. 6. At a high level, Type A evaluations come from statistical analysis, and Type B evaluations are determined from other methods. Both Type A and Type B evaluations are required to determine the total measurement uncertainty. These terms refer to the method of evaluation and not the source of uncertainty associated with them, although it is commonplace to see uncertainties referred to as "Type A" or "Type B." These are not directly equivalent to the older terminology of "random error" and "systematic error," although in some instances they are used interchangeably.

Figure 2.3 depicts the relationships of measurement uncertainty. The true value is unknown, and therefore could fall anywhere within the estimated uncertainty, or even outside of the uncertainty estimate if the level of confidence is less than 100%.

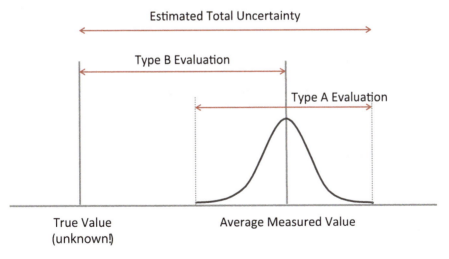

Fig. 2.3 Illustration summarizing terms for the preferred Uncertainty Approach to measurement terminology. The measured value will fall within the total uncertainty above or (not shown) below the true value

Fig. 2.4 Typical chart representation of measurement uncertainty using error bars

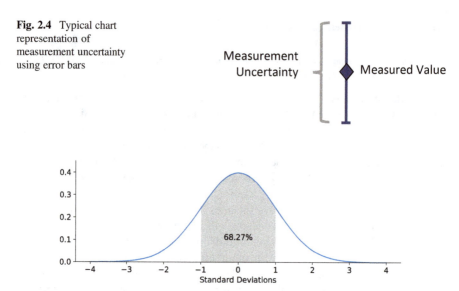

Fig. 2.5 One standard deviation under the normal curve (gray shaded region), corresponding to a level of confidence of 68.27% (often rounded to 68%)

Figure 2.4 depicts the common visualization of a measured value and its uncertainty using error bars.

2.2.3.2 Level of Confidence (Coverage Probability)

> *Probability that the set of true quantity values of a measurand is contained within a specified coverage interval* (VIM 2.37, p. 28).

In measurement terminology, the level of confidence is an important part of the measurement uncertainty. It represents the probability that the constructed uncertainty interval will contain the true value. The width of the uncertainty interval will change depending on the desired level of confidence, which should be reported along with the uncertainty interval.

The levels of confidence (coverage probabilities) for a normal random variable are often associated with multiples of the standard deviation, as illustrated in Figs. 2.5, 2.6, and 2.7.

2.2.3.3 Coverage Interval

> *Interval containing the set of true quantity values of a measurand with a stated probability, based on the information available* (VIM 2.36, p. 27).

2.2 Measurement Terminology

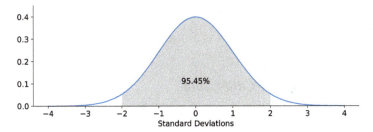

Fig. 2.6 Two standard deviations under the normal curve, corresponding to a level of confidence of 95.45% (often rounded to 95%). This is the typical value used on calibration certificates

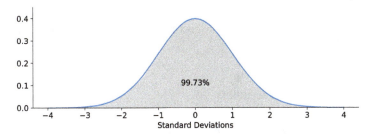

Fig. 2.7 Three standard deviations under the normal curve, corresponding to a level of confidence of 99.73% (often rounded to 99.7%)

The coverage interval is a range of possible values that includes the true value of the measurement with the specified level of confidence. The level of confidence is associated with the choice of coverage factor, discussed in later sections for both normal and non-normal distributions. Sometimes the term "confidence interval" is used; however, this term is not recommended, to avoid possible confusion with the statistical term of the same name.

2.2.3.4 Measurement Model

Mathematical relation among all quantities known to be involved in a measurement (VIM 2.48, p. 32).

A measurement model is the formula used to combine measurements of different quantities in order to obtain the quantity of interest. Measurement models can be classified as direct or indirect (Chap. 6). A direct measurement refers to a measurement in which a single value is read directly from the measuring instrument (for example, a measurement of length using a ruler). An indirect measurement refers to a measurement obtained by measuring other quantities functionally related to the measurand. For example, when measuring electrical resistance, R, by making measurements of voltage, V, and current, I, the measurement model is Ohm's law

$R = V/I$. In this case, the value R is determined indirectly by first measuring two other quantities (V and I).

2.2.4 Terminology of Calibration

Relying on the uncertainty approach terminology described in Sect. 2.2.3, calibration has its own set of definitions.

2.2.4.1 Measuring and Test Equipment (M&TE)

Device or system to certify, test, or inspect items for conformance to requirements (DOE-STD-1054-96 1995).

2.2.4.2 Metrological Traceability

> *Property of a measurement result whereby the result can be related to a reference through a documented unbroken chain of calibrations, each contributing to the measurement uncertainty* (VIM 2.41, p. 29).

"Measurement traceability" refers to an unbroken chain of comparisons relating a device's measurements to a known standard. The top of the traceability chain should relate to one or more of the seven the SI base units. For details on traceability and the SI units, see Chap. 3.

2.2.4.3 Calibration

> *Operation that, under specified conditions, in a first step, establishes a relation between the quantity values with measurement uncertainties provided by measurement standards and corresponding indications with associated measurement uncertainties and, in a second step, uses this information to establish a relation for obtaining a measurement result from an indication* (VIM 2.39, p. 28).

Calibration is a periodic comparison of measurement values delivered by a device under test to a known standard or reference. The calibration results in a statement of the device's value, the associated uncertainty, and the level of confidence in that uncertainty. Although the calibration process is purely a comparison, the concept of measurement uncertainty is introduced by relating the accuracies of the device under test and the standard. Even though calibration and adjustments are often performed at the same time, calibration must not be confused with adjustment; an adjustment is a modification of the device's properties to ensure performance within specified limits. See Chap. 3 for more details on calibration.

2.2.4.4 Tolerance Test

A tolerance test is a pass or fail determination based on whether a measurement falls within required limits. A tolerance test may be considered a calibration if traceable reference standards are used in the measurement. See Chap. 5 for guidance on determining whether a particular measurement is adequate to perform a tolerance test.

2.2.4.5 Certification Uncertainty

The certification uncertainty is the uncertainty consisting of the measurement uncertainty and any uncertainties due to use, environment, handling, or variation with time along with the expiration criteria for the calibration.

Certification uncertainty is infrequently reported by commercial calibration laboratories; in fact, ISO/IEC 17025:2017 discourages reporting anything but uncertainty due to the measurement itself. However, certain industries require reporting a certification uncertainty and associated expiration date for all calibrations to appropriately account for uncertainty due to drift and handling between calibrations.

2.3 Types of Measurements

Measurements of a device can generally be divided into physical measurements and electrical measurements. A short overview of basic measurement types is provided here.

2.3.1 Physical Measurements

Physical measurements seek to estimate measurands such as dimension, mass, force, temperature, pressure, acceleration, material hardness, yield strength, fluid viscosity, thermal energy, and others. Such measurements may be direct or indirect (refer to Chap. 6). For example, using a Coordinate Measuring Machine (CMM) to measure the dimensions of a part constitutes a physical measurement (Fig. 2.8). Refer to Appendix C for a case study involving a CMM to determine the radius of curvature on a part. Another example of a physical measurement is the measurement of velocity using photometric methods.

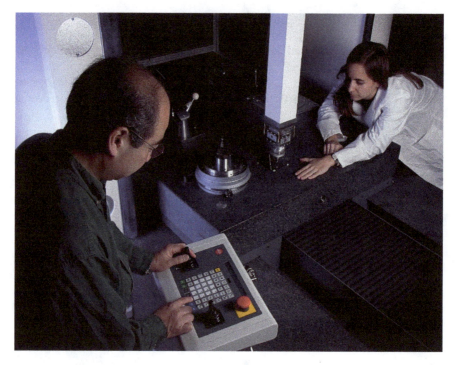

Fig. 2.8 Metrologists setting up a dimensional measurement of a reference standard using a CMM

2.3.2 Electrical Measurements

Electrical measurements seek to estimate measurands such as voltage, current, resistance, capacitance, inductance, and others. Electrical parameters may be continuous (DC), pulsed, or time-varying (AC). The measurement of a high-current shunt is shown in Fig. 2.9. Voltage and current measurements using digital multifunction meters are commonplace in industry.

2.3.3 Other Types of Measurements

Chemical, radiation, and spectroscopic measurements all have associated uncertainty, but sometimes determining uncertainties is not straightforward. Some surface science measurements may be used to describe qualitative results (e.g., energy-dispersive x-ray spectroscopy [EDS] or X-ray photoelectron spectroscopy [XPS]) but can also be used to determine a quantitative result. Figure 2.10 illustrates a quantitative method called angle-resolved X-ray photoelectron spectroscopy, which is used to determine film thickness.

2.4 Sources of Uncertainty

Fig. 2.9 Measurement of a high-current shunt

2.4 Sources of Uncertainty

All measurements, regardless of the type, have associated uncertainty. Primary contributors to measurement uncertainty include equipment, operators, environment, measurement models, etc. These types of measurement uncertainty will be highlighted here.

M&TE Equipment and instrumentation used to perform measurements is inherently inaccurate, regardless of whether it is calibrated or not. All equipment used to perform the measurement will contribute to the overall uncertainty. A calibration certificate seeks to quantify this component of uncertainty for the user of the equipment.

Standards and Calibration Calibrating M&TE against known reference standards, ultimately traceable back to the SI base units, is necessary to ensure equipment is functioning within its specifications, ensure equipment is traceable to known quantities, and establish uncertainties. Reference standards are not perfect, and every time a calibration is performed, the uncertainty of the reference standard becomes part of the uncertainty of the M&TE being calibrated. In general, the uncertainty of the M&TE can never be lower than the uncertainty of the standard

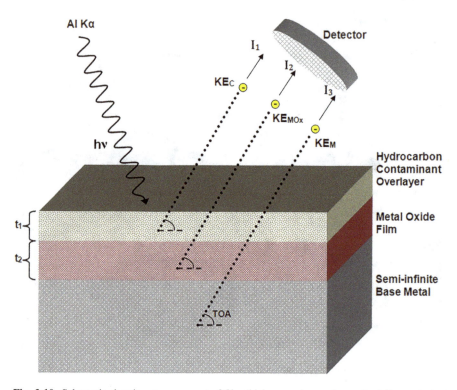

Fig. 2.10 Schematic showing measurement of film thickness using angle-resolved X-ray photoelectron spectroscopy (AR-XPS) (Forrest et al. 2015). © 2015 Elsevier Ltd., used with permission

against which it was calibrated. In fact, it will usually be larger due to additional uncertainty introduced from the calibration/measurement process, as well as inherent uncertainty associated with the M&TE.

Operator An operator may introduce additional uncertainty into a measurement process through imperfect execution or use of the M&TE. This becomes especially important for complex measurements, where extensive on-the-job training (OJT) is required due to intricacies of the measurement process. Operator-to-operator variability can become relevant to the reproducibility of the measurement.

Procedure Operating procedures are unable to eliminate all sources of variability during a measurement. Poorly written procedures may be ambiguous and introduce additional uncertainty into the measurement process.

Environment Environmental factors play a huge role in most measurement processes and can dominate uncertainty if not properly corrected or considered. For instance, accurate dimensional measurement requires careful control of temperature due to thermal expansion. Per ISO 1:2016 (ISO 1 2016), dimensional calibrations are always performed at 20 °C with compensations made when used at different temperatures. Leak rates from permeation-type leak elements typically vary ~3% per °C, meaning that a leak rate measured 10 °C away from ambient could have a 30% offset. Piezoelectric transducers, such as those used to measure pressure or acceleration, have large temperature coefficients, and a correction (with associated uncertainty) must be applied to achieve reasonable measurements. Failure to apply such corrections results in gross measurement uncertainties. Electrical resistors drift with temperature, and many electrical instruments must be used within narrow operating ranges of humidity. M&TE typically has recommended shock and acceleration limitations that must be observed for proper function. Consistency in environmental conditions is critical and must be carefully considered.

Random Disturbances Measurements are subject to disturbances, which it may not be possible to eliminate, thus contributing to measurement uncertainty. For example, electrical instruments may be subject to electromagnetic interference, leading to undesired variability in the observed signal that is not attributable to the measurand of interest. All electronics generate some intrinsic electrical noise that cannot be eliminated even with proper shielding and grounding. Radiation counting experiments are subject to a "counting uncertainty" due to the probabilistic nature of radioactive decay and emission processes. Random disturbances may be modeled as additive noise which, when significant, must be accounted for in the uncertainty of the measurement.

Measurement Model Uncertainty In general, measurements make assumptions about the measurand being investigated. For instance, measurement of a linear dimension on a part may assume the part is perfectly straight, when it may not be perfectly straight. In this case, the model assumes perfection. More complex measurement models typically require more assumptions, leading to additional sources of uncertainty. For example, determining the velocity of an object by measuring the distance traversed over a certain duration assumes that the object was traveling at constant velocity. What if it were accelerating or decelerating? This would introduce additional uncertainty into the measurement due to the model assumption of constant velocity. Sometimes the model can be expanded to include additional factors, such as a temperature coefficient in a resistor measurement model, that eliminate some of the model uncertainty.

2.4.1 Evaluating Sources of Uncertainty

Ultimately, it is nearly impossible to determine all potential sources of uncertainty in a measurement process. However, the most significant uncertainty components can be captured by considering the sources above and using engineering judgement, experience, and published recommendations about a measurement.

2.5 Summary

Consistent, well-defined terminology is a critical component of metrology. This chapter has introduced common terms used when discussing measurement uncertainty to prevent confusion later in the book. While most readers are likely familiar with older terms used in the error approach, this book advocates usage of the uncertainty approach to terminology for compliance with modern standards and the GUM. It is important to also understand older terminology as it is still widely encountered in textbooks, journals, reports, and other documents.

Basic measurement concepts and an understanding of uncertainty sources are foundational elements for the practitioner of metrology. Understanding these concepts is also important for those who analyze measurement data, in order to realize inherent limitations of different techniques. Ultimately, experience performing a particular type of measurement provides the greatest understanding of uncertainty contributors to the measurement process. While this chapter does not delve into specific measurement disciplines, Appendix B presents related reading supporting achievement of valid measurements in a variety of disciplines.

2.6 Related Reading

The definitive reference for terminology used in metrology and uncertainty analysis is the International Vocabulary of Metrology, or VIM (JCGM 200 2012). Written in a dictionary-style format, it is an important reference but may not easily convey concepts to those without prior experience in metrology. Appendix D1 of NIST Technical Note 1297 (Taylor and Kuyatt 1994) provides further explanation and interpretation of the VIM terminology and serves as an important companion for those seeking a conceptual-level understanding of VIM terminology.

The direct and indirect measurement models are covered in greater detail in Chap. 6 of this book and in a draft GUM supplement (JCGM 103 2018) that is still in development. Also discussed is the concept of independence of measurements (or observations).

Kline (1985) provides a concise pre-VIM summary of terminology used in uncertainty analysis and explains important underlying concepts and the need for uncertainty analysis with any measurement. The paper also offers a number of examples and historical cases stressing the importance of uncertainty analysis in measurements.

An Introduction to Error Analysis: The Study of Uncertainties in Physical Measurements by Taylor (1997) provides coverage of important concepts and topics for those new to the field of metrology and uncertainty analysis. The Error Approach and Uncertainty Approach terminology is interspersed and used somewhat interchangeably throughout, as the first addition was written prior to the VIM. The text is especially useful for those with limited prior experience in statistics, data analysis, and uncertainty concepts.

John Mandel, a distinguished scientist and statistician at the National Bureau of Standards (renamed the National Institute of Standards and Technology in 1988), originally authored *The Statistical Analysis of Experimental Data* (Mandel 1984) in 1964. In addition to introducing key concepts such as measurement models and sources of variability, the book presents statistical concepts that are relevant to the experimentalist and metrologist, such as independence of measurement observations. Although utilizing the terminology of the Error Approach, the text serves as a good companion to this book.

While proper understanding of discipline-specific measurement processes and instrumentation selection are critical to obtaining good measurement data, these areas are not discussed at length in this book. Appendix B offers additional related reading for several metrology disciplines along with related reading for each area. For topical coverage of discipline-specific instrumentation and selection for the experimentalist, Holman (2012) provides an overview of various measurement principles and discussion of instrumentation options in the framework of reducing measurement uncertainty. With an emphasis on mechanical engineering, the text also provides coverage of basic electrical measurements utilized in engineering experiments.

Measurement uncertainty concepts discussed here are applicable to testing laboratories as well as chemical metrology. For additional information in these areas, see Kimothi (2002).

2.7 Exercises

1. Repeatability and reproducibility can both be used to assess measurement variability.
    a. Describe how repeatability differs from reproducibility.
    b. For the same type of measurement, which will represent larger variability, and why?
    c. Is reproducibility or repeatability a better representation of measurement variability? Justify your response.

2. A digital multimeter (DMM) with microvolt resolution reads to six decimal places, such that the measurement of a fully charged automotive battery (nominal 12.6 V) is 12.623172 V.
    a. Does the reading on the multimeter mean the measurement is accurate to a microvolt? Why or why not?
    b. What else could contribute to the accuracy of the automotive battery voltage measurement?

3. Consider the following scenario: You are at the pump filling your vehicle with gasoline. The gasoline is sold by volume at a price of $2.74 per gallon (1 US gallon = 3.785411784 L), as displayed on the pump. The total fill amount is 14.462 gallons as shown on the digital pump display.

a. What is the resolution of the gasoline volume measurement?
b. Do you think the total uncertainty is the same as the displayed resolution displayed?
c. How much would you be willing to overpay for the entire tank of gasoline due to measurement uncertainty?
d. Using your value from (b), calculate a maximum acceptable uncertainty on the fuel volume measurement, per gallon (provide as an uncertainty in $\pm U$/gallon and $\pm\%$/gallon).
e. Will temperature and other environmental factors affect the measurement uncertainty? Why or why not?

4. You are watching a stock car race and want to estimate the speed of your favorite driver, Zip McFly, in car #7. You brought a stopwatch to the event and know that the length of the oval superspeedway is 2.0 miles.

   a. How would you determine Zip's speed? Is this a direct or indirect measurement?
   b. Write down all potential sources of uncertainty in trying to estimate Zip's speed using this method.
   c. Which source of uncertainty do you think will be largest, and why?
   d. Are there any sources of uncertainty that you can ignore? Why?

5. A colleague asks you to assess the repeatability of a DC electrical current measurement using a multifunction meter with high-speed data acquisition system. You have been told to take at least 30 independent measurements. You know that more measurements are better in assessing repeatability, so you set up you high-speed data acquisition system to record one thousand measurements over one second. Are the results valid for assessing measurement repeatability? Why or why not?

6. A laboratory environment is monitored using a thermometer and hygrometer with a digital readout, as seen in Fig. 2.11.

   a. What is the resolution of the temperature measurement shown?
   b. What is the resolution of the relative humidity measurement?

Fig. 2.11 Digital readout showing temperature and relative humidity measurement

2.7 Exercises

c. For the highest accuracy measurements in a laboratory environment, why is monitoring of environmental conditions (such as temperature and humidity) important?

7. Identify the following sources of uncertainty as either Type A or Type B. Provide your reasoning.
    a. Stated uncertainty on the calibration certificate of a pressure gauge.
    b. Uncertainty arising from ambiguous measurement instruction in a laboratory procedure for the measurement of flow rate in metal tubing.
    c. Variability in a measurement result when the M&TE is used by different operators.
    d. A subject matter expert's estimate of the worst case uncertainty associated with the measurement of bullet velocity using a high-speed camera.
    e. A shift in the sensitivity of an accelerometer due to $\pm 5.0\ °C$ variation in temperature.
    f. Uncertainty due to the limited resolution (0.01 mg) of a mass balance.
    g. Fluctuating readings of a microvolt-level signal due to electromagnetic interference.
    h. Counting uncertainty in the measurement of alpha particles from a radioactive source.
    i. The change in what should be a nominally identical measurement result when making the measurement two weeks apart.

8. What is the resolution of the DC electrical current measurement shown in Fig. 2.12? Can you think of any advantages an analog readout might provide over a digital one?

9. On February 25, 2008, a B-2 bomber (*the Spirit of Kansas*) crashed shortly after takeoff from Anderson Air Force Base in Guam. Fortunately, both the pilot and mission commander ejected from the aircraft and survived the crash, although with injuries. However, as seen in Fig. 2.13, the accident resulted in a total loss of the aircraft, costing over one billion US dollars. It is the only B-2 aircraft lost to date.

Fig. 2.12 Analog gauge showing measurement of DC electrical current

Fig. 2.13 Emergency teams responding to the B-2 crash site

Investigation of the crash after the fact revealed that an error in calibrating certain Port Transducer Units (PTUs) resulted in a measurement bias of air pressure in excess of -0.40 in Hg (-1355 Pa).

Research the crash (both NASA and the U.S. Air Force offer documentation online) and discuss how the calibration error happened. Explain how such a small error in air pressure measurement led to the crash. What sources of uncertainty were not considered in the calibration of the PTUs?

References

Abernathy, R.B., Benedict, R.P., Dowdell, R.B.: ASME measurement uncertainty. J. Fluids Eng. **107**, 161–164 (1985)

ANSI/ASME PTC 19.1: Measurement uncertainty (1983)

DOE-STD-1054-96: Guideline to good practices for control and calibration of measuring and test equipment (M&TE) at DOE nuclear facilities (1995)

Forrest, E., Schulze, R., Liu, C., Dombrowski, D.: Influence of surface contamination on the wettability of heat transfer surfaces. Int. J. Heat Mass Transf. **91**, 311–317 (2015)

Holman, J.: Experimental Methods for Engineers, 8th edn. McGraw-Hill Companies, Inc., New York (2012)

ISO 1: Geometrical product specifications (GPS) — standard reference temperature for the specification of geometrical and dimensional properties (2016)

ISO/IEC Guide 98: Guide to the expression of uncertainty in measurement (1993)

JCGM 100: Evaluation of measurement data - guide to the expression of uncertainty in measurement (2008)

JCGM 103: Guide to the expression of uncertainty in measurement — developing and using measurement models (committee draft) (2018)

References

JCGM 200: International Vocabulary of Metrology – basic and general concepts and terms (VIM), 3rd Edition, 2008 Version with Minor Corrections (2012)

Kimothi, S.K.: The Uncertainty of Measurements: Physical and Chemical Metrology: Impact and Analysis. ASQ Press, Milwaukee (2002)

Kline, S.J.: The purposes of uncertainty analysis. J. Fluids Eng. **107**, 153–160 (1985)

Mandel, J.: The Statistical Analysis of Experimental Data. Dover Publications, Inc., New York (1984)

Taylor, J.R.: An Introduction to Error Analysis: the Study of Uncertainties in Physical Measurements, 2nd edn. University Science Books, Sausalito, CA (1997)

Taylor, B.N., Kuyatt, C.E.: Guidelines for Evaluating and Expressing the Uncertainty of NIST Measurement Results. NIST Technical Note 1297. National Institute of Standards and Technology, Gaithersburg, MD (1994)

Chapter 3
The International System of Units, Traceability, and Calibration

Consistent and well-defined units, much like terminology, are critical in the comparison of measurement results and to the understanding of measurement uncertainty. This chapter provides a history of the International System of Units, or "SI," and a modern description of its seven base units utilized throughout the majority of the world. These units include the second, the meter, the kilogram, the ampere, the kelvin, the mole, and the candela. Recent advancements in unit definitions are discussed, with an emphasis on redefinition using fundamental constants to achieve lower uncertainties. This chapter also describes calibration and traceability, important metrological concepts, along with calibration policies and standards.

3.1 History of the SI and Base Units

The International System of Units, abbreviated "SI" from the French name, Le Système International d'Unités, is the modern form of the metric system. In 1790, Louis XVI of France authorized a scientific investigation to reform the country's weights and measures, which led to the development of the first metric system in 1795. In 1875, 17 countries signed the Metre Convention, which established the Bureau International des Poids et Mesures (BIPM). The BIPM provided a single international system of measurement standards, officially instituting unit definitions for length and mass: the meter and kilogram. The Conference Générale des Poids et Mesures (CGPM), a conference of delegates from around the world, was simultaneously established under the BIPM to periodically evaluate and update the unit definitions and provide guidance on their dissemination to the measurement community. Subsequent meetings of the CGPM added definitions of other units, terminology, the metric prefixes, and measurement guidance. In 1954, the 10th meeting of GCPM declared that a complete measurement system would be derived from 6 base units directly related to mechanical, electromagnetic, temperature, and optical radiation. These six base units—meter, kilogram, second, ampere, Kelvin, and candela—would form the basis of the SI. In 1960, the SI was formally enacted by the 11th CGPM.

The SI units continue to evolve based on the decisions of the CGPM. For example, the mole was added as a 7th base unit by the 14th CGPM in 1971 (Bureau International des Poids et Measures (BIPM) 2019). The meter was redefined in 1983 based on the speed of light, rather than the length of a physical artifact. The latest update, approved by the CGPM in 2018, effective in 2019, redefines the kilogram based on the value of Planck's constant rather than the mass of a physical piece of metal known as the International Prototype of the Kilogram (IPK). This latest update changed the definitions of the kilogram, ampere, mole, and Kelvin; all SI units are now defined based on physical constants such as the speed of light and electron charge, rather than on artifacts that are prone to damage or drift over time. This redefinition also established some additional physical constants, such as the electron charge, as fixed values with no uncertainty. In other words, the value of these constants is determined based on the best consensus value between metrologists. Prior to the redefinition, physical constants were empirical values, with uncertainty, based on measurements using the unit definitions at the time.

The base units can be thought of as the fundamental set of seven units from which all other SI units can be derived. The SI and base units are part of the same basic measurement language used internationally. The seven base units are defined below (by their 2019 definitions) along with the seven fundamental physical constants that define the units. A list of some common derived units is also provided. The definitions in italics are quoted from the 9th edition of the SI Brochure (Bureau International des Poids et Measures (BIPM) 2019).

3.1.1 SI Constants

In defining the seven base units, the SI relies on several physical constants that are fixed with no uncertainty, as shown in Table 3.1. These constants are used to define the SI units.

Table 3.1 Physical constants defining the SI (Bureau International des Poids et Measures (BIPM) 2019)

Defining Constant	Symbol	Value	Unit
Hyperfine transition frequency of Cs-133	$\Delta\nu_{Cs}$	9,192,631,770	s^{-1}
Speed of light in vacuum	c	299,792,458	$m\ s^{-1}$
Planck constant	h	$6.62607015 \times 10^{-34}$	$kg\ m^2\ s^{-1}$
Elementary charge	e	$1.602176634 \times 10^{-19}$	$A\ s$
Boltzmann constant	k	1.380649×10^{-23}	$kg\ m^2\ s^{-2}\ K^{-1}$
Avogadro constant	N_A	$6.02214076 \times 10^{23}$	mol^{-1}
Luminous efficacy	K_{cd}	683	$cd\ kg^{-1}\ m^{-2}\ s^3$

3.1 History of the SI and Base Units

3.1.2 Time: Second (s)

The second is the duration of 9,192,631,770 ($\Delta\nu_{Cs}$) periods of radiation corresponding to the transition between two hyperfine levels of the cesium-133 atom ground state.

Many definitions of the second have been used or proposed throughout metrological history, ranging from measuring the period of a one-meter pendulum to dividing the time between full moons. Currently based on atomic vibrations, atomic clocks are now in use with an uncertainty of less than 1 s in 100 million years (Ball 2013).

3.1.3 Length: Meter (m)

The meter is the length of the path travelled by light in vacuum during a time interval of 1/299,792,458 (1/c) of a second.

The first meter was defined in 1795 to be one ten-millionth the circumference of the Earth. A metal bar was created a few years later based on this length and the meter was redefined to be the length of the bar. Some form of bar was used as the standard until the 1960s. In 1983, the current definition based on the speed of light was adopted. Devices called interferometers can convert the speed of light into an actual meter of length.

3.1.4 Mass: Kilogram (kg)

The kilogram is the mass found by taking the value of the Planck constant, h, of $6.62607015 \times 10^{-34}$ kg m^2 s^{-1}, given the definition of the meter and second [given above].

The first definition of a kilogram was the mass of one liter of pure water. Because it was difficult to replicate this definition under different conditions, in 1799 the kilogram was defined based on a platinum artifact. This artifact was formally replaced in 1889 with a platinum-iridium cylinder, known as the IPK and maintained by the BIPM in a vault near Paris. When the IPK was made, many duplicates were also manufactured and sent to different National Metrology Institutes (NMIs). However, measurements on the IPK and its siblings now show microgram-level discrepancies in their mass values, prompting the BIPM's 2019 redefinition of the kilogram to its current form based on Planck's constant. A Kibble balance (see Sect. 3.4.1) can convert this definition into a kilogram of mass.

For historical reasons, the kilogram is the only base unit that includes a metric prefix: "kilo." The original base unit was the gram, but the small mass of one gram was difficult to fabricate and work with, so the defining artifact was made one kilogram. Eventually, the definition was updated to consider a kilogram the base unit, so the kilogram artifact could be used directly.

3.1.5 Electric Current: Ampere (A)

The ampere is that constant current corresponding to the flow of $1/1.602176634 \times 10^{-19}$ elementary charges (1/e) per second.

Prior to 2019, the elementary charge was an empirical value with some uncertainty. The 2019 SI definition fixes the elementary charge as an exact quantity and bases the Ampere on that value.

3.1.6 Temperature: Kelvin (K)

The kelvin is equal to the change of thermodynamic temperature that results in a change of thermal energy kT by 1.380649×10^{-23} J (k).

Previous definitions of the Kelvin were based on the triple-point of water: the temperature and pressure at which water can exist as a solid, liquid, and gas simultaneously. Obtaining pure water to realize this definition was difficult, so the 2019 change redefines the Kelvin based on a value of the Boltzmann constant, k.

3.1.7 Quantity of Substance: Mole (mol)

The mole is the amount of substance of a system that contains $6.02214076 \times 10^{23}$ (N_A) elementary entities.

The mole is a measure of the amount of a substance and is mostly used in chemistry. Unlike the other units, it is simply a number (Avogadro's constant) based on the number of atoms in 12 g of carbon-12. The mole has been an SI base unit since 1971, although there was and remains controversy in the metrology community over whether the mole is truly a base unit or simply a scaling factor (Josephson 1962).

3.1.8 Luminous Intensity: Candela (cd)

The candela is the luminous intensity, in a given direction, of a source that emits monochromatic radiation of frequency 540×10^{12} Hz and that has a radiant intensity in that direction of 1/683 watt per steradian ($1/K_{cd}$).

The original definition of the candela relied on the intensity of black body radiation emitted from platinum at its melting point. The current definition was adopted in 1979 to avoid reliance on a specific material. One candela is approximately the brightness of a typical candle flame.

3.3 Unit Realizations

Table 3.2 Examples of derived units (Bureau International des Poids et Measures (BIPM) 2019)

Derived quantity	Special SI name	Base units
Area		m^2
Volume		m^3
Speed, velocity		m/s^{-1}
Acceleration		m/s^{-2}
Density		$kg\, m^{-3}$
Frequency	Hertz (Hz)	s^{-1}
Force	Newton (N)	$kg\, m\, s^{-2}$
Pressure	Pascal (Pa)	$kg\, m^{-1}\, s^{-2}$
Energy	Joule (J)	$kg\, m^2\, s^{-2}$
Voltage	Volt (V)	$kg\, m^2\, s^{-3}\, A^{-1}$
Dose equivalent	Sievert (Sv)	$m^2\, s^{-2}$

3.2 Derived Units

All other measurement units can be formed by combining the base units stated in Sect. 3.1. Derived units are products of powers of the base units. For example, the SI unit for area is meters squared, m^2. The SI also defines the prefixes kilo-, mega-, milli-, micro-, etc., so that km^2 is also an SI unit. Some derived units are given special names by the SI; for example, the newton is an SI unit of force defined as $kg\, m\, s^{-2}$ in base units. Examples of common derived units, including special names given by the SI, if applicable, appear in Table 3.2.

While discouraged by the BIPM, use of imperial and United States customary units (feet, inches, pounds) is still common in USA manufacturing and industry. In 1959, the few countries still using this unit system signed an agreement to define the yard and pound based on an exact number of meters and kilograms, respectively (Mueller and Astin 1959), effectively making the imperial system an SI-derived system.

3.3 Unit Realizations

The units described in Sect. 3.1 are given abstract definitions based on fundamental constants. To be useful for measurement, the units must be "realized" by converting the abstract definition into a physical value that can be used for comparison. The BIPM offers guidance on experimental setups that can realize the units. Based on the current definition of units, anyone can, in theory, construct one of these experiments to realize a unit without reliance on a higher organization such as BIPM or the National Institute of Standards and Technology (NIST) to provide traceability. As technology progresses, new experimental methods may be developed that realize the units with lower uncertainty or reduced cost.

This section provides two examples of unit realization for length and voltage.

3.3.1 Gauge Block Interferometer

The meter is based on the distance light travels in a specific amount of time. To be useful for calibration, this definition must be transferred to a set of master gauge blocks. Gauge blocks are standards used to obtain SI traceability in dimensional metrology. A Gauge Block Interferometer (GBI) is a device used to realize the meter. GBIs are instruments used worldwide for the highest accuracy calibration of gauge blocks and similar metrology standards. Two frequency-stabilized lasers at different wavelengths are reflected off the gauge block and a reference mirror. The interference of the two lasers creates a fringe pattern that is used to determine phase difference and thus length difference of the two beam paths. This difference can determine the length of the gauge block and thus realize the meter.

3.3.2 Josephson Volt

The SI unit for electric potential, the volt, can be realized using the Josephson effect. In 1962, Brian Josephson predicted quantum effects that occur when electron pairs tunnel between two superconductors separated by a thin insulating barrier now known as a Josephson Junction (Josephson 1962). The I–V curve predicted by Josephson includes regions of constant voltage V_n occurring at values $nhf/(2e)$, where n is an integer representing the voltage step, h is Planck's constant, and e is the elementary charge of the electron. e and h are both fundamental constants defined by the SI (see Table 3.1). The ratio $2e/h$ is referred to as the Josephson constant, K_J. Prior to 2019, the Josephson constant was an empirical value with some uncertainty. Under the 2019 SI redefinition, K_J is an exact value with no uncertainty. To differentiate between the two values, K_J is used to represent the value under the 2019 SI redefinition, while $K_{J\text{-}90}$ is used to represent the conventional value defined by CGPM in 1990.

A Josephson voltage standard uses a superconductive integrated circuit chip containing thousands of Josephson junctions in a series–parallel configuration. When cooled to 4 K and biased with microwave frequencies, the chip becomes superconducting and will generate stable voltages that depend only on the applied frequency and K_J constant. The Josephson voltage standard is the most accurate method to generate or measure voltage and is the universal basis for voltage measurements.

3.4 Advancements in Unit Definitions

Ongoing research in the field of metrology is continually improving measurement techniques which, in turn, affect the BIPM's unit definitions. This section gives two examples of advancement in measurement science. First, the Kibble Balance was key in redefining the kilogram to be based on Planck's constant. Second, the intrinsic pressure standard is a new method of realizing the pascal over a large pressure range with much lower uncertainty.

3.4.1 Kibble (Watt) Balance

In 1999, the CGPM formally declared that the kilogram should be redefined without need for a physical artifact (the IPK). Several competing methods were considered for this redefinition, but eventually the Kibble Balance proved the method of choice. The Kibble Balance provides a method of realizing the value of the kilogram in terms of current and voltage. Since current and voltage can be defined in terms of fundamental physical constants, such as the speed of light and Planck's constant, they provided a way to define the kilogram in terms of constants rather than the IPK, which is vulnerable to deterioration or damage and can drift. In 2018, the CGPM approved the new definition of the kilogram, with use of the Kibble Balance to realize the unit (Bureau International des Poids et Measures (BIPM) 2019).

The Kibble Balance was originally called a "Watt" balance since it performs measurements of both current and voltage in the coil, the product of which is expressed in watts, the SI unit of power. This product equals the mechanical power of the test mass in motion. The balance was renamed in honor of its inventor, Bryan Kibble, after he passed away in 2016.

The Kibble Balance is an extremely accurate device for measuring mass. Like conventional balances, it is designed to equalize one force with another, but on a Kibble Balance, the weight of the test mass is offset by the force produced by an electromagnetic field. An upward force is exerted on the electromagnet coil when its field interacts with the surrounding magnetic field. By adjusting the applied current, the magnitude of the force can be controlled and adjusted as necessary to support the weight of the test mass. While similar electromagnetic balances have been used for many years, the Kibble Balance adds a second measurement: using the balance at constant velocity. By swinging the balance arms at a constant velocity, the electrical power can be related to mechanical power. When combined with the static measurement of electrical power required to equalize the balance, the mass of the object can be accurately determined (Kilogram: The Kibble Balance 2018).

The first step to measure a mass with a Kibble Balance is to use the "weighing" or "force" mode. In this mode, a test mass is placed on the mass pan. It exerts a downward force equal to its mass (m) times the local gravitational field(g); $F = mg$. Current applied to the electromagnetic coil creates an upward force that balances the downward force of the mass. The force (F) on the coil is equal to the current (I) times the magnetic field (B) times the length of the wire in the coil (L); in other words, $F = IBL$. Combining these two forces, one can see that $mg = IBL$.

The second measurement, using the "velocity" mode, is necessary because BL is extremely difficult to measure. Adding this measurement allows the measurement of BL to be circumvented. In this mode, the test mass is removed from the mass stage and the current removed from the coil. The coil is then moved through the surrounding field at a controlled velocity. The resulting voltage (V) is measured. Here, $V = vBL$ where v is the velocity, and B and L are the same field strength and wire length as in weighing mode.

Combining the equations from weighing mode and velocity mode, the mass can be calculated as $m = IV/gv$. Here, current and voltage can be determined using quantum-electrical effects that are measurable on laboratory instruments. Both current and voltage are defined in terms of h, Planck's constant, and e, the charge of the electron.

3.4.2 Intrinsic Pressure Standard

The current method of realizing the pascal, the derived SI unit for pressure, uses mercury manometers. To improve uncertainties, reduce measurement time, and eliminate the need for hazardous mercury, an intrinsic pressure standard is being developed based on highly accurate optical interferometry to link the pascal to the refractive index of helium (Ahmed et al. 2016). The refractive index of a gas depends on its density, which is a function of temperature and pressure. Using the relationship between refractive index, temperature, and pressure for atomic helium, it is possible to determine the refractive index to an accuracy better than one part in 10^6. This will significantly decrease measurement uncertainties and could drastically improve uncertainties in pressure measurement. The same method could improve the realization of Kelvin.

To implement this pressure realization, it is necessary to measure the refractive index with extremely high accuracy, which is possible using laser interferometry. A Fixed-Length Optical Cavity is being developed by NIST to measure the difference in light passing through two channels: one under vacuum and the other filled with gas at the pressure to be measured. The presence of gas in the one cavity will slow the light by an amount relative to the pressure and refractive index. After the light passes through the two cavities it is combined to generate an interference pattern that can be used to calculate the difference in optical length and then calculate the pressure of the gas (Lee 2019).

This new mercury-free intrinsic pressure standard promises to improve pressure and temperature uncertainties. The optical pressure standard is expected to become a commercial device that is comparably priced to commercial pressure standards. The optical pressure standard should provide a large pressure range from a planned 1 Pa to 360 kPa along with lower uncertainties (Hendricks et al. 2014).

3.5 Metrological Traceability

Metrological traceability is obtained through an unbroken chain of calibrations, with uncertainty evaluated at each step, linking them to a relevant primary standard of the SI unit of measurement. The link to SI units may be achieved by reference to national measurement standards. National measurement standards may be primary standards, which are primary realizations of the SI units, or representations of SI units based on fundamental physical constants, or secondary standards that are calibrated by another NMI. Traceability to the SI between the standards used to perform calibrations and the unit being calibrated must be documented. Note that metrological traceability establishes the units of measurement that express a measurement result. Metrological traceability requires an established calibration hierarchy, and each step in the hierarchy chain has its own contribution to the measurement uncertainty.

When there is a measurement model with more than one input quantity, each of these input quantities should be metrologically traceable and the calibration hierarchy may form a network rather than a chain (JCGM 200 2012). When a calibration is performed using imperial units, traceability can be established using the fixed conversion factors between the imperial units and the SI units.

3.6 Measurement Standards

Measurement standards are artifacts used to define and realize measurement units and are the basis for all lower-echelon measurements. Standards are the fundamental reference for any system of weights and measures against which all other measuring devices are compared. There is a three-level hierarchy of physical measurement standards that are typically used to realize an SI unit.

1. "Primary standards" are at the top level. Primary standards have the highest metrological quality and are the realization for the unit of measurement.
2. "Secondary standards" are at the second level. Secondary standards are calibrated with reference to a primary standard through an unbroken chain to achieve metrological traceability to the SI.
3. "Working standards" are the lowest tier of a standard. Working standards are typically used to calibrate commercial and industrial measurement equipment (de Silva 2002).

An example of a measurement standard used in a calibration laboratory is a Thomas-type 1-ohm resistor commonly used as a primary standard for calibrating secondary resistances.

3.6.1 Certified Reference Materials

Certified Reference Material (CRM) is a type of measurement standard made of discrete quantities of a known substance or minor artifact that has been certified based on its composition, purity, concentration, or some other characteristic. CRMs are essential in stoichiometry, the metrology of chemistry, and are also finding much use in the growing field of nanotechnology. Commercially available CRMs include ores, pure metals and alloys, gases and gas mixtures, nanoscale particles and structures, and many biochemical substances and organic compounds.

Purchased CRMs have their properties certified by a procedure that establishes traceability to an accurate realization of the appropriate unit. The certificate should also include an uncertainty at a stated level of confidence. Certified reference materials that are provided by a supplier are considered traceable when the supplier complies with ISO Guide 34, as accredited by an organization that currently holds mutual recognition signatory status under the International Laboratory Accreditation Cooperation (ILAC) Mutual Recognition Arrangement (MRA) (Sharp 1999). The term "Standard Reference Material" is a trademark applied to CRMs sold by NIST that meet additional documentation requirements (May et al. 2000).

An example of a CRM is a film thickness standard consisting of a quartz substrate with precise and traceable 180-nm deep grooves etched into the surface. The step height of the grooves has been calibrated by the manufacturer and is traceable through NIST. The standard is used to calibrate the thickness of a thin film on a substrate used as a secondary standard to calibrate equipment used for measuring film thickness with X-ray fluorescence.

3.6.2 Check Standards

A check standard is a reference artifact that can be used to closely monitor the status of a system or piece of M&TE. A check standard should be periodically measured using the system and the results should be recorded in a control chart to evaluate the health of a system. This is not the same as a calibration; a check standard needs to be stable over time with little drift, but not necessarily calibrated. The check standard ensures that a measurement is being carried out correctly and that the system is working as intended.

Check standards are a great way to expose errors that afflict a process over time. Once a baseline for the quantities has been established based on the check standard's historical data, they can be used to control the bias and long-term variability of the process.

Check standards must be similar to the test items measured on the system and ideally should be actual test items from the system. Check standards should always be stable and available to the measurement process. If the check standard is measured regularly and at regular intervals, it can help ascertain the true output of the measurement process by determining the part-to-part variation of each measurement sequence (NIST/SEMATECH 2013).

An example of a check standard is a set of microwave attenuators, 3 dB, 40 dB, and 80 dB, maintained by the Microwave Lab at the Sandia Primary Standards Laboratory. These attenuators are measured before every calibration to ensure the measurement equipment is configured correctly and that cables and connectors are functioning as expected.

3.7 Calibration

Calibration is the process of comparing an unknown value of interest to a known traceable value. Calibration is critical because the accuracy of all measuring devices degrades over time. Calibration ensures that equipment is operating in accordance with its certified specifications throughout its operating life. Ensuring that the equipment can measure accurately and agree with measurements around the world ensures product quality.

In a calibration, a measurement standard provides a known traceable value and the device under test (DUT) provides the unknown value. Calibration allows users to determine the uncertainty of a DUT and ensures that the measurements performed using that DUT are metrologically traceable. The accepted method for determining uncertainty in a calibration is defined by the GUM (JCGM 200 2012) and is described in detail in Chaps. 6, 7, and 8.

A proper calibration results in a report of the DUT's value, uncertainty, and the confidence level in that uncertainty. For example, a calibration report for a resistor may state that the resistor value is $1.000\ \Omega \pm 35\ \mu\Omega/\Omega$ at 95% confidence. The report should list any measurement standards used for the calibration and provide a statement establishing metrological traceability to the SI. This is typically ensured by periodically calibrating the measurement standards using an accredited calibration provider or NMI, first assuring the provider is accredited for the desired calibration.

The calibration report from commercial calibration laboratories typically provides a "time-of-test uncertainty," including only uncertainty in the measurement made on the DUT the day it was calibrated. Some labs, including the Primary Standards Laboratory at Sandia, provide a calibration certificate with a "certification uncertainty." This value includes additional uncertainty for aspects such as drift, shipping, and handling to account for use of the device during the upcoming interval.

Tolerance testing can be a form of calibration. The result of a tolerance test is a pass or fail determination based on whether the device value falls within an acceptable range. No uncertainty statement is provided in the report. To perform a tolerance test, the measurement standard must have uncertainty at least four times better than the DUT's required tolerance. The mathematical basis for this requirement comes from a risk evaluation and is described in Chap. 5.

3.7.1 The Calibration Cycle

Calibration must be completed periodically to ensure a DUT is operating as expected throughout its lifetime. The recalibration interval, often one year, is determined by the calibration lab with input from the customer and ensures that drift or degradation in the DUT is accounted for before it becomes problematic in use. If results of a calibration indicate the DUT value has changed by more than the previous calibration's expanded uncertainty, or if a tolerance test results in a fail condition, the DUT is considered out of tolerance (OOT). An OOT indicates that at some point during the last interval, the DUT drifted outside of its certification uncertainty. When this occurs, all measurements made using that DUT during the past interval are questionable and should be analyzed for the possible impact of a bad measurement.

Some DUTs can be adjusted at the time of calibration, but adjustment is not the same as calibration. Before an adjustment is made, a calibration must be performed to determine the "as-found" condition of the DUT. Then, the adjustment is made to bring the DUT closer to the desired nominal value. Finally, a second calibration is performed to determine the "as-left" condition, which can be used until the next calibration date.

At the end of a DUT's life cycle, when it is no longer needed for metrologically traceable measurements, a close-out calibration is performed. This ensures that the equipment functioned as expected during the final calibration interval. If the as-found close-out calibration is OOT, an impact analysis should be performed to determine if the failure impacted the equipment's measurements during the calibration interval.

3.7.2 Legal Aspects of Calibration

"Legal metrology" refers to legal requirements related to metrology used to control the daily transactions of trade and commerce around the world. Legal metrology exists to ensure fair trade and the safety and health of the public and the environment. Domestic uniformity in legal metrology is the responsibility of NIST through its Office of Weights and Measures. Internationally, it is handled by an international agreement and a quasi-official body, the International Organization of Legal Metrology (OIML). Examples of legal metrology include ensuring that fuel purchased by volume is measured using calibrated pumps through all parts of the distribution chain, emissions from power plants are monitored using calibrated equipment to ensure compliance with environmental laws, and municipal water sources use calibrated measurements to evaluate contaminant levels in the water supply. Recall the "Deflategate" case study from Chap. 1 and consider how calibration (or lack thereof) could affect the outcome of civil and criminal investigations when measurements are used as evidence.

3.7.3 Technical Aspects of Calibration

M&TE must be calibrated with consideration of customer requirements and needs. If a calibration is not performed in a similar fashion to how the equipment is used, the results of the calibration are not valid, and the certification does not apply to the measurements being performed. An example is the difference between pulsed measurements and AC or the difference between DC and AC. If the customer is using the equipment at AC frequencies, the equipment shall be calibrated at the same AC frequencies for the range of use.

Always consider measurement uncertainty, both when performing a calibration and when using the M&TE. When using the DUT to determine compliance to a specification, the DUT uncertainty must always be at least 4 times better than the required tolerance of the measurement. In some cases, a DUT uncertainty as low as 1.5 times better than the requirement may be acceptable if proper guardbanding techniques are used to mitigate the risk of a lower-quality measurement (see Chap. 5.) The quality of the measurements may affect the related risk associated with decisions made regarding the measurement.

3.7.4 Calibration Policies and Requirements

The calibration certificate provided by an outside lab must be trustworthy and the measurements certified in the calibration must be metrologically traceable. Accreditation gives equipment owners greater confidence that the calibration has been performed correctly. Accreditation helps ensure that the calibration process has been reviewed and complies with internationally accepted technical and quality metrology requirements. Calibration laboratories are accredited to the international metrology standard ISO/IEC 17025:2017 (ISO/IEC 17025 2017). In addition, the American National Standard ANSI/NCSL Z540.3:2006 (ANSI/NCSLI 2006) is the American standard listing general requirements for the competency of calibration laboratories.

All accredited calibration laboratories must be periodically assessed by an independent accreditation body such as the National Voluntary Laboratory Accreditation Program (NVLAP) or American Association for Laboratory Accreditation (A2LA). To obtain accreditation status, a calibration lab must demonstrate the ability to perform acceptable calibrations, have a defined scope that documents their measurement capabilities, and have documented uncertainty analyses of their approved measurement systems to ensure they can meet their defined scope. Accreditation assessments require that the laboratory can demonstrate they can make traceable measurements.

3.7.4.1 ISO 17025

ISO/IEC 17025: *General Requirements for the Competence of Testing and Calibration Laboratories* is the recognized worldwide ISO standard for testing and calibration laboratories. For a laboratory to be deemed technically competent in calibration, it must either hold accreditation to ISO/IEC 17025 or at least comply with the requirements in the document. ISO/IEC 17025 gives calibration laboratories one standard to demonstrate that they can operate competently and provide valid results that are metrologically traceable. Having one international standard enables a wider acceptance of results between countries, ensuring that when a measurement is performed in one part of the world, it can be repeated accurately in another.

A revision to ISO/IEC 17025 was released in 2017. Calibration labs have until November 2020 to fully implement the new standard, which emphasizes taking a risk-based approach to calibration, to maintain their accreditation status.

3.7.4.2 ANSI Z540.1 and ANSI/NCSL Z540.3:2006

ANSI/NCSL Z540.1 and ANSI/NCSL Z540.3:2006 are calibration standards facilitated by the ANSI and the National Conference of Standards Laboratories International (NCSLI). ANSI Z540.1 was originally used to accredit calibration laboratories and provides guidance for general competency of a calibration lab, including requirements on maintaining records, quality assurance, and personnel. ANSI/NCSL Z540.3:2006 provides the requirements for the calibration of an organization's equipment when used in the manufacturing, modification, or testing of products. This promotes confidence in calibration laboratories and ensures that M&TE can be operated in compliance with its certified specifications. One of the ANSI/NCSL Z540.3:2006's major requirements is that decisions based on measurements from calibrated equipment must result in a false acceptance rate of less than 2%. Chapter 5 provides more information on determining false acceptance rates.

3.8 Summary

The International System of Units, or SI, is central to metrology and essential for making credible estimates of measurement uncertainty. The SI constitutes what we define as *real* in the world of measurement. Traceability of measurements to the SI is necessary for making meaningful estimates of measurement uncertainty. Calibration by an accredited laboratory is a means of providing this traceability. While any vendor, laboratory, or individual can claim traceability, without accreditation from a reputable body, it is up to the end-user of M&TE to prove it.

Calibration has associated technical, quality, and legal aspects making it an industry in and of itself. Expertise in many aspects of calibration takes years of hands-on experience. Calibration technicians constitute a specialized field in both military services and the commercial sector.

The SI is not static, as seen by its long history. Modernizations have been necessary to reduce measurement uncertainties for advanced manufacturing and technology requirements. While many in the USA may claim they do not use the SI, for metrology purposes modern U.S. customary units are effectively derived from the SI.

3.9 Related Reading

As seen in this chapter, the SI is a constantly evolving system. With redefinition of four (out of seven) base units in 2019 alone, it is clear that improvements to the system of base units are at the forefront of research conducted by National Metrology Institutes in the quest for reducing uncertainties in measurement. For this reason, we recommended that the reader check the BIPM website (https://www.bipm.org) for the latest version of the SI brochure. NIST also offers additional documentation on the fundamental system of units (Newell and Tiesinga 2019).

The research and development efforts that went into the replacement of the artifact defining the kilogram (IPK) were tremendous and cannot be adequately described here. For further reading on the development and theory behind the Kibble balance, Robinson and Schlamminger (2016) offer a comprehensive review of the new SI definition for the unit of mass and include historical perspectives.

Other modernizations of SI units are equally as interesting, and in some cases, controversial. In modern times, temperature has been realized using the International Temperature Scale. Preston-Thomas (1990) provides an English translation of the currently used ITS-90. While ITS-90 is still used to realize temperature, the redefinition of the Kelvin in 2019 is considered by some to be the most radical change in the SI to date. Machin (2018) provides a backdrop to the unit's redefinition, concisely describing more detailed technical aspects motivating the redefinition.

The full history of the International System of Units cannot be adequately covered in this chapter. Previously introduced in Chap. 1, Klein (1988) provides a well-written, in-depth history with the aforementioned gaps in recent developments.

Interlaboratory comparisons (ILCs) and proficiency tests (PTs) are an important part of metrology and crucial to ensure the validity and quality of accredited calibration laboratories. For more detail on how ILCs and PTs are conducted, ISO/IEC 17043:2010 (ISO/IEC 17043 2010) provides an overview of requirements. NIST also provides more specific general guidelines in the NVLAP accreditation program (Merkel and White 2016) along with discipline-specific guidance in conducting PTs and ILCs (Gust and Harris 2005).

While this book serves as the perfect guide for the metrologist in applying statistics and conducting uncertainty analyses, it does not serve as a reference for training in calibration fundamentals for different disciplines. *The Metrology Handbook* (Bucher 2012), mentioned in Chap. 1, is a good reference for the calibration technician or calibration laboratory manager, in covering detailed quality requirements and principles. *Calibration: A Technician's Guide* (Cable 2005) offers foundational instruction for either the benchtop or field technician performing calibrations on M&TE. For discipline-specific calibration references, the second edition of *Calibration*: *Philosophy and Practice* (Spang 1994) from Fluke provides comprehensive information regarding calibration of electrical instrumentation, especially instrumentation operating in the DC and low frequency regimes. For dimensional calibration, Mitutoyo's *Metrology Handbook: The Science of Measurement* delivers coverage of numerous dimensional metrology topics for those performing calibration.

3.10 Exercises

1. How did the ancient Egyptians define the unit of length? List three potential issues with using and applying this unit in practice.

2. List three units still in modern usage derived from measures of the human body. Why do you think these are still used?

3. The *cord* is a unit of measure of volume used for firewood sales in many countries around the world.

   a. How is this unit standardized in the USA?
   b. What is its equivalent to in cubic meters?
   c. What are some inherent sources of variability in this unit?

4. The time constant for an electrical circuit is an important parameter in determining the dynamic response of the system. For the following cases, show how the units on the left-hand side and right-hand side of the equation are equivalent (hint, derive in terms of base units).

   a. In a simple RC circuit, the time constant is $\tau = RC$. Here τ is in seconds, R is in ohms, and C is in farads.
   b. In a simple RL circuit, the time constant is $\tau = \frac{L}{R}$. Here τ is in seconds, R is in ohms, and L is in henrys.

5. The British thermal unit (Btu) is a non-SI unit of energy in common usage in the USA. How is the Btu defined? Has it always been defined this way? What issues do you see with having a unit whose definition has changed over time?

6. The typical heat flux in a nuclear reactor is 1595 Btu/(minute $\times$ ft^2). Express this in terms of SI units.

7. The SI unit for force is the newton. Show how this can be expressed in terms of base units.

8. The electron volt (eV) is the amount of kinetic energy gained by an electron accelerating from rest through a constant potential difference of 1 V in vacuum. Based on the 2019 redefinition of base units, derive the value of the electron volt in terms of it's SI equivalent, the joule (J).

9. The sale of many produce items at grocery stores and supermarkets is by mass. In most localities, these scales are required to be calibrated. Draw a traceability chain, going back to the fundamental definition of the kilogram, for the measured mass of bananas purchased at a grocery store.

10. The previous definition of the kilogram was set as the mass of the International Prototype of the Kilogram (IPK). 40 prototype kilograms in total were manufactured in 1882 from a platinum-iridium alloy (90% platinum, 10% iridium) and distributed to National Metrology Institutes around the world. Calibration of these prototypes against the IPK showed an average relative

mass drift of 50 μg from 1889 to 1989, with some units drifting by more than 100 μg over this period.

a. Why is this drift concerning? Discuss the implications for daily measurements of mass in commerce or in a laboratory environment.
b. Research the cause of this drift and propose three possible explanations. What could be done to fix the issue(s)?
c. Discuss advantages and disadvantages of redefining the kilogram in terms of fundamental constants.

References

Ahmed, Z., Klimov, N., Douglass, K., Fedchak, J., Scherschligt, J., Hendricks, J, Strouse, G.: Towards photonics enabled quantum metrology of temperature, pressure, and vacuum (2016). https://arxiv.org/abs/1603.07690v1. Accessed 29 Apr 2020

ANSI/NCSLI: Requirements for the Calibration of Measuring and Test Equipment. ANSI/NCSL Z540.3:2006 (2006)

Ball, P.: Precise atomic clock may redefine time. Nature (2013). https://www.nature.com/news/precise-atomic-clock-may-redefine-time-1.13363. Accessed 29 Apr 2020

Bucher, J.L.: The Metrology Handbook, 2nd edn. ASQ Quality Press, Milwaukee (2012)

Bureau International des Poids et Measures (BIPM): The international system of units (SI), 9th ed. (2019). https://www.bipm.org/en/publications/si-brochure (2019). Accessed 29 Apr 2020

Cable, M.: Calibration: A Technician's Guide. ISA, Research Triangle Park, NC (2005)

De Silva, G.M.S.: Basic Metrology for ISO 9000 Certification. Routledge, London (2002)

Gust, J.C., Harris, G.L.: Weights and measures division quality manual for proficiency testing and interlaboratory comparisons. In: NISTIR 7214. National Institute of Standards and Technology, Gaithersburg, MD (2005)

Hendricks, J.H., Ricker, J.E., Egan, P.F., Strouse, G.F.: Search of better pressure standards. Phys. World. **2014**, 67 (2014)

ISO/IEC 17025: General requirements for the competence of testing and calibration laboratories (2017)

ISO/IEC 17043: Conformity assessments — general requirements for proficiency testing (2010)

JCGM 200: International Vocabulary of Metrology – Basic and General Concepts and Terms (VIM), 3rd edn, 2008 Version with Minor Corrections (2012)

Josephson, B.D.: Possible new effects in superconductive tunneling. Phys. Lett. **1**(7), 251–253 (1962)

Kilogram: The Kibble Balance: National Institute of Standards and Technology (2018). https://www.nist.gov/si-redefinition/kilogram-kibble-balance. Accessed 29 Apr 2020

Klein, H.A.: The Science of Measurement: A Historical Survey. Dover Publications, Inc., New York (1988)

Lee, J.L.: FLOC takes flight: First portable prototype of photonic pressure sensor (National Institute of Standards and Technology, 2019). https://www.nist.gov/news-events/news/2019/02/floc-takes-flight-first-portable-prototype-photonic-pressure-sensor. Accessed 29 Apr 2020

Machin, G.: The Kelvin redefined. Meas. Sci. Technol. **29**(2), 022001 (2018)

May, W., Parris, R., Beck, C., Fassett, J., Greenberg, R., Guenther, F., Kramer, G., Wise, S.: Definitions of terms and modes used at NIST for value-assignment of reference materials for chemical measurements. In: NIST Special Publication 260-136, Standard Reference Materials. National Institute of Standards and Technology, Gaithersburg, MD (2000)

Merkel, W.R., White, V.R.: National Voluntary Laboratory Accreditation Program: procedures and general requirements. In: NIST Handbook 150. National Institute of Standards and Technology, Gaithersburg, MD (2016)

Mueller, F.H., Astin, A.V.: Research Highlights of the National Bureau of Standards Annual Report, Fiscal Year 1959. Miscellaneous Publication 229. National Bureau of Standards, Washington, D.C. (1959)

Newell, D.B., Tiesinga, E.: NIST Special Publication 330 - The International System of Units (SI). National Institute of Standards and Technology, Gaithersburg, MD (2019)

NIST/SEMATECH: e-Handbook of Statistical Methods (2013). https://www.itl.nist.gov/div898/handbook/. Accessed 29 Apr 2020

Preston-Thomas, H.: The International Temperature Scale of 1990 (ITS-90). Metrologia. **27**(1), 3–10 (1990)

Robinson, I.A., Schlamminger, S.: The Watt or Kibble balance: A technique for implementing the new SI definition of the unit of mass. Metrologia. **53**(5), A46–A74 (2016)

Sharp, D.B.: Measurement standards. In: Webster, J.G. (ed.) The Measurement, Instrumentation, and Sensors Handbook, p. 51. CRC Press, Boca Raton (1999)

Spang, S.: Calibration: Philosophy in Practice, 2nd edn. Fluke Corporation, Everett, WA (1994)

Suga, N.: The Metrology Handbook: The Science of Measurement, 2nd edn. Mitutoyo America Corporation, Aurora, IL (2016)

Chapter 4
Introduction to Statistics and Probability

This chapter provides an overview of statistics and probability concepts that are used in metrology. It begins with a discussion on data types along with Exploratory Data Analysis techniques that can be used to help understand and visualize data. It then gives an overview of probability distributions common to metrology with guidance on choosing probability distributions based on available information. This chapter concludes with information on estimating parameters and assessing goodness-of-fit of probability models.

4.1 Introduction

The practice of metrology relies heavily on a variety of statistical methods used to design, characterize, and quantify the uncertainty of measurement systems. In this section we provide an overview of basic probability and statistics concepts that are often used to this end.

In statistics for metrology, the primary goal is to draw conclusions (e.g., estimate parameters) about a target population (measurement system) using a finite sample from that population. In metrology the parameters of concern are usually uncertainty terms, expressed as standard deviations. Once uncertainty terms have been estimated, probability statements can be made regarding future measurements. The probability statement will usually be in the form of an interval that is expected to contain the true value of a measurand.

This chapter will first introduce different types of data and will then discuss statistical methods that can be used to describe sample data. It will end with an overview of common probability distributions used in metrology, as well as guidance for estimating parameters of those distributions and assessing the goodness-of-fit of probability models.

4.2 Types of Data

A first step before analysis is to identify the type of data that was collected and the methods that may be appropriate for that type of data. Data can be divided into two broad categories: continuous and discrete. Continuous variables, such as temperature and voltage, can take on any value in a continuous interval. Alternatively, discrete variables can only take on distinct, separate values. There are several types of discrete data, including:

- Binary data, which can fall into one of two categories. Commonly, binary data are described as "Pass/Fail", "0/1," or "Go/No Go." In metrology, it may be of interest to evaluate the performance of a "Go/No Go" gauge used in the inspection of manufactured parts.
- Count data, in which the data are nonnegative integers. The integer values come from a counting process rather than from measuring some characteristic on a continuous scale. An example is using a neutron detector to count the number of neutrons produced by a manufactured neutron source.
- Categorical data, in which the data fall into one of several distinct, nonnumeric categories. Categorical data can be further divided as ordinal or nominal. Ordinal data has an intrinsic ordering (such as low, medium, and high), while nominal data is nonordered (such as Tester A vs. Tester B).

In metrology, the data are most likely to be continuous variables, although we will present examples of assessing uncertainty using both binary (Sect. 11.3) and count data (Sect. 6.4.1).

It is important to identify the variable type as it affects the exploratory data analysis (Sect. 4.3) and choice of probability distribution (Sect. 4.4). Note that continuous data provide much more information than discrete data, so it is advantageous to use continuous measures whenever possible. For example, if instead of categorizing temperature using ratings such as "low," "medium," and "high," the measurements are on a continuous scale, then more information is available to estimate population parameters such as the mean or standard deviation.

4.3 Exploratory Data Analysis

Once the data type has been determined, Exploratory Data Analysis (EDA) can be used to better understand the overall characteristics of the data. In particular, EDA can be helpful for

- Identifying potential outliers and trends present in the data. Outliers can inflate the uncertainty standard deviations. If an "assignable cause" of an outlier can be identified and removed from the measurement process, the outlier can be discounted. Trends over time that are identified in the measurement process must be accounted for when estimating uncertainty intervals that are expected to be valid over time.

4.3 Exploratory Data Analysis

- Determining an appropriate probability distribution. The validity of uncertainty intervals will depend on correctly specifying the underlying probability distribution.
- Detecting subpopulations in the data. Subpopulations (i.e., a subset of measurements taken under very different conditions than the bulk of the data) can have the effect of inflating the measurement uncertainty. Subpopulations may need to be discounted or analyzed separately.
- Generating summary statistics. Summary statistics (e.g., sample means and standard deviations) are inputs to uncertainty analyses for both direct and indirect measurements (Chap. 6).

Commonly, EDA consists of two main parts: calculating summary statistics and creating graphical displays of the sampled data.

4.3.1 Calculating Summary Statistics

A statistic is simply a measure of some attribute of sampled data. There are many summary statistics that can be used to concisely describe features of data. Generally, for continuous data, measures of the distribution's center (e.g., mean, median or mode) and spread (e.g., range, standard deviation, inter-quartile range) are used. For data sets with multiple continuous variables, the relationship between variables (measured by the correlation coefficient) is often of interest in metrology studies. The following sections describe each of these measures in more detail. A discussion of summary statistics for discrete data follow below.

4.3.1.1 Summary Statistics for Continuous Data

Measures of Central Tendency Measures of central tendency are often used to quantify where the mid-point or center of sampled data falls. The most commonly used measure is the sample mean, also known as the sample average, which is simply the sum of the data values divided by the number of observations n. Formally, for observations $x_1, x_2, \ldots, x_n$ it is defined as

$$\bar{x} = \frac{1}{n} \sum_{i=1}^{n} x_i. \tag{4.1}$$

Another common measure of central tendency is the median, defined as the 50th percentile of the data (i.e., the point below which 50% of the observations fall). The mode is the most frequently observed value in the sampled dataset and can also be used as a measure of central tendency.

While the sample mean is most commonly used, it is sensitive to outliers, meaning it can be heavily skewed in the direction of outliers if they are present. Where there are extreme outliers, the median can be a more informative measure of central tendency. In metrology, the sample mean is the most frequently used measure of central tendency.

Measures of Spread Measures of spread can be used to quantify the range of the data. The sample variance is a popular measure of spread and is defined as

$$s^2 = \frac{1}{n-1} \sum_{i=1}^{n} (x_i - \bar{x})^2. \tag{4.2}$$

The sample standard deviation, denoted s, is the positive square root of the sample variance. The standard deviation is most commonly used as a measure of spread because it has the same units as the observed data. Other measures of spread include the range (the difference between the maximum and minimum values in the sample) and the interquartile range (the difference between the 25th and 75th percentiles of the sample). Each of these sample statistics provides an estimate of the population standard deviation (often denoted by σ), the true but unknown standard deviation.

Another measure of spread frequently used in metrology applications is the standard error of the mean (referred to in the GUM as the "experimental standard deviation of the mean") and is given by the formula

$$\frac{s}{\sqrt{n}}. \tag{4.3}$$

This statistic is a measure of the spread in sample means about the true but unknown population mean (often denoted by μ), based on n observations.

Correlation The correlation between two continuous variables can be thought of as the strength of the relationship between the variables. In metrology studies with multiple continuous variables, evaluation of the pairwise correlation between variables should be part of the exploratory data analysis when possible. The correlation coefficient, a number between -1 and $+1$, quantifies the strength of the linear dependence between the two variables.

The most commonly used correlation coefficient is Pearson's correlation coefficient (r_{xy}), defined as:

$$r_{xy} = \frac{\sum_{i=1}^{n}(x_i - \bar{x})(y_i - \bar{y})}{\sqrt{\sum_{i=1}^{n}(x_i - \bar{x})^2}\sqrt{\sum_{i=1}^{n}(y_i - \bar{y})^2}}, \tag{4.4}$$

4.3 Exploratory Data Analysis

where n is the sample size, and x_i, y_i are the individual measurements of variable x and y. The formula assumes a natural pairing of the x and y variables. Values of r_{xy} close to zero indicate little to no correlation, while values of $|r_{xy}|$ close to one suggest strong correlation (positive or negative) between the variables. If variables that appear in a measurement equation (Eq. (6.27)) are highly correlated, the total variance (Eq. (7.25)) associated with the measurement equation must include additional terms to account for the correlation. This component of the EDA is thus an important consideration in any complex measurement system uncertainty analysis.

4.3.1.2 Summary Statistics for Discrete Data

For discrete categorical data, common summary statistics are the frequency and relative frequency of each category in which the data fall. Frequency is defined as the number of values in a category, while relative frequency is defined as the frequency divided by the total number of data points.

The values in a frequency table provide a rough estimate of how the data are distributed. Table 4.1 provides an example of the frequencies and relative frequencies for binary data. In this example most of the data are classified as Pass. Again, discrete data inherently provide less information than continuous data, so it is preferable to use a continuous measure if possible.

For discrete count data, the summary statistics described above for continuous variables data still apply, but the interpretation of the sample means and standard deviations may be different depending on the specific distribution used to describe the data. For example, the sample mean (Eq. (4.1)) of count data assumed to be from a Poisson distribution (Sect. 4.4.1.2) estimates both the mean and variance of the distribution. Interpretation of the summary statistics in these cases must thus consider the particular discrete probability distribution.

4.3.2 Graphical Displays of Data

In addition to summary statistics, it can be useful to visualize data to assess if there are significant outliers, trends, correlations, subpopulations, or unusual variation present. Graphical tools can also be used to suggest which probability distribution may be appropriate for the measurement variables. The choice of graphical display also depends on the type of data.

Table 4.1 Frequency and relative frequency for discrete data

Category	Frequency	Relative frequency
Pass	15	$15/20 = 0.75$
Fail	5	$5/20 = 0.25$
Total	20	$20/20 = 1.00$

4.3.2.1 Graphical Displays for Continuous Data

Histograms Histograms are generated by first dividing the range of the variable into equally spaced intervals, and then tabulating the number of data points that fall within each interval. The height of the histogram bar is equal to the proportion of data in its corresponding interval. Histograms provide a visualization of the distribution of continuous data including its central tendency and spread. The histogram can also be used to identify outliers and subpopulations and help determine an appropriate probability distribution.

Note that histograms can be sensitive to the choice of bin width; that is, the size of the interval spanned by the bar. Smaller bin widths will produce a more-refined version of the histogram, while larger bin widths will produce a less-refined histogram. The choice of bin width can be particularly important when there are subtle characteristics in the data distribution. The left and right plots of Fig. 4.1 show a histogram with small and large bin widths, respectively. The data are centered around zero and they appear to be normally distributed. Also included in these plots is a smoothed estimate of the histogram, shown as a black line. This smoothed estimate was created using Kernel Density Estimation (KDE), which is a nonparametric approach to estimating a distribution of data. KDE can be performed in most statistical software packages; for more information, see Silverman (1986).

Boxplots Boxplots provide a simple way to visualize the central tendency and spread of a dataset. Four characteristics can be assessed from a boxplot:

- The location of the median.
- The spread of the data.
- Whether the distribution of the data is skewed or symmetric.
- Whether there are potential outliers.

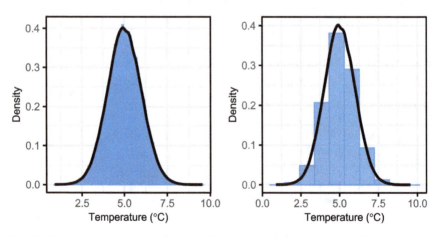

Fig. 4.1 Histograms with varying bin-widths. The left plot shows a smaller bin width (and a more-refined histogram), while the right plot shows a larger bin width (and a less-refined histogram)

4.3 Exploratory Data Analysis

Figure 4.2 shows an example boxplot. Here, the center line represents the median (Q2) of the sample, i.e., the point where 50% of the data fall above and 50% of the data fall below. The bottom and top edges of the box are the 1st and 3rd quartiles (Q1 and Q3), respectively, which are the 25th and 75th percentiles of the data. The difference between the 1st and 3rd quartiles is called the interquartile range. The "whiskers" (i.e., the lines extending from the box) show the location of the effective minimum and maximum of the data. Finally, an outlier is shown above the effective maximum. Boxplot points are generally considered outliers if they fall more than 1.5 times the interquartile range below or above the 1st or 3rd quartiles, respectively. Note that this is simply a rule-of-thumb for identifying outliers and expert judgement should always be used to determine if a point is truly an outlier.

Scatterplots Scatterplots plot individual values of data versus an independent variable. For example, if measurements are taken on multiple units, it may be useful to plot the data by unit to see if there is significant unit-to-unit variability. Additionally, if the time of measurement is provided with the data, it can be helpful to plot data across time to see if there are any visible trends. Figure 4.3 provides two examples of scatterplots; the top plot shows data with no discernable trend, while the bottom plot shows data with variation that decreases over time. This change in variation may be an indication that there are two subpopulations present in the data that need to be analyzed separately. Plots like these can be particularly helpful in identifying relationships and trends in data.

Correlation Plots Correlation plots are a simple way to assess the magnitude of the correlation (linear dependence) between two variables. The plots usually include the computed Pearson correlation coefficient r_{xy}.

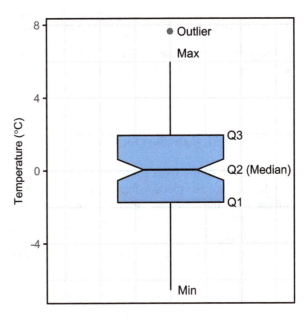

Fig. 4.2 Example boxplot

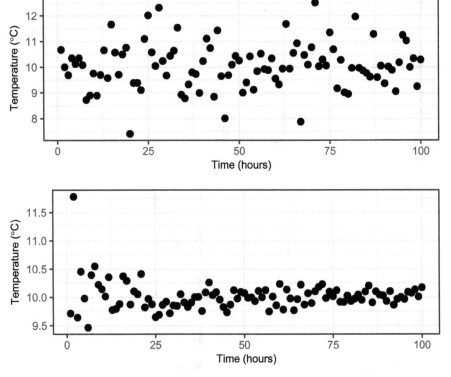

Fig. 4.3 Example of scatterplots of temperature data across time. The top plot shows no observable trend, while the bottom plot shows data with decreasing variance over time

The graphs below (each plotting a y variable versus an x variable) show examples of both low correlation (r_{xy} near zero) and high correlation ($|r_{xy}|$ near 1) (Fig. 4.4).

This component of the EDA can be used to help determine what terms should be included in the expression for the total variance (Eq. (7.25)) in an uncertainty analysis.

4.3.2.2 Graphical Displays for Discrete Data

For discrete data, the most commonly used graphical display is a frequency plot (also known as a bar plot). A frequency plot is similar to a histogram; the primary difference is that bars are given for each category or level of the discrete variable. The height of the bar is equal to the number of data points that fall into that category. Figure 4.5 gives a frequency plot with the discrete variable (temperature) on the x-axis and the count on the y-axis.

4.3 Exploratory Data Analysis

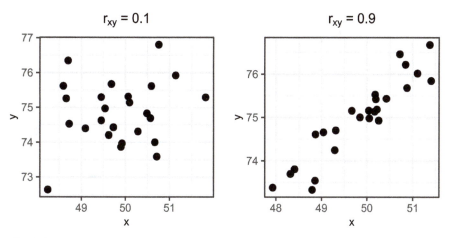

Fig. 4.4 Example correlation plots showing low (left) and high (right) correlations between variables

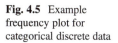

Fig. 4.5 Example frequency plot for categorical discrete data

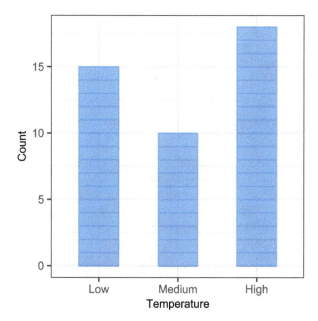

A frequency plot can also be used to explore the shape of a discrete probability distribution involving count data or other discrete ordinal data. Figure 4.6 shows the shape of a discrete frequency distribution based on particle count data. The shape of the distribution can be used, as with the histogram of continuous variables, to suggest a probability distribution for the particle count data.

The previous sections discussed how to calculate summary statistics and generate graphical displays of a data sample. The next section provides guidance on choosing and fitting a probability distribution that represents the population from which the

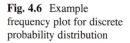

Fig. 4.6 Example frequency plot for discrete probability distribution

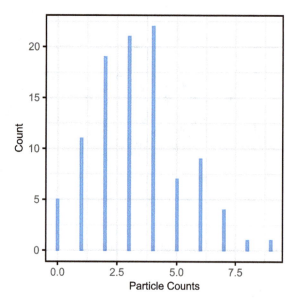

data were sampled. This distribution predicts future events based on the parameters estimated for the population.

4.4 Probability Distributions

Probability distributions are used in metrology to express the uncertainty in both direct measurements and in the input variables of an indirect measurement equation (Chap. 6). These input variables can be considered random variables, defined in ISO 3534-1 1993 as "a variable that may take any of the values of a specified set of values and with which is associated a probability distribution." A probability distribution is the function that assigns probabilities to the various outcomes of a random variable. Probability distributions are defined differently for discrete and continuous random variables. For a continuous random variable, a Probability Density Function (PDF) is used to assign probabilities to intervals of potential outcomes, while for a discrete random variable, a Probability Mass Function (PMF) assigns probabilities to potential outcomes that are exact values (Casella and Berger 2002).

The characteristics of a probability distribution are determined by its parameters. Most distributions have between one and three parameters. For example, the location and spread of a normal distribution are determined by its two parameters: μ, the mean, and σ^2, the variance. For a normal distribution, these parameters can be estimated directly from the data using the sample mean and the sample variance (as defined above). For other distributions, parameters can be estimated using a variety of techniques, as discussed in Sect. 4.4.2.

The three key steps of fitting a probability distribution are:

4.4 Probability Distributions

1. Determine a set of potential distributions,
2. Estimate the parameters for the potential distributions, and
3. Assess the distributional fits.

The following sections describe each of these steps in greater detail.

4.4.1 Identification of Probability Distributions

There are several factors to consider when identifying a probability distribution, including:

- *Type of data*: As discussed above, different distributional forms should be considered depending on whether the data are discrete or continuous.
- *Amount of available data*: When limited data are available, additional information (such as historical data, manufacturer's specifications, or expert opinion) may be required before choosing an appropriate distribution. If a large amount of data is available, the results from the EDA (histogram, probability plots) can help inform which distribution might be appropriate.
- *Information about the location, spread, and shape of the distribution*: Using EDA and/or expert judgement to determine whether the data are bounded or if certain values are more probable than others can help inform the choice of distribution.
- *Accuracy required for the distribution*: Distributions for variables that greatly affect the analysis should be chosen with great care. In many metrology applications, accuracy in the tails of the distribution is most important.

Continuous and discrete distributions that are commonly used in metrology are given below.

4.4.1.1 Continuous Distributions

Normal Distribution The normal distribution is an unbounded (in its argument x), symmetric, bell-shaped distribution that is regularly used in metrology. The PDF of the normal (also known as Gaussian) distribution is defined as

$$f(x) = \frac{1}{\sigma\sqrt{2\pi}} \exp\left[-\frac{1}{2}\left(\frac{x-\mu}{\sigma}\right)^2\right] \quad -\infty < x < \infty, \quad (4.5)$$

where μ is the mean and σ is the standard deviation. Figure 4.7 provides a visual depiction of a normal distribution with $\mu = 0$ and $\sigma = 1$. A normal distribution with these parameter values is called a standard normal distribution.

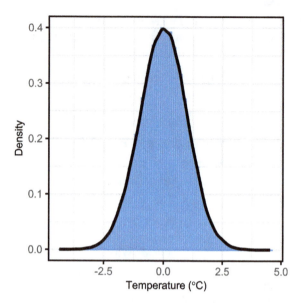

Fig. 4.7 Example of a normal distribution with mean 0 and standard deviation 1. The bars show a histogram of the data, while the black line is a KDE fit

Many statistical methods in metrology have an underlying assumption of normality, but when performing an uncertainty analysis, distributional assumptions such as this must be checked to validate the specific analysis. More information on choosing probability distributions and assessing distributional fits can be found below.

Rectangular Distribution The rectangular distribution, also known as the uniform distribution, is used when the data are known to fall within an interval, but there is no specific knowledge about the probability of possible values within that interval. In other words, data are bounded, but all values are assumed to be equally probable within those bounds. This distribution is often used in metrology when measures of uncertainty are given in the form of a simple interval on a calibration report or a manufacturer's specification, and no additional detail is given. The PDF of the rectangular distribution, given a lower and upper bound of $[a, b]$, is given as

$$f(x) = \begin{cases} \dfrac{1}{(b-a)}, & b \leq x \leq a \\ 0, & \text{otherwise.} \end{cases} \quad (4.6)$$

With these bounds, the mean is $\frac{a+b}{2}$ and the standard deviation is $\frac{(b-a)}{2\sqrt{3}}$. Figure 4.8 shows a histogram of data with a fitted rectangular distribution (shown in black) that is bounded on the interval $[-3, 3]$. Note that the distribution is not perfectly rectangular because the data were simulated to produce an estimate of the rectangular distribution.

Triangular Distribution The triangular distribution differs from the rectangular distribution in that some knowledge may exist as to which values are more probable within its bounds. It is a bounded, symmetric distribution that can be used when

4.4 Probability Distributions

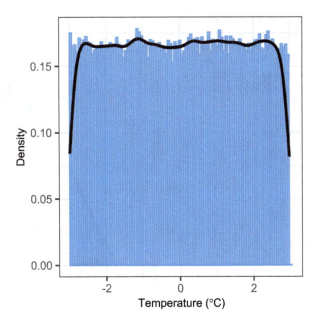

Fig. 4.8 Example of a uniform distribution bounded on the interval [−3, 3]. The bars show a histogram of the data, while the black line is a KDE fit

values are more likely to fall somewhere near the center of the distribution. The PDF for a triangular distribution bounded on the interval $[-a, a]$ is given as

$$f(x) = \begin{cases} \dfrac{(x+a)}{a^2}, & -a \leq x \leq 0 \\ \dfrac{(a-x)}{a^2}, & 0 < x \leq a. \end{cases} \qquad (4.7)$$

This distribution has a mean of 0 and a standard deviation of $a/\sqrt{6}$. An example of a triangular distribution with parameter $a = 3$ is given in Fig. 4.9. In metrology, the triangular distribution is used less frequently than the normal and rectangular distributions.

t-distribution The t-distribution (also known as Student's t-distribution) is generally used when the sample size is small (e.g., less than 30) and the population standard deviation is unknown. The PDF of the t-distribution with parameter ν (known as the degrees of freedom) is

$$p(t, \nu) = \frac{1}{\sqrt{\pi\nu}} \frac{\Gamma\left(\frac{\nu+1}{2}\right)}{\Gamma\left(\frac{\nu}{2}\right)} \left(1 + \frac{t^2}{\nu}\right)^{-\frac{(\nu+1)}{2}} \qquad -\infty < t < \infty, \ \nu > 0. \qquad (4.8)$$

Here, Γ is the gamma function. The mean for the t-distribution is 0 and the standard deviation for $\nu > 2$ is defined as

Fig. 4.9 Example of a triangle distribution bounded on the interval [−3, 3]. The bars show a histogram of the data, while the black line is a KDE fit

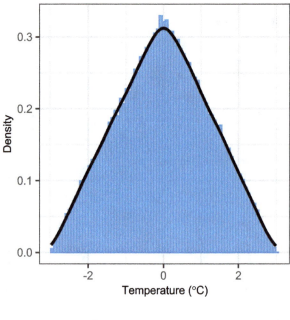

$$\left(\frac{\nu}{\nu-2}\right)^{\frac{1}{2}}. \tag{4.9}$$

In metrology, the degrees of freedom ν can be thought of as the number of independent data values used to estimate an uncertainty term. For example, the degrees of freedom associated with estimating the variance of a probability distribution using the sample variance (Eq. (4.2)) is $n - 1$. The t-distribution, with the appropriate number of degrees of freedom, is used to compute the expanded uncertainty of a measurand (Sect. 6.2.1).

An example of a t-distribution with $\nu = 3$ is given in Fig. 4.10.

For small samples, the t-distribution resembles a heavy-tailed version of the standard normal distribution, as shown in Fig. 4.11. In the left plot of this figure, the light red density represents the t-distribution with three degrees of freedom, while the light green density represents the standard normal distribution. When the degrees of freedom increase to 30, as shown in the right plot, the t-distribution converges to the standard normal distribution. This implies that for large samples, using a t-distribution is equivalent to using a standard normal distribution.

Comparison of distributions. Table 4.2 provides an overview of when each type of distribution may be most appropriate.

Figure 4.12 shows a visual comparison of the normal, rectangular, triangular, and t-distributions.

While there are many continuous distributions, the ones described above are the most commonly used in metrology. Most measurement data are continuous, but when the data are discrete, the most commonly used distribution is the Poisson distribution, described below.

4.4 Probability Distributions

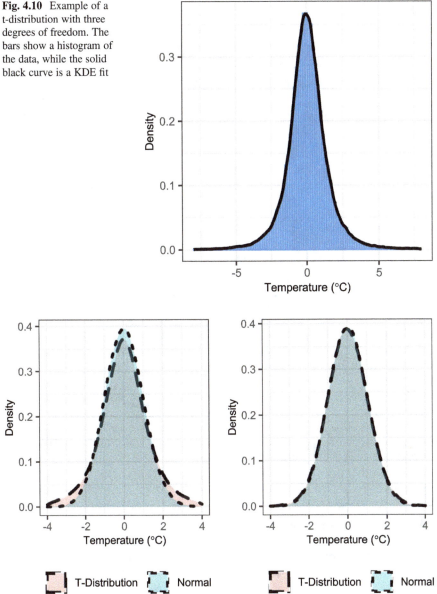

Fig. 4.10 Example of a t-distribution with three degrees of freedom. The bars show a histogram of the data, while the solid black curve is a KDE fit

Fig. 4.11 Comparison of a t-distribution with three (left) and 30 (right) degrees of freedom to a normal distribution

Table 4.2 Characteristics of probability distributions commonly used in metrology

Distribution	Bounded data	Knowledge of probability of values	When to use
Normal	No	Yes	Continuous variables such as voltage, temperature
Rectangular	Yes	No	Calibration interval without additional details regarding probability density
Triangular	Yes	Yes	Bounded interval with data more likely near the midpoint
t-Distribution ($n < 30$)	No	Yes	Computing expanded uncertainties

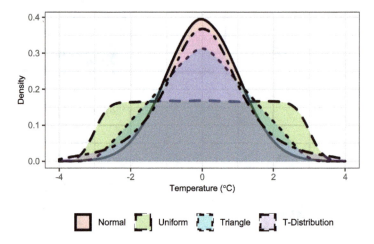

Fig. 4.12 Comparison of normal, uniform, triangular, and t-distributions. Here, the normal distribution has mean 0 and standard deviation 1, the uniform and triangular distributions are bounded on $[-3, 3]$, and the t-distribution has three degrees of freedom

4.4.1.2 Discrete Distributions

Poisson Distribution The Poisson distribution is often used when the data are counts of some event or process occurrence. For example, when counting the number of defects that occur per manufactured part, the Poisson distribution would likely be appropriate (Montgomery 2013). The count of neutrons produced by a laboratory neutron source, analyzed in Sect. 6.4.1, was also modelled using a Poisson distribution.

The Poisson distribution is defined as

4.4 Probability Distributions

$$p(x) = \frac{e^{-\lambda}\lambda^x}{x!}, \quad \text{with } x = 0, 1, 2, 3, \ldots \tag{4.10}$$

The parameter λ, known as the event rate, is both the mean and variance of the distribution. Figure 4.13 provides an example of the Poisson distribution with λ equal to 3. As the value of λ increases, the shape of the distribution becomes more symmetric, and can be approximated by a normal distribution in uncertainty analyses.

4.4.2 Estimating Distribution Parameters

EDA and expert judgement should be used to identify potential distributions. After identification of reasonable distributions for the data, parameters for those distributions are estimated. A commonly used method for estimating parameters of a distribution with moderate to large sample sizes is Maximum Likelihood Estimation (MLE). MLE chooses parameters that are most probable (likely) for a distribution based on the observed data. Many statistical software packages perform MLE. Note that there are many other ways to estimate parameters (particularly if the samples are small), such as the method of moments and Bayesian inference (See Sect. 11.6). For more information on parameter estimation, see Casella and Berger (2002).

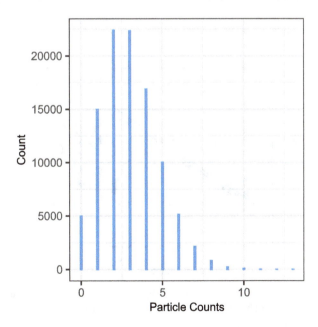

Fig. 4.13 Poisson distribution with mean and variance equal to 3

4.4.3 Assessing Distributional Fit

Goodness of fit diagnostics can be used to compare empirical distributions to different hypothesized distributions after parameters have been estimated. A common diagnostic tool is a probability plot. A probability plot is a visual tool used to assess whether data come from a particular probability distribution. A common probability plot is known as the QQ plot, where the empirical quantiles are plotted against the theoretical quantiles of the hypothesized probability distribution. If the data come from the hypothesized distribution, the points should be approximately linear. If the probability plot is markedly nonlinear, it is evidence that the hypothesized probability distribution is not a good fit.

Figure 4.14 shows how to use the QQ plot as a diagnostic tool. The left and right plots show simulated data generated from normal and Weibull distributions, respectively. The quantiles of the data are compared to the quantiles of a normal distribution. Confidence bounds are also given to show the maximum deviation from the line that should encompass most of the data. In this case, the normal-generated data falls within the confidence bounds, providing evidence that the data are normally distributed. Alternatively, the simulated Weibull data deviate from the fitted line and several points fall outside the confidence bounds, evidence that a normal distribution may not be appropriate.

In metrology applications, assessment of the distributional fit will often be most important in the tails of the distribution. For small sample sizes, it is often difficult to determine which distribution best fits the data, particularly in the tails. Caution should be exercised before choosing a distribution with limited data. Expert judgement or additional data may be needed to properly identify a distribution.

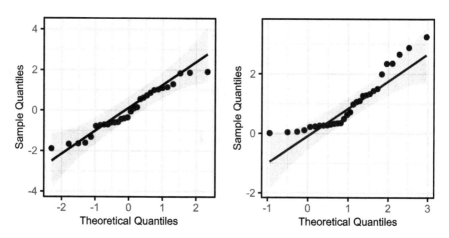

Fig. 4.14 Examples of normal QQ plots used to assess whether data are normally distributed. The left and right plots show data that were generated from normal and Weibull distributions, respectively

When the data are clearly non-normal, the standard uncertainty analysis (Chap. 6) will need to be modified. Monte Carlo (simulation) methods (Chap. 8) may be more appropriate in this case.

4.5 Related Reading

Annex C of the JCGM 100 (2008) (GUM) provides definitions for basic statistical concepts and terms. It gives formal mathematical definitions for some of these concepts along with brief comments on how they might be used in metrology applications. The definitions in the GUM are taken from ISO 3534-1 1993. This ISO provides additional probability and general statistical definitions that are not included in the GUM, including a list of symbols and abbreviations.

The JCGM 101 (2008) (GUM-S1) expands on probability density functions (PDFs) in metrology, including a brief description of how Bayesian methods can be used to estimate a PDF (see Sect. 11.6). The GUM-S1 discusses the principle of maximum entropy for choosing a PDF among a candidate of possible PDFs. Table 1 on p. 20 provides high level guidance on choosing the appropriate PDF based on the type of available information. It includes less-common PDFs such as the curvilinear trapezoid, trapezoidal, and arc sine distributions. It also includes mention of the multivariate Gaussian distribution. The multivariate t-distribution is discussed in the JCGM 102 (2011) (GUM-S2), along with a brief mention of methods for the construction of multivariate PDFs.

Heumann et al. (2016) introduces statistics at an undergraduate level. It provides additional discussion on types of data, methods for analyzing variables using graphs, and summary statistics. It also includes methods for multivariate data including joint and marginal distributions and graphical representations of bi-variate data. Later it discusses elements of probability theory, random variables, and PDFs, with applications of these methods in R.

Additional publications provide an overview of statistical methods in metrology. Chapter 2 of Vardeman and Jobe (2016) begins with basic concepts of statistics along with an introduction to probability modeling. The authors then discuss how one-sample and two-sample statistical methods can be applied to measurement data. The remainder of the chapter briefly discusses several of the concepts covered later in this book, including linear regression (Chap. 9) and Gauge R&R studies (Chap. 10). The National Bureau of Standard's Special Publication 747 (Ku 1988) offers an overview of statistical concepts in metrology along with a postscript on statistical graphics used to check models and their assumptions (e.g. residual plots).

Willink (2006) provides a discussion of principles for probability and statistics in metrology that aid in good decision making. These principles are given with the intent of helping practitioners of metrology produce a statistically defensible analysis. This paper provides high-level considerations for both uncertainty analysis and interlaboratory comparisons.

4.6 Exercises

The data in Table 4.3 provide 15 measurements of temperature, pressure, and number of defects. It will be used for Exercises 4.1–4.8 below.

4.1 Categorize the temperature, pressure, and defects data as continuous or discrete. Further categorize any discrete variables as binary, count or categorical.

4.2 Calculate the following summary statistics for the sample of temperature and pressure data—mean, median, mode, variance, standard deviation.

4.3 What is the standard error of the mean of the pressure data? How can the standard error be interpreted?

4.4 What graphical display(s) are most appropriate for continuous data? For categorical data?

4.5 Generate a histogram of the temperature data. What characteristics do you notice about the spread of these data?

4.6 Generate a boxplot for the pressure data. Are there any outliers present according to the boxplot? If so, what should be done about these outliers?

4.7 Plot the defects data. What distribution would be most appropriate for this type of data?

4.8 Generate a scatterplot for pressure and temperature where pressure is the dependent variable. What do you notice? Now calculate the correlation coefficient for pressure and temperature. How can the correlation coefficient be interpreted?

4.9 Provide an example of when a uniform distribution might be more appropriate than a normal distribution.

4.10 Choose the appropriate probability distribution(s) for the following cases.

a. A manufacturer's specification states that the radius of a cylinder is expected to fall between $\pm 10\%$ of the nominal value.

Table 4.3 Data for Exercises 4.1–4.8

Measurement	Temperature (°F)	Pressure (mmHg)	Defects
1	64.69	645.22	2
2	73.67	728.04	3
3	71.61	717.41	1
4	68.26	678.02	5
5	76.42	761.36	4
6	64.93	644.97	0
7	67.45	678.63	2
8	69.16	687.6	2
9	68.49	689.78	4
10	80.13	814.8	1
11	74.5	744.32	0
12	70.68	706.27	0
13	68.99	683.18	1
14	65.97	661.92	3
15	73.22	737.01	2

Table 4.4 Data for Exercise 4.11

Measurement	Length (m)	Measurement	Length (m)	Measurement	Length (m)
1	1.118	13	0.861	25	1.036
2	1.262	14	1.055	26	1.618
3	1.118	15	0.834	27	1.187
4	0.480	16	1.106	28	0.884
5	1.014	17	0.912	29	0.858
6	0.955	18	0.773	30	1.269
7	0.978	19	0.958	31	0.811
8	0.972	20	1.152	32	0.811
9	1.190	21	1.324	33	0.636
10	1.055	22	2.192	34	0.710
11	0.886	23	0.779	35	0.675
12	1.071	24	0.692	36	0.830

b. The angle of a weld is known to be between 30° and 60°, though angles near 45° are most likely.

c. 50 measurements are taken of the internal pressure of a tank. A plot of the data shows that these measurements are relatively symmetric around the mean value of 50 MPa.

4.11 The data in Table 4.4 show 36 length measurements of a steel plate. Use these data and the statistical software of your choice to complete the following:

    a. Calculate the sample mean and standard deviation.

    b. Plot a histogram of the data.

    c. Use the mean and standard deviation calculated in (a) to fit a normal distribution to the data. Overlay a curve of the normal distribution on the plotted histogram in (b).

    d. Generate a QQ plot that compares the data in Table 4.4 to the quantiles of the normal distribution. Does the normal distribution seem appropriate in this case? Why or why not?

References

Casella, G., Berger, R.L.: Statistical Inference, 2nd edn. Duxbury, Pacific Grove, CA (2002)

Heumann, C., Schomaker, M., Shalabh, A.: Introduction to Statistics and Data Analysis. Springer, New York (2016)

ISO 3534-1: Statistics vocabulary and symbols – Part 1: Probability and general statistical terms (1993)

JCGM 100: E valuation of measurement data - Guide to the expression of uncertainty in measurement (2008)

JCGM 101 Evaluation of measurement data - Supplement 1 to the guide to the expression of uncertainty in measurement - Propagation of distributions using a Monte Carlo method (2008)

JCGM 102 Evaluation of measurement data - Supplement 2 to the -guide to the expression of uncertainty in measurement - extension to any number of output quantities (2011)

Ku, H.: NBS Special Publication 747. National Bureau of Standards, Gaithersburg (1988)

Montgomery, D.: Introduction to Statistical Quality Control. John Wiley & Sons, New York (2013)

Silverman, B.: Density Estimation for Statistics and Data Analysis. Chapman & Hall, Boca Raton, FL (1986)

Vardeman, S., Jobe, J.: Statistical Methods for Quality Assurance. Springer-Verlag, New York (2016)

Willink, R.: On using the Monte Carlo method to calculate uncertainty intervals. Metrologia. **43**(6), 39–42 (2006)

Chapter 5
Measurement Uncertainty in Decision Making

5.1 Introduction

Measurements are used to make important decisions ranging from accepting or rejecting product (Smith 1990) to deciding on R&D proposals (Cividino et al. 2019) to assessing the outcome of medical diagnostics (Bhise et al. 2018). However, because all measurements have an associated uncertainty, there is inherent risk that incorrect decisions will be made. Decisions regarding the air pressure of footballs (Sect. 1.4) may be of minor consequence, but decisions based on the pandemic fatality rate (Sect. 1.5) can have profound consequences. In cases like these, failure to address measurement uncertainty leaves the risk of incorrect decisions unknown and unmanageable.

This chapter provides a statistical framework for using measurement uncertainty to quantify risk in measurement related decisions. Test uncertainty ratios, measurement decisions, false accept and false reject probabilities, guardbanding, and risk associated with biased measurements are covered. These topics will first be presented in the context of a manufacturing environment in which the decision is to accept or reject manufactured product. The same topics will then be discussed in the context of a calibration lab in which a decision must be made regarding the state of a unit under test (UUT).

5.2 Measurement Uncertainty and Risk

In metrology, risk is defined as the probability that a measurement result will lead to an incorrect decision. Since the actual value of a measurand can never be known, there is always some probability that a measured value falls within an acceptable region when the actual value lies outside that region. This results in a *False Accept* (FA) condition. There is also some probability that a measured value falls outside an acceptable region when the actual value lies inside that region. This results in a *False Reject* (FR) condition. National standard ANSI/NCSLI Z540.3 2006 (see Sect. 3.7.4) recommends that compliance testing should control the false accept risk to be no more than 2%.

False accept and false reject risks are only one part of a full risk assessment. Another factor in the assessment, not considered in detail here, is the *consequence* of making an incorrect decision. These consequences may include the cost of rejecting a good part (inflating rework and scrap costs) or the cost of accepting a bad part (inflating future costs and risks). Depending on the consequences of an incorrect decision, false accept and false reject risks may need to be controlled at less than the typical 2%. However, controlling these risks at lower levels comes with its own cost, as more sampling, testing, and scrapping of marginal parts may be required.

5.2.1 Measurement Uncertainty and Risk in Manufacturing

In a manufacturing setting, a measurement is performed on each part (or a suitable sample of parts) at product acceptance to determine whether it meets requirements. Ideally, all parts that meet specifications are accepted and all parts that do not meet specifications are rejected. In the presence of measurement uncertainty, this ideal outcome is not possible. In the sections that follow, the focus is on the quantification and management of risks associated with uncertainty in the measurement of manufactured parts.

5.2.1.1 Test Uncertainty Ratio

An accept or reject decision is usually based on whether a part's measurement falls within its specification limits (SL). For example, if the voltage of a battery must fall within 10.0 V $\pm$ 1.0 V for its intended application, then the specification limit is $SL = 1.0$ V. Absolute limits may also be used: in this case the lower limit is $SL_L = 9.0$ V and the upper limit is $SL_U = 11.0$. Most equations in this chapter use absolute limits to avoid ambiguity. The JCGM (2012) uses the term "tolerance limits" in place of "specification limits," and defines the tolerance interval to be the range of acceptable values between the limits. Tolerance limits and specification limits are used interchangeably in this chapter.

Historically, the Test Uncertainty Ratio (TUR) has been used as an indicator of a measurement's suitability to make an accept or reject decision. The TUR is defined as TUR = [$\pm$Specification Limit]/[$\pm$Measurement Uncertainty at 95% confidence]. For example, if a part's length must be within $\pm$ 0.1 inch, and the caliper used to measure this length has total uncertainty of ± 0.005 inch, the TUR is (0.1 inch/ 0.005 inch) = 20. When the specification limits are defined in terms of absolute limits, the TUR formula can be expressed as

$$\text{TUR} = \frac{SL_U - SL_L}{2U_c}, \tag{5.1}$$

where U_c is the expanded uncertainty (see Sect. 6.2.5), evaluated with a 95% confidence level.

Requiring a TUR of at least 4 (referred to as "meeting a 4:1 ratio") is often used to ensure the adequacy of a measurement to make a particular accept/reject decision, although sometimes a TUR of at least 10 is required for more conservative work. There are, however, important assumptions associated with using TUR as a metric and the requirement of a TUR of 4 or 10. Using a TUR assumes that all measurement biases have been removed from the measurement process and the measurements involved follow a normal distribution. If there are significant biases that cannot be removed, the TUR will not account for the increased risk (see Sect. 5.2.1.5). The TUR was originally introduced in the 1950s as a shorthand way to evaluate a measurement process without a full assessment of false accept and false reject probabilities (Harben and Reese 2012). With modern computers now readily available to evaluate these probabilities, the use of TUR alone becomes questionable. As a result, computing the actual risk is the recommended way to ensure false accept and false reject rates are acceptable. All of the risk computations presented in this chapter are supported by the uncertainty calculator that is available at https://sandiapsl.github.io/.

The TUR is still commonly used to determine if a measurement is adequate or if guardbanding (Sect. 5.2.1.4) is necessary to reduce the risk of false accepts. Within the NSE production agencies, whenever the TUR ratio is greater than or equal to 4:1, a measurement result can be directly compared to the specification limits for purposes of accept or reject. Otherwise, guardbanding is required (see Sect. 5.2.1.4).

Occasionally, the Test Accuracy Ratio (TAR) is used in place of the TUR. The TAR is similar to the TUR in its structure, but differs in that its denominator includes the uncertainty of only the test equipment, rather than including the additional Type A uncertainties, environmental effects, and any other known uncertainties in the measurement. The TAR can be expressed as:

$$\text{TAR} = \frac{\pm \text{Specification Limit}}{\pm \text{Accuracy of Test Equipment}}. \tag{5.2}$$

The TUR as an indicator of measurement capability is recommended as it includes all known uncertainties in the measurement. The TAR, while simpler to determine, could significantly understate the total uncertainty and associated risks.

5.2.1.2 Measurement Decisions

Any measurement performed on a manufactured part will have uncertainty, and any decision based on that measurement has inherent risk that should be considered. Figure 5.1 illustrates this concept, with y representing the measured

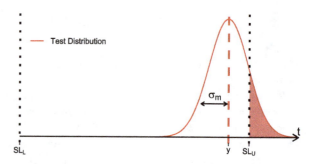

Fig. 5.1 Measurement uncertainty distribution. Dotted lines are shown at the specification limits. The dashed line represents the measurement result, y value and (SL_L, SL_U) representing the specification limits. In this particular case, the measured value falls within the specification interval, however, when the measurement uncertainty σ_m is considered, there is some probability that the true value falls outside the limits. That probability is represented by the shaded region in the figure. When the measured value is close to the specification limit SL_U, there is some non-negligible probability that the true value falls outside the limit.

The measurement uncertainty is commonly assumed to be normally distributed with mean zero (if unbiased) and standard deviation σ_m. If the measured value is y and the actual value is t, then y given t has a $N(t, \sigma_m^2)$ distribution, and t given y has a $N(y, \sigma_m^2)$ distribution.

The probability of a false accept decision *given* the measurement y falls inside the specification limits is found by integrating the conditional probability density of t given y outside the specification limits. From Fig. 5.1, the probability of t being less than SL_L is considered negligible, so the integration is only over the region where t is greater than SL_U.

In this case the probability of a false accept (shaded area), given measurement y, is calculated by evaluating the integral

$$Pr(t > SL_U \mid y) = \int_{SL_U}^{+\infty} \frac{1}{\sigma_m \sqrt{2\pi}} e^{-\frac{(t-y)^2}{2\sigma_m^2}} dt. \quad (5.3)$$

To illustrate this concept more clearly, consider a voltage measurement on a sample of batteries. The results are displayed in Fig. 5.2, with specification limits drawn at 9.0 and 11.0 V. The uncertainty associated with the test measurements is ±0.25 V (95% confidence). Sample numbers 1, 3, 4, and 5 all represent an accept decision with negligible risk of an incorrect decision. For sample number 7, the measurement value would lead to a reject decision with negligible risk of an incorrect decision. Sample number 2 presents a considerable amount of risk that an accept decision could be incorrect. Since there is a non-negligible risk that the actual value is beyond the specification limit of 11 V, caution should be exercised before accepting based on this measurement. Typical practices dictate that this product be accepted as meeting specifications, but the risk and consequences of accepting a battery outside of specifications should be taken into account. Sample number 6 results in a situation where the accept or reject decision based on this measurement has a 50% probability of being incorrect.

5.2 Measurement Uncertainty and Risk

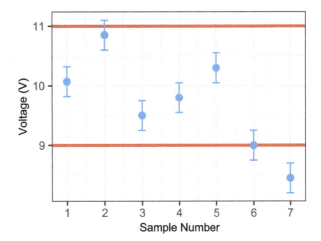

Fig. 5.2 Voltage measurement results in decision making. The horizontal lines represent the specification limits

While a single measurement with up to 50% probability of an incorrect decision may seem troubling, the probability that a part will exhibit a value close to the specification limit should be relatively small. The manufactured parts will themselves have a distribution of actual values described by a probability density function called the product (or process) distribution. It is commonly assumed that these values have a $N\left(\mu_p, \sigma_p^2\right)$ distribution, where μ_p represents the mean and σ_p the standard deviation of the distribution we will call the *product* distribution.

For instance, batteries may have a nominal value of 10.0 V, but due to variations in manufacturing and operating conditions, the actual voltages of all manufactured batteries may follow a normal distribution with mean 10.0 V and standard deviation 0.5 V.

The unconditional probability that any part is out of tolerance (OOT) can be determined by evaluating the integral

$$Pr(\text{OOT}) = 1 - \frac{1}{\sqrt{2\pi}\sigma_p} \int_{SL_L}^{SL_U} e^{-\frac{(t-\mu_p)^2}{2\sigma_p^2}} dt. \tag{5.4}$$

A graphical illustration of this integration is given in Fig. 5.3, with $Pr(\text{OOT})$ represented by the sum of the shaded areas.

The product distribution parameters μ_p and σ_p are typically estimated using historical production data. In-tolerance probability (*itp*) is sometimes used as a metric of process quality, describing the probability that any part is manufactured within specifications, and is simply $1 - Pr(\text{OOT})$.

Fig. 5.3 Product distribution with specification limits. Dotted lines are shown at the specification limits

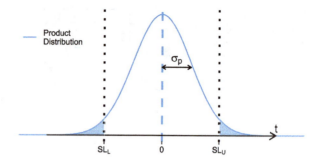

5.2.1.3 False Accept and False Reject Risks

The overall false accept and false reject risks associated with a manufacturing process can be determined by jointly considering the product and measurement uncertainty distributions (Smith 1990). The unconditional probability of false accept (*PFA*), sometimes called consumer's risk, is the probability that the measurement y is in tolerance ($SL_L < y < SL_U$) when the actual value t is out of tolerance ($t < SL_L$ or $t > SL_U$). This probability can be expressed as:

$$PFA = Pr(SL_L \leq y \leq SL_U \text{ and } t < SL_L) \\ + Pr(SL_L \leq y \leq SL_U \text{ and } t > SL_U). \tag{5.6}$$

Assuming normal distributions, this probability can be calculated as:

$$PFA = \int_{-\infty}^{SL_L} \left(\int_{SL_L}^{SL_U} \frac{1}{\sigma_m \sqrt{2\pi}} e^{-\frac{1}{2\sigma_m^2}(y-t)^2} dy \right) \frac{1}{\sigma_p \sqrt{2\pi}} e^{-\frac{1}{2\sigma_p^2}(t-\mu_p)^2} dt + \\ \int_{SL_U}^{\infty} \left(\int_{SL_L}^{SL_U} \frac{1}{\sigma_m \sqrt{2\pi}} e^{-\frac{1}{2\sigma_m^2}(y-t)^2} dy \right) \frac{1}{\sigma_p \sqrt{2\pi}} e^{-\frac{1}{2\sigma_p^2}(t-\mu_p)^2} dt. \tag{5.7}$$

In this equation, the inner integral can be thought of as the conditional probability of accepting the part given a particular value of t. The outer integral averages over all values of t that result in the false accept condition.

Rather than requiring specific knowledge of σ_m, σ_p, and (SL_L, SL_U), it can be useful to express the *PFA* (Eq. (5.7)) in terms of TUR and *itp* using Eq. (5.1) and an estimate of σ_p based on *itp* and (SL_L, SL_U) (see Eq. (5.20)).

5.2 Measurement Uncertainty and Risk

Figure 5.4 shows the *PFA* as a function of in-tolerance probability for various TURs. This plot justifies the use of 4:1 TUR as a benchmark for measurement quality. With process distributions exceeding 80% *itp*, a TUR of 4:1 guarantees the *PFA* remains less than 2%, meeting the Z540.3 requirement.

Alternatively, the unconditional probability of false reject (PFR), also called producer's risk, is the probability that the measurement y is out of tolerance ($y < SL_L$ or $y > SL_U$) when the actual value t is in tolerance ($SL_L < t < SL_U$). This probability can be expressed as:

$$\text{PFR} = Pr(y < SL_L \text{ and } SL_L < t < SL_U) \\ + Pr(y > SL_U \text{ and } SL_L < t < SL_U). \quad (5.8)$$

In terms of normal distributions:

$$PFR = \int_{SL_L}^{SL_U} \left(\int_{-\infty}^{SL_L} \frac{1}{\sigma_m \sqrt{2\pi}} e^{-\frac{1}{2\sigma_m^2}(y-t)^2} dy \right) \frac{1}{\sigma_p \sqrt{2\pi}} e^{-\frac{1}{2\sigma_p^2}(t-\mu_p)^2} dt \\ + \int_{SL_L}^{SL_U} \left(\int_{SL_U}^{+\infty} \frac{1}{\sigma_m \sqrt{2\pi}} e^{-\frac{1}{2\sigma_m^2}(y-t)^2} dy \right) \frac{1}{\sigma_p \sqrt{2\pi}} e^{-\frac{1}{2\sigma_p^2}(t-\mu_p)^2} dt. \quad (5.9)$$

Figure 5.5 shows the PFR as a function of *itp* and TUR. A TUR of 4:1 guarantees the PFR will remain below 3% for any *itp*.

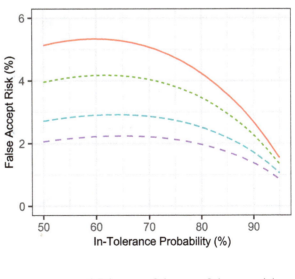

Fig. 5.4 False Accept Risk curve against in-tolerance probability. The curves represent various TURs

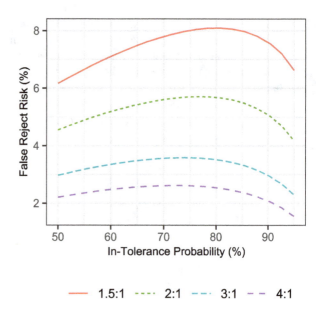

Fig. 5.5 False reject risk curve against in-tolerance probability. The curves represent various TURs

5.2.1.4 Guardbanding

Guardbanding modifies the specification limits to reduce the level of risk (*PFA*) associated with measurement uncertainty. The original specification limit, *SL*, is reduced by the guardband factor (GBF):

$$AL = SL \cdot \text{GBF} \quad (5.10)$$

to result in new *acceptance* limits *AL*. GBF is a multiplier between 0 and 1, with a value of 1 meaning no guardband is applied. In terms of absolute specification limits, absolute acceptance limits become

$$AL_L = SL_L + (1 - \text{GBF}) \frac{SL_U - SL_L}{2} \quad (5.11)$$

$$AL_U = SL_U - (1 - \text{GBF}) \frac{SL_U - SL_L}{2}. \quad (5.12)$$

The most popular method for calculating GBF is the root-sum-of-squares (RSS) method. The GBF for this method is

$$\text{GBF} = \sqrt{1 - \frac{1}{\text{TUR}^2}}. \quad (5.13)$$

5.2 Measurement Uncertainty and Risk

As an example, consider a product requirement of 60.0 µV ± 6.0 µV. The acceptance range for this requirement is 54.0–66.0 µV (the center region of Fig. 5.6).

The measurement process used to verify this measurement requirement has an uncertainty of ±2.0 µV, resulting in a TUR of 6 µV/2 µV = 3. Using Eq. 5.12, a guardband factor of 0.943 is calculated and used to establish the new acceptance limits: AL = ±6.0 µV × 0.943 = ±5.7 µV, illustrated in Fig. 5.7. The new acceptance range is 60.0 ± 5.7 µV = 54.3–65.7 µV. A measured value in the intermediate regions (54.0–54.3 and 65.7–66.0), while technically appearing within product requirements, would be rejected due to the high level of risk that the true value falls outside the limits.

Although the guardbanded acceptance limits reduce the PFA involved with the measurement decision, narrower limits also result in more scrapped product. The benefits of reducing the PFA would be weighed against the cost of more scrapped product. Applying the guardband, the PFA and PFR calculations become

$$\text{PFA} = \int_{-\infty}^{SL_L} \left(\int_{AL_L}^{AL_U} \frac{1}{\sigma_m\sqrt{2\pi}} e^{-\frac{1}{2\sigma_m^2}(y-t)^2} dy \right) \frac{1}{\sigma_p\sqrt{2\pi}} e^{-\frac{1}{2\sigma_p^2}(t-\mu_p)^2} dt$$
$$+ \int_{SL_U}^{+\infty} \left(\int_{AL_L}^{AL_U} \frac{1}{\sigma_m\sqrt{2\pi}} e^{-\frac{1}{2\sigma_m^2}(y-t)^2} dy \right) \frac{1}{\sigma_p\sqrt{2\pi}} e^{-\frac{1}{2\sigma_p^2}(t-\mu_p)^2} dt \tag{5.14}$$

$$\text{PFR} = \int_{SL_L}^{SL_U} \left(\int_{-\infty}^{AL_L} \frac{1}{\sigma_m\sqrt{2\pi}} e^{-\frac{1}{2\sigma_m^2}(y-t)^2} dy \right) \frac{1}{\sigma_p\sqrt{2\pi}} e^{-\frac{1}{2\sigma_p^2}(t-\mu_p)^2} dt$$
$$+ \int_{SL_L}^{SL_U} \left(\int_{AL_U}^{+\infty} \frac{1}{\sigma_m\sqrt{2\pi}} e^{-\frac{1}{2\sigma_m^2}(y-t)^2} dy \right) \frac{1}{\sigma_p\sqrt{2\pi}} e^{-\frac{1}{2\sigma_p^2}(t-\mu_p)^2} dt. \tag{5.15}$$

Plotting the PFA and PFR curves with guardbanding results in Fig. 5.8. While guardbanding decreases the PFA to below 2% for all cases, it also increases the PFR.

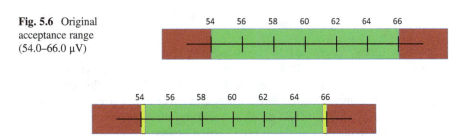

Fig. 5.6 Original acceptance range (54.0–66.0 µV)

Fig. 5.7 Guardbanded acceptance range (54.3–65.7 µV)

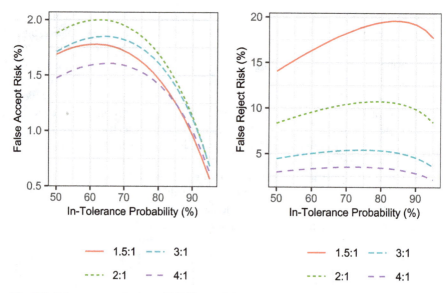

Fig. 5.8 False accept and reject with RSS guardbanding. The curves represent various TURs

Applying too much guardband results in rejecting many good parts. Guardbanding is often used in practice because the consequence of accepting a bad part is typically more serious than the consequence of rejecting a good part. Even with guardbanding, a TUR of at least 1.5 is recommended. Although guardbanding can reduce PFA to below 2%, a PFR exceeding 20% with a TUR of less than 1.5 is deemed unacceptable.

The RSS guardbanding method (Eq. (5.13)) has found favor because it offers acceptable risks for a wide range of TURs as well as the ease of calculating GBF. Other methods for calculating an appropriate GBF for specific measurement problems have been proposed and are summarized in Deaver (1994).

One such method is based on Dobbert (2008). The GBF for this method is

$$\text{GBF} = 1 - \frac{M}{\text{TUR}}, \quad (5.16)$$

where

$$M = 1.04 - e^{0.38 \ln (\text{TUR}) - 0.54}. \quad (5.17)$$

Dobbert (2008) developed this guardbanding equation first by finding the maximum PFA for a given TUR as a function of *itp*, then finding the M value at each maximum point that results in 2% PFA. After performing this calculation at various TUR values, Dobbert found the best fit curve to the M vs. TUR values, resulting in Eq. (5.17). This process could be re-evaluated with different maximum PFA to obtain a modified Eq. (5.16), if desired.

5.2 Measurement Uncertainty and Risk

Figure 5.9 shows the value of the GBF as a function of TUR for the RSS and Dobbert guardbanding methodologies. Note that the GBF for the RSS method approaches a value of one (no guardbanding) as the TUR increases. Meanwhile, the GBF for the Dobbert method surpasses a value of one when the TUR is around 4.6. Thus, when using the Dobbert guardbanding method beyond a TUR of 4.6:1, the acceptance limits are set beyond the specification limits, which is not practically feasible. Another proposed method for calculating guardband factors involves minimizing the total cost (economic or otherwise) of false accept and false reject decisions (Easterling et al. 1991).

If a specification limit is one-sided, guardbanding can still be used and risk calculations can still be performed. Examples of one-sided requirements include a connector whose resistance must be less than 1 Ω or a battery whose voltage must be greater than 10 V. When working with one-sided limits, TUR cannot be calculated in the usual way. The recommended practice is to guardband the acceptance limit by the measurement uncertainty. In other words, with an upper specification limit of 1 Ω and a measurement uncertainty of 0.05 Ω (95% confidence), the guardband limit would be set at 0.95 Ω. Risk for one-sided requirements can be calculated in the manner described previously, with the region of integration modified for the single specification limit.

Occasionally, specification limits are asymmetric; for example, a voltage requirement of 10.0 V–0.0 V/+2.0 V. In these cases, the range of acceptable values should be considered: acceptable voltage occurs anywhere between 10.0 V and 12.0 V. This is equivalent to a symmetric specification limit of 11.0 V $\pm$ 1.0 V. This shifted, symmetric specification limit can be used to determine the TUR and the absolute guardbanded acceptance limits. If the TUR is 3, resulting in a GBF of 0.945 (RSS method), the new acceptance limit is ± 1.0 V $\times$ 0.945, resulting in an acceptance range of 11.0 V $\pm$ 0.945 V. In other words, accept parts measuring

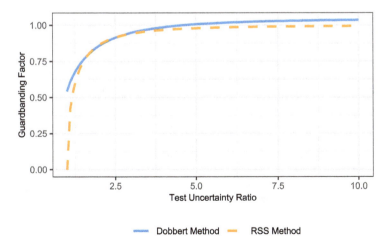

Fig. 5.9 The value of the GBF as a function of TUR for the RSS (dashed) and Dobbert (solid) guardbanding methodologies

between 10.055 V and 11.945 V. Risk for asymmetric requirements can be calculated in the manner described previously, with the region of integration modified for the asymmetric limits.

It should be pointed out that the need for guardbanding can be reduced by measuring marginal parts multiple times. Multiple independent measurements reduce the standard uncertainty, improving the ability to discriminate between good and bad parts. Cost models such as given by Smith (1990) consider the cost of additional measurements, along with the cost of false accept and false reject decisions.

5.2.1.5 Risk with Biased Measurements

While the 4:1 TUR requirement is commonly used to ensure a measurement is adequate for making an accept/reject determination, this metric assumes that the process distribution is centered between the specification limits, that is $\mu_p = (SL_U + SL_L)/2$. If this is not the case, TUR cannot be reliably used as an indicator of risk, however, the PFA and PFR equations are still valid assuming the correct μ_p is used.

The measurement uncertainty distribution is also assumed to be centered about the actual value t when calculating TUR. The measurement process is said to be biased if it is not centered about t and systematically overstates or understates the true value of the measurement. Properly accounting for measurement bias provides a more accurate risk evaluation. If bias is ignored, the risk might be understated, perhaps significantly.

In the presence of bias, the distribution of the measurement y, given the actual value t, shifts from a $N(t, \sigma_m^2)$ distribution to a $N(t - b_m, \sigma_m^2)$ distribution, where b_m is the measurement bias.

With bias b_m, the expressions for the PFA and PFR (without guardbanding) become

$$\begin{aligned} \text{PFA} &= \int_{-\infty}^{SL_L} \left(\int_{SL_L}^{SL_U} \frac{1}{\sigma_m \sqrt{2\pi}} e^{-\frac{1}{2\sigma_m^2}(y-(t-b_m))^2} dy \right) \frac{1}{\sigma_p \sqrt{2\pi}} e^{-\frac{1}{2\sigma_p^2}(t-\mu_p)^2} dt \\ &+ \int_{SL_U}^{+\infty} \left(\int_{SL_L}^{SL_U} \frac{1}{\sigma_m \sqrt{2\pi}} e^{-\frac{1}{2\sigma_m^2}(y-(t-b_m))^2} dy \right) \frac{1}{\sigma_p \sqrt{2\pi}} e^{-\frac{1}{2\sigma_p^2}(t-\mu_p)^2} dt. \end{aligned} \qquad (5.18)$$

5.2 Measurement Uncertainty and Risk

$$\text{PFR} = \int_{SL_L}^{SL_U} \left(\int_{-\infty}^{SL_L} \frac{1}{\sigma_m \sqrt{2\pi}} e^{-\frac{1}{2\sigma_m^2}(y-(t-b_m))^2} dy \right) \frac{1}{\sigma_p \sqrt{2\pi}} e^{-\frac{1}{2\sigma_p^2}(t-\mu_p)^2} dt$$

$$+ \int_{SL_L}^{SL_U} \left(\int_{SL_U}^{+\infty} \frac{1}{\sigma_m \sqrt{2\pi}} e^{-\frac{1}{2\sigma_m^2}(y-(t-b_m))^2} dy \right) \frac{1}{\sigma_p \sqrt{2\pi}} e^{-\frac{1}{2\sigma_p^2}(t-\mu_p)^2} dt. \qquad (5.19)$$

Figure 5.10 shows PFA and PFR when μ_p is shifted by 60% of the specification limit $\left(\frac{SL_U - SL_L}{2}\right)$ without guardbanding (GBF = 1), and with $b_m = 0$. Figure 5.11 shows PFA and PFR when the measurement is biased by 20% of the specification limit ($b_m = 0.2 \left(\frac{SL_U - SL_L}{2}\right)$).

As discussed earlier, if either shift is ignored and the 4:1 TUR rule is blindly applied, then the true PFA may exceed the traditionally accepted risk rate of 2%, even with a 4:1 TUR. For this reason, if it is not feasible to correct for measurement bias, the risk for the measurement process with bias needs to be explicitly quantified and deemed acceptable.

5.2.2 Measurement Uncertainty and Risk in Calibration

Calibration labs frequently perform tolerance tests on measurement equipment to ensure the equipment meets its specifications. Tolerance tests result in a pass or fail determination regarding whether the M&TE is in tolerance. Evaluating the

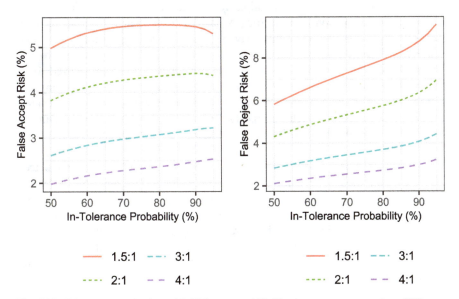

Fig. 5.10 False accept and reject with 60% process shift. The curves represent various TURs

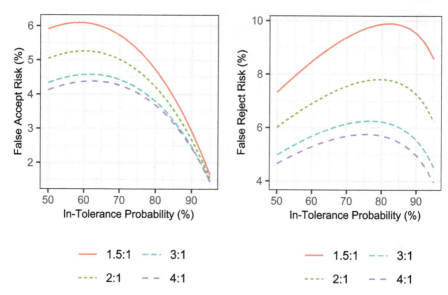

Fig. 5.11 False accept and reject with 20% measurement bias. The curves represent various TURs

risk of falsely accepting or rejecting M&TE in a calibration lab is similar to that of the manufacturing setting. In fact, the mathematics are identical, but some of the terminology, especially encountered in literature written for the calibration community, may be slightly different (see Deaver 1995).

The M&TE being measured is often referred to as the unit under test (UUT) because the unit may contain multiple measurement functions, as in a digital multimeter with functions for electrical voltage, current, and resistance. In these cases, multiple pass/fail decisions may be made on a single M&TE for each function and range.

The probability distribution of the measurement is sometimes referred to as the distribution of the standard or reference, because the M&TE is often compared to a known reference standard. However, a proper evaluation of risk includes all known uncertainties, such as repeatability and environmental factors, in the measurement uncertainty distribution, in addition to the calibration uncertainty in the reference standard itself.

The manufacturing product distribution as defined in Sect. 5.2.1.2 is replaced with the distribution of all equivalent M&TEs calibrated by the lab. This could be all devices of the same model number or series of similar model numbers. Depending on how many devices are in the lab's database, the data may be very sparse for calculating the distribution parameters μ_p and σ_p. A frequent assumption in the presence of little data is to set the mean of the distribution μ_p exactly centered between specification limits and to estimate the standard deviation by (Dobbert 2008):

5.2 Measurement Uncertainty and Risk

$$\sigma_p \approx \frac{SL}{F^{-1}\left(\frac{1+itp}{2}\right)}. \qquad (5.20)$$

The *itp* is again in-tolerance probability, sometimes referred to as end-of-period reliability (EOPR), based on the *observed* fraction of equivalent devices that were determined in-tolerance when submitted for recalibration. The function F^{-1} is the inverse normal cumulative distribution function, also known as the quantile function or percent-point-function. Most statistical software have functions to calculate this quantity, which assumes the population of equivalent M&TE follows a normal distribution centered at nominal.

When a calibration lab performs a true calibration, as opposed to a tolerance test (Sect. 2.2.4), no pass or fail determination is being made; only a value and measurement uncertainty are reported on the certificate. However, the calibration lab's accreditation statement lists a maximum uncertainty for a particular measurement, and the lab should verify the risk of exceeding their accredited uncertainty is minimal.

Finally, when a calibration lab makes an incorrect decision when testing M&TE, the consequences and costs of such a decision are different from a manufacturing setting. Depending on what the M&TE is used for, any measurement made with that equipment during its calibration cycle should be considered suspect. If the lab calibrates a reference standard for use in another calibration lab down the traceability chain, the incorrect decision can affect many downstream calibrations.

5.2.2.1 Decision Rules in Calibration

ISO 14253-1 2017 defines requirements for calibration labs, including a requirement to state a "decision rule" on calibration certificates. The decision rule defines the action(s) to take given a measurement result. "Simple acceptance," sometimes called "as-read," is the most common decision rule, where the outcome is a pass if the measurement as-read from the M&TE indicates a passing condition. This decision rule is commonly employed when the TUR is greater than 4:1. When the TUR is less than 4:1, a "guarded acceptance" decision rule is used, where the as-read measurement must be compared to a reduced acceptance interval.

Some calibration labs have begun using "indeterminate acceptance" decision rules, where a measured value within the acceptance limits is considered a pass, a value outside the specification limits is a fail, but measured values falling within the guardband (between the acceptance limit and specification limit) is an "indeterminate pass" condition. In this situation, the lab's measurement indicates a pass condition, but with a high risk of being OOT. "Indeterminate fail" conditions are also occasionally used, when the measured value is outside the specification limits, but only by some small tolerance. The indeterminate outcomes effectively shift the pass/fail decision on to the end-user of the equipment.

Sometimes, non-binary decision rules are used. For example, when manufacturing a part with a maximum diameter requirement of 1.0 cm, all components measuring less than 1.0 cm will be accepted, parts measuring greater than 1.2 cm will be rejected, but parts measuring between 1.0 cm and 1.2 cm can be sent for rework, and not rejected outright. The same concepts from this chapter apply to non-binary decision rules assuming care is taken to appropriately set the integration regions in the risk calculations.

5.3 Summary

Measurements are used to make conformity decisions, yet because all measurements have uncertainty, there is inherent uncertainty in the decision outcome and risk of making an incorrect decision. Historically, a TUR of at least 4:1 has been used to ensure a measurement process is adequate for ensuring compliance with a specification. When a 4:1 ratio cannot be achieved, guardbanding should be applied to reduce the risk of false accept decisions. However, guardbanding also increases the risk of false reject decisions. Any non-normal behavior or bias in the distributions will affect the false accept and reject levels and is not accounted for in the TUR 4:1 rule of thumb. In this case, a full analysis of risk, including evaluation of *PFA* and *PFR*, is the best option to ensure adequate *PFA* rates.

5.4 Related Reading

Additional details and more motivation for the risk calculations as applied to manufacturing are given by Smith (1990). He includes an evaluation of the expected cost per part based on the cost of a false accept, the cost of a false reject, and the cost of making additional measurements. Basnet and Case (1992) and Sankle and Singh (2012) discuss evaluation of manufacturing risk based on sampling from a batch process rather than testing every single part. The process capability index, *Cpk*, closely related to *itp* defined in this chapter, is a commonly used measure of the ability of a manufacturing process to meet requirements. As discussed in Mottonen et al. (2008), if the *Cpk* is high enough, risk is minimal regardless of the test measurement.

The discussion in this chapter calculates risk based on a single feature of a part. In many cases, a part will have numerous specifications that must all be met in order to meet requirements. Jackson (2019) discusses combined false accept rates and when multiple individual attributes are measured on one component.

Much of the background work on consumer's and producer's risk as applied to calibration was done by Deaver (1993, 1995) and Dobbert (2007, 2008). This work was incorporated into the NCSLI's guidance for application of standard Z540.3 in National Conference of Calibration and Standards Laboratories International (2009). Mimbs (2007) gives an overview of measurement risk with some interesting history on the confusing inconsistent definitions of TAR and TUR and NASA's decision to change from a 10:1 to a 4:1 TUR requirement.

JCGM (2012) provides general statistical guidance in calculating risk as well as setting acceptance and tolerance limits. To avoid confusion between different and often ambiguous definitions of TUR and TAR found in literature, it uses the term "measurement capability index," the same quantity as TUR given in this chapter. ISO 10576-1 (2013) is the international standard, referenced by JCGM 106, on conformity assessment practices. ISO 14253-1 considers conformity assessment on geometrical product specifications. The NCSLI also offers recommended practices for applying the concepts in these standards with NCSLI RP-18 (2014), which includes a discussion of risk assessment using a Bayesian approach. Interpreting manufacturer's specifications with regard to risk assessment is discussed in National Conference of Calibration and Standards Laboratories International (2016). Finally, IEC 31010 2009 covers risk management from a general perspective, not specific to risk only in measurements.

Several other methods of setting acceptance limits to mitigate measurement risk have been proposed. Deaver (1994) compares several common methods based on TUR. Easterling et al. (1991) and Kim et al. (2007) consider total cost minimization methods, and Hund et al. (2017) proposes a data-driven probabilistic approach.

Mimbs (2011) discusses the usefulness of end-of-period reliability (EOPR) as a means for satisfying the 2% PFA metric required by Z540.3, differentiating observed EOPR from true EOPR. Mimbs, along with Dobbert (2008), introduce a Monte Carlo approach to calculating false accept and false reject risk that requires no calculus but does require a suitable random number generator (Chap. 8). Singh (1996) considered non-normal distributions in risk analysis, using an Edgeworth series expansion to approximate non-normal distributions. Castrup (2001) mentions the importance of accounting for bias in normal distributions, while Delker (2020) directly considers the effect of bias and non-normal behavior, as quantified by skewness and kurtosis, on false accept and false reject probabilities.

5.5 Exercises

1. Calculate the test uncertainty ratio (TUR) of a resistance measurement used to verify a resistor falls within requirements $1000 \, \Omega \pm 4 \, \Omega$, measured with an ohmmeter having $\pm 1 \, \Omega$ (95% confidence) uncertainty.
2. A caliper is used to measure a component and ensure it has a diameter of less than 0.5 cm. If the caliper has total (95% confidence) uncertainty of 0.1 mm, calculate an appropriate acceptance limit for this measurement.
3. A caliper is used to measure a component and ensure its width falls between 9.9 cm and 10.1 cm. The caliper has total (95% confidence) uncertainty of 0.05 mm.

a. Calculate the TUR of this measurement.
b. Historical data shows that 80% of the calipers are received in tolerance. Use Fig. 5.4 to estimate the probability of false accept for this measurement.
c. Use the RSS guardbanding method to calculate appropriate acceptance limits for this measurement.
d. Use Fig. 5.8 to estimate the false accept risk of the guardbanded measurement.

4. Using your favorite programming language, write a function to compute an estimate of conditional false accept risk (Eq. (5.3)), sometimes called "specific risk"—the probability that a specific test measurement result is actually out of tolerance—given a test result y. You may use a numerical (trapezoidal) approximation for the integral. The function should take the measured value y, measurement uncertainty σ_m, and the specification limits as arguments. Validate the function by checking the specific risk values for measurements given in Table 5.1 having limits of [9, 11].

5. Use the function from exercise 4 to make a "probability of pass" plot. This plot shows the probability that the device's true value is within the specification limits as a function of the measured value y. For an upper specification limit of 10 and measurement standard uncertainty of 0.25, calculate and plot 1 minus the specific risk as a function of measured value in the range of 8–12. Set the lower limit to something much smaller than 10.

6. One conservative guardbanding approach is to use the probability of pass plot (exercise 5) to find the measured value at which the passing probability falls below some threshold. Using the plot in exercise 5, find an approximate acceptance limit that results in 90% probability of pass.

7. A pressure gauge is calibrated annually at a set value of 180 psi. While in-tolerance at the most recent calibration, the gauge has been slowly drifting closer to its tolerance limit of 185 psi. The gauge is predicted to drift to a reading of 184.7 psi with standard uncertainty ±0.22 psi by the end of its calibration interval (see Chap. 10 for techniques on predicting drift). What is the probability that the gauge will be out of tolerance when it comes back for calibration?

8. Using your favorite programming language, write a function to compute probability of false accept and reject given a product distribution and test distribution. Rather than attempting to implement the double integral numerically, you may use a Monte Carlo approach. First generate N random samples from the process distribution, then generate N random samples from the test distribution with mean values set by the sampled process distribution.

Table 5.1 Specific risk validation data

Measured value	Measurement standard deviation	Specific risk (%)
10.0	0.25	0.006
10.5	0.25	2.27
11.0	0.25	50.0

5.5 Exercises

Then the PFA is the fraction of samples whose test value indicates it falls within limits but whose process value indicates it is out of limits. The function should take sigma_p, sigma_m, mu_p, upper and lower specification limits (LL, UL), and lower and upper guardbanded limits (GBL, GBU) as arguments. Use $N = 10^5$.

Pseudocode:

```
process = random.normal(mean=mu_p, sigma=sigma_p, size=N)
test = random.normal(mean=process, sigma=sigma_m, size=N)
PFA = count(process out-of-tolerance & test in-tolerance) / N
PFR = count(process in-tolerance & test out-of-tolerance) / N
```

Validate the function by checking that PFA for sigma_p=1, sigma_m=0.25, LL=-2, UL=2 results in ~0.8% PFA and ~1.5% PFR. The remaining exercises can use these functions for their calculations.

9. Write a function that computes PFA and PFR given values for *itp*, TUR, and GBF. This can wrap your function written in exercise 8. Fixing $SL = \pm 1$, use Eq. (5.20) to approximate the process distribution standard deviation, and Eq. (5.1) to calculate sigma_m from the TUR. Validate the function by ensuring that inputs of ITP = 0.95, TUR = 4, and GBF = 1 result in PFA of ~0.8% and PFR of ~1.5%.
10. Use the function developed in exercise 9 to make a plot of PFA vs ITP for TURs of 1.5, 2, 3, and 4, with ITP ranging from 50 to 95%. Compare with Fig. 5.4.
11. Historical data analysis shows that 95% of the manufactured components fall within the specification limits of 9.9–10.1 cm. When each new component is measured using the caliper with 0.05 mm uncertainty (95% confidence), what is the total probability of false accept?
12. A voltmeter is being calibrated against manufacturer specifications. When a 10 V reference voltage is applied, the meter must measure $10\,V \pm 0.05\,V$ to pass the tolerance test. The total 95% confidence uncertainty in the reference voltage is 0.020 V.

    a. What is the TUR of this measurement?
    b. Use the RSS guardbanding method to determine an appropriate guardbanding factor and acceptance limits for the voltmeter.
    c. After calibrating many voltmeters of the same model, it was determined that 90% of them return for calibration in-tolerance. Estimate the product distribution σ_p for these meters.
    d. Calculate the total probability of false accept for the voltmeters with and without the guardbanding found in part (b).

13. In addition to the RSS method (Eq. (5.12)) and Dobbert method (Eq. (5.16)) of guardbanding, two other approaches are calculating $GBF = 1 - 1/TUR$ and $GBF = 1.25 - 1/TUR$.

    a. For these four methods, plot GBF as a function of TUR up to TUR = 10.

b. Typically, guardbanding is applied when TUR < 4. Which of these methods result in a reduced guardband factor even above TUR = 4?

c. Which methods may result in a GBF greater than one (acceptance limits set outside the specification limits)?

14. Another guardbanding strategy involves minimizing the total cost of false accepts and false rejects. Although they are usually quite different, assume the cost of a false accept is the same as the cost of a false reject. In this case, an optimal guardband can be determined by minimizing the sum PFA + PFR.

15. Given an in-tolerance probability of 90% and TUR of 3, plot PFA + PFR as a function of guardband factor ranging from 0 to 2. It can be helpful to plot the probability on a log scale.

    a. What guardband factor minimizes PFA + PFR? (Approximate the minimum graphically using the plot.) What is interesting about this GBF?

    b. Now assume a false accept costs 100 times more than a false reject. What guardband factor minimizes the total cost PFA + PFR/100?

16. Speed limit enforcement on public highways is an example of a single-sided conformance limit. With a speed limit of 65 miles/h, a sample of traffic was observed to follow a normal distribution with mean of 68 mph and standard deviation of 3 mph (note: the authors do not condone speeding). The radar guns have a 95% confidence uncertainty of ± 1 mph.

    a. If a police officer issues a ticket to any vehicle traveling above 65 mph as indicated by radar, what is the probability that a ticket is issued incorrectly?

    b. Most districts apply a tolerance to speed limits. If the officer only issues tickets to drivers when measured above 70 mph, what is the probability that a ticket is issued incorrectly?

17. Given a TUR of 3, use the RSS method (Eq. (5.13)) and the Dobbert method (Eqs. (5.16) and (5.17)) to calculate an appropriate guardband factor. Assuming an in-tolerance probability of 95%, what is the PFA when each guardbanding method is used?

18. Plot the false accept probability as a function of TUR ranging from 1 to 4 assuming (1) no guardband; (2) RSS guardband; (3) Dobbert's guardband; (4) constant GBF of 0.9.

References

ANSI/NCSLI Z540.3: Requirements for the calibration of measuring and test equipment (2006)

Basnet, C., Case, K.E.: The effect of measurement error on accept/reject probabilities for homogeneous products. Qual. Eng. **4**(3), 383–397 (1992)

Bhise, V., Rajan, S., Sittig, D., Morgan, R., Chaudhary, P., Singh, H.: Defining and measuring diagnostic uncertainty in medicine: a systematic review. J. Gen. Intern. Med. **33**(1), 103–115 (2018)

Castrup, H.: Estimating bias uncertainty. NCSL Workshop and Symposium, Washington D.C. (2001)

Cividino, S., Egidi, G., Zambon, I., Colantoni, A.: Evaluating the degree of uncertainty of research activities in Industry 4.0. Future Internet. **11**, 196 (2019)

Deaver, D.: How to maintain your confidence. Proceedings of the NCSL Workshop and Symposium, Washington D.C. (1993)

Deaver, D.: Guardbanding with confidence. In: Proceedings of the NCSL Workshop and Symposium, Albuquerque, NM (1994)

Deaver, D.: Managing calibration confidence in the real world. In: Proceedings of the NCSL Workshop and Symposium, Dallas, TX (1995)

Delker, C. J.: Evaluating risk in an abnormal world: how arbitrary probability distributions affect false accept and false reject. In: Proceedings of the NCSLI Workshop and Symposium, Aurora, CO (2020)

Dobbert, M.: Understanding measurement risk. Proceedings of the NCSL International Workshop and Symposium, St. Paul, MN (2007)

Dobbert, M.: A guard-band strategy for managing false-accept risk. In: Proceedings of the NCSL International Workshop and Symposium, Orlando, FL (2008)

Easterling, R., Johnson, M., Bement, T., Nachtsheim, J.: Statistical tolerancing based on consumer's risk considerations. J. Qual. Technol. **23**(1), 1–11 (1991)

Harben, J., Reese, P.: Risk mitigation strategies for compliance testing. NCSLI Meas. **7**(1), 38–49 (2012)

Hund, L.B., Campbell, D.L., Newcomer, J.T.: Statistical guidance for setting product specification limits. In: Proceedings of the IEEE Annual Reliability and Maintainability Symposium (2017)

IEC 31010: Risk management – Risk assessment techniques (2009)

ISO 10576-1: Statistical methods – Guidelines for the evaluation of conformity with specified requirements (2013)

ISO 14253-1: Geometrical product specifications (GPS) – Inspection by measurement of workpieces and measuring equipment (2017)

Jackson, D.: A surprising link between measurement decision risk and calibration interval analysis. In: Proceedings of the NCSLI Workshop and Symposium, Cleveland, OH (2019)

JCGM 106: Evaluation of measurement data - The role of measurement uncertainty in conformity assessment (2012)

Kim, J.Y., Byung, R.C., Kim, N.: Economic design of inspection procedures using guard band when measurement errors are present. Appl. Math. Model. **31**(5), 805–816 (2007)

Mimbs, S.M.: Measurement decision risk - The importance of definitions. In: Proceedings of the NCSLI Workshop and Symposium, St. Paul, MN (2007)

Mimbs, S.M.: Using reliability to meet Z540.3's 2% rule. In: Proceedings of the NCSL Workshop and Symposium, National Harbor, MD (2011)

Mottonen, M., Belt, P., Harkonen, J., Haapasalo, H., Kess, P.: Manufacturing process capability and specification limits. Open Ind. Manuf. Eng. J. **1**, 29–36 (2008)

National Conference of Calibration and Standards Laboratories International: Handbook for Application of ANSI/NCSL Z540.3-2006 (2009)

National Conference of Calibration and Standards Laboratories International: Recommended Practice 18 - Estimation and Evaluation of Measurement Risk (2014)

National Conference of Calibration and Standards Laboratories International: Recommended Practice 5 - Measuring and Test Equipment Specifications (2016)

Sankle, R., Singh, J.R.: Single sampling plans for variables indexed by AQL and AOQL with measuremetn error. J. Mod. Appl. Stat. Methods. **11**(2), 12 (2012)

Singh, H.R.: Producer and consumer risks in non-normal populations. Am. Soc. Qual. **8**(2), 335–343 (1996)

Smith, J.R.: Statistical aspects of measurement and calibration. Comput. Ind. Eng. **18**(3), 365–371 (1990)

Chapter 6
The Measurement Model and Uncertainty

This chapter introduces the basic terms and models used to quantify the uncertainty of both direct and indirect measurements. Basic definitions such as Type A and Type B evaluation of uncertainty are introduced, and the GUM approach to the propagation of uncertainties is explained. Case studies are used to illustrate the approaches to uncertainty analyses for both types of measurements.

6.1 Introduction

Measurements can be categorized as either direct or indirect, depending on how the measurement is taken. The basic definitions that are used throughout the chapter for both types of measurements are given first, followed by the two measurement models and their respective uncertainty analyses. The direct measurement model has received less attention in metrology guidelines, while the indirect measurement model has been emphasized in guidelines such as the JCGM 100 (GUM). We believe that direct measurement models, however, can involve subtle features that warrant separate attention. These subtleties appear in a case study involving direct voltage measurements. A second case study involving measurement of neutron yield is used to illustrate indirect measurements and their associated uncertainty analyses. Much of the terminology and methodologies for indirect measurements parallel the presentation in the GUM.

6.2 Uncertainty Analysis Framework

This section presents an overlying framework for uncertainty analysis based on the development of the appropriate measurement model and analysis for both direct and indirect measurements. The key terms are defined followed by presentation of the conceptual models and analysis approaches.

Sections 6.3 and 6.4 present and illustrate with case studies the estimation techniques that can be applied to both types of measurements.

6.2.1 Standard Uncertainty

Standard uncertainty is defined as the uncertainty of a measurement expressed in terms of the standard deviation of an uncertainty distribution. If the uncertainty distribution is symmetric and approximately normal, this value can be used in a straightforward way to construct uncertainty intervals with specified confidence levels. If the uncertainty distribution is highly non-normal, the standard deviation may not be as useful, and other measures of dispersion should be used.

6.2.2 Type A Uncertainty Evaluation

The Type A uncertainty method of evaluation is defined in the GUM as the "method of evaluation of uncertainty by the statistical analysis of series of observations." These data typically come from a series of tests performed as part of a planned measurement experiment. Examples include data from Gauge R&R experiments performed on the factory floor, small experiments to confirm the performance of secondary standards, or larger controlled experiments used to evaluate high-precision equipment in a primary standards laboratory. Sources of uncertainty that are not part of a statistical analysis are handled by the Type B evaluation of uncertainty.

6.2.3 Type B Uncertainty Evaluation

The Type B uncertainty method of evaluation is defined in the GUM as the "method of evaluation of uncertainty by means other than the statistical analysis of series of observations." Sources of Type B uncertainties may include historical data, calibration reports, manufacturer's specifications, and reference data from handbooks. Type B uncertainty evaluations attempt to account for variation that is not observable in Type A uncertainty evaluations. For example, long-term tester drift would

6.2 Uncertainty Analysis Framework

typically not affect the uncertainty observed in a short-term experiment. Systematic bias would also not affect the variation observed in a short-term experiment. Type B uncertainty evaluations are thus included as part of the overall uncertainty to give validity to the reported uncertainty interval over an extended period of time. In practice, both Type A and Type B evaluations may involve statistical analysis of series of observations. The distinction is that Type A evaluations are based on immediate time-of-test data, while Type B evaluations are not. The values reported as Type B uncertainties, however, have often been derived from statistical analysis of historical data. As such, caution must be used to ensure that Type B uncertainty evaluations do not "double count" uncertainties present in Type A uncertainty evaluations.

6.2.4 Combined Standard Uncertainty

The combined standard uncertainty is expressed in terms of the standard deviation of an uncertainty distribution obtained by combining the individual variances and covariances of each Type A and Type B evaluation that contribute to the overall uncertainty. The combined standard uncertainty is determined by the root sum of squares (RSS) of those variance and covariance terms.

6.2.5 Confidence Level and Expanded Uncertainty

The conceptual form of a measurement uncertainty interval for either direct or indirect measurements is given by the following:

$$\text{Mean} \pm \text{Measurement Uncertainty, or} \\ \text{Mean} \pm k \text{ (Combined Standard Uncertainty)}. \tag{6.1}$$

The measurement uncertainty in this expression is referred to as the expanded uncertainty, defined in more detail below. In terms of the mean of a measurand, coverage factor k, and combined standard uncertainty, this interval can be expressed as

$$\bar{y} \pm t_p(v_{\text{eff}}) u_c(y). \tag{6.2}$$

The remainder of this section will provide additional definitions and show how to calculate the sample mean $\bar{y}$, the multiplier $t_p(v_{\text{eff}})$, and the combined standard uncertainty $u_c(y)$. In Eq. (6.2) $\bar{y}$ is an average of a Type A series of measurements. For direct measurements it is usually the simple average of a series of measurements, while for indirect measurements it is evaluated using a measurement equation.

The combined standard uncertainty $u_c(y)$ is a combination (RSS) of the Type A standard uncertainty $u_A(y)$ and the Type B standard uncertainty $u_B(y)$:

$$u_c(y) = \sqrt{u_A^2(y) + u_B^2(y)}. \tag{6.3}$$

In the case of multiple Type A evaluations and multiple Type B evaluations, the terms are combined as the positive square root of the sum of the squares (RSS) of the standard uncertainties:

$$u_c(y) = \sqrt{\sum_{i=1}^{N_A} u_{A_i}^2(y) + \sum_{i=1}^{N_B} u_{B_i}^2(y)}. \tag{6.4}$$

The multiplier $t_p(v_{\text{eff}})$ in Eq. (6.2) is the coverage factor, computed from the t-distribution with effective degrees of freedom v_{eff} and confidence level $p \cdot 100\%$. The coverage factor is the percentile from the t-distribution that makes the uncertainty interval a $p \cdot 100\%$ confidence interval. It is typically in the range of two to three. Percentiles of the t-distribution are tabled in elementary statistics textbooks such as Montgomery (2013) and appear in Appendix D.

The effective degrees of freedom, v_{eff}, is computed using the Welch–Satterthwaite (W-S) approximation (see Satterthwaite 1946). The effective degrees of freedom are the degrees of freedom associated with the chi-square distribution that is used to approximate the distribution of the combined standard variance $u_c^2(y)$:

$$v_{\text{eff}} = \frac{u_c^4(y)}{\frac{u_A^4(y)}{v_A} + \frac{u_B^4(y)}{v_B}} \tag{6.5}$$

The terms v_A and v_B are the Type A and Type B degrees of freedom. The formula extends in the case of multiple Type A and Type B terms:

$$v_{\text{eff}} = \frac{u_c^4(y)}{\sum_{i=1}^{N_A} \frac{u_{A_i}^4(y)}{v_{A_i}} + \sum_{i=1}^{N_B} \frac{u_{B_i}^4(y)}{v_{B_i}}} \tag{6.6}$$

In the formulas above, N_A and N_B are the number of Type A and Type B terms, respectively.

The expanded uncertainty is a quantity used to construct an interval about a measurement result that is expected to contain a large fraction of the distribution of values that could reasonably be attributed to the measurand. It is denoted by U. The expanded uncertainty is calculated from the combined standard uncertainty using

$$U = t_p(v_{\text{eff}}) \cdot u_c(y)$$
$$= t_p(v_{\text{eff}})\sqrt{u_A^2(y) + u_B^2(y)}. \qquad (6.7)$$

The formulas introduced in Sect. 6.2 apply to both the direct measurement case and the indirect measurement case, although the Type A and Type B standard uncertainties are computed in different ways. The direct measurement problem, the direct measurement model, and the uncertainty analysis of a direct voltage measurement are demonstrated in Sect. 6.3.

For both the direct measurement and indirect measurement uncertainty analyses, care must be taken to distinguish between the uncertainty of a single measurement and the uncertainty of an average of multiple measurements. Both cases are important, depending on the measurement problem under consideration. To accept or reject manufactured parts based on a single measurement, for example, the uncertainty of that single measurement is of interest. To quantify the uncertainty of a device such as a voltage standard, typically multiple measurements of the standard are made and the uncertainty of the average measurement will be of interest.

6.3 Direct Measurements and the Basic Measurement Model

A *direct measurement* is defined as one in which the value of the measurand is read directly from the measuring instrument. Examples include measurements of weight, length, and time, where the value of the measurand is determined with a simple measuring device. In some cases, a direct measurement may be adjusted by environmental or other time-of-test factors, making it an effectively indirect measurement.

In practice, the two types of measurements are not always easily distinguished from one another but will depend on one's point of view. For example, an ohmmeter measures resistance by applying a current, measuring the resulting voltage, and dividing the two. From the perspective of the ohmmeter manufacturer, the resistance measurement is indirect, but from the perspective of the ohmmeter user, the resistance measurement is direct.

The simple conceptual model proposed for direct measurements is as follows, with y representing the measured value of the measurand:

$$y = \text{Actual} + \text{Bias} + \varepsilon_A + \varepsilon_B. \qquad (6.8)$$

In this expression, the measured value (y) of the measurand is equal to the actual value plus a potential constant bias term, plus ε_A and ε_B, the Type A and Type B random error terms, respectively.

In the presence of a known bias, the preferred approach is to adjust the measurement device to remove the effect of bias. If the measurement device cannot be corrected, any resulting uncertainty interval should be adjusted by the amount of bias, determined from an independent assessment. This approach is illustrated below in a case study involving direct measurements of voltage.

In terms of a formal statistical model for direct measurements, the ith measurement, y_i, in a series of n measurements can be expressed as

$$y_i = \text{Actual} + \text{Bias} + \varepsilon_{A,i} + \varepsilon_B. \tag{6.9}$$

In this expression, $\varepsilon_{A,i}$ is the Type A error term associated with the ith measurement and ε_B is the Type B random uncertainty term. Note with this model that the Type B term does not increase variation in the time-of-test data. The actual value, bias, and error terms are not known, but bounds can be constructed that are expected to contain the actual value. The approach is illustrated in the voltage case study in Sect. 6.3.1.

For the direct measurement case, the Type A evaluation typically depends on the sample standard deviation

$$s_A(y) = \sqrt{\frac{1}{(n-1)} \sum_{i=1}^{n} (y_i - \bar{y})^2}. \tag{6.10}$$

The standard uncertainty $u_A(y)$ associated with a *single* determination of the measurand is

$$u_A(y) = s_A(y), \tag{6.11}$$

and the standard uncertainty of the *average* of n determinations (sometimes called the "standard error") is

$$u_A(\bar{y}) = \frac{s_A(y)}{\sqrt{n}}. \tag{6.12}$$

The degrees of freedom associated with the Type A standard uncertainty calculated this way is $(n-1)$. In general, the degrees of freedom will depend on the total number of measurements and how the standard uncertainty is estimated.

For the direct measurement case, the Type B standard uncertainty evaluation will depend on statements of uncertainty and engineering judgement using all available information about sources of variation affecting the measurand over time.

If the uncertainty interval for a Type B source of uncertainty is stated in a calibration report, for example, as $\pm U$ with 95% coverage ($k = 2$), the standard uncertainty $u_B(y)$, assuming an underlying Normal distribution, is simply

6.3 Direct Measurements and the Basic Measurement Model

$$u_B(y) = \frac{U}{2}. \tag{6.13}$$

A more conservative value for the Type B standard uncertainty assumes a Rectangular (Uniform) distribution Eq. (4.6) for the Type B source of uncertainty, resulting in the standard uncertainty:

$$u_B(y) = \frac{U}{\sqrt{3}}. \tag{6.14}$$

This distribution may be appropriate if little is known about the shape of the distribution. In this case, $\frac{U}{\sqrt{3}}$ estimates the standard deviation of the Uniform distribution on the interval $(-U, U)$.

A less conservative value for the Type B standard uncertainty assumes an underlying Triangular distribution Eq. (4.7) for the Type B source of uncertainty, resulting in the standard uncertainty:

$$u_B(y) = \frac{U}{\sqrt{6}}. \tag{6.15}$$

The triangular distribution has been used for resolution uncertainty when reading an analog gauge or ruler, but in practice the Uniform distribution is most often used when no other information is provided because of its conservatism. Fig. 4.12 compares the normal, uniform, and triangular distributions used for this problem.

If the Type B value U is treated as exactly known, then $u_B(y)$ can be treated as exactly known and its associated degrees of freedom can be set to $+\infty$. If this assumption cannot be made, a subjective estimate of the relative uncertainty in $u_B(y)$ can be used to define the Type B degrees of freedom as

$$v_B \cong \frac{1}{2} \left[\frac{\Delta u_B(y)}{u_B(y)} \right]^{-2}. \tag{6.16}$$

In this expression, the quantity inside the brackets is the relative uncertainty in $u_B(y)$. The rationale for this calculation is discussed in the GUM (G.4.2). The use of this equation will be discussed further in Sect. 11.5.

The methodology recommended for direct measurements is presented next with a case study involving the direct measurement of voltage.

6.3.1 Case Study: Voltage Measurement

A voltage measurement uncertainty study was performed on a National Instruments PXI-5124 digitizer (the tester) using a Tektronix TDS 3024 digitizer as the standard. The purpose of the study was to quantify the uncertainty of the standard digitizer and

Table 6.1 Results of digitizer uncertainty experiment at 350 V

Sample	Standard (V)	Tester (V)	Delta = standard-tester (V)
1	354.0	338.2	15.8
2	350.1	336.6	13.5
3	354.0	337.1	16.9
4	352.2	336.3	15.9
5	354.0	338.0	16.0
6	352.0	337.1	14.9
7	350.0	338.5	11.5
8	352.2	338.2	14.0
9	352.2	337.7	14.5
10	358.0	337.4	20.6
11	356.0	335.4	20.6
12	354.0	337.9	16.1
13	358.0	336.8	21.2
14	354.0	337.1	16.9
15	358.0	336.8	21.2
16	356.0	336.0	20.0
17	356.0	337.9	18.1
18	354.0	337.6	16.4
19	356.0	337.9	18.1
20	356.0	337.4	18.6
21	354.0	338.6	15.4
22	353.0	338.3	14.7
23	352.0	336.9	15.1
24	350.0	339.1	10.9
25	352.0	338.6	13.4
26	352.0	339.1	12.9
27	352.0	336.9	15.1
28	350.0	338.0	12.0
29	350.0	338.3	11.7
30	354.0	338.6	15.4
Avg	353.5	337.6	15.9
Std Dev	2.43	0.91	2.90

to quantify the uncertainty and bias of the tester digitizer. The study consisted of 30 measurements of output voltage from a trigger voltage tester with the output voltage set at 60 V, 150 V, and 350 V. The output trigger voltage was measured by each digitizer. The difference between the two digitizer measurements provides an estimate of the bias in the PXI-5124 tester. The combined standard uncertainty for the voltage measurement, applying Eq. (6.9), consists of Type A, Type B, and bias uncertainties. The uncertainty analysis for direct measurements illustrated below uses the data collected at 350 V. The data were collected over a short period of time. Table 6.1 lists the results, along with summary statistics, for the voltage measurements made at 350 V.

6.3 Direct Measurements and the Basic Measurement Model

Table 6.2 Calibration records providing Type B standard uncertainties

Digitizer	Assumed PDF	$u_B\ (k = 1)$
NI PXI-5124 (tester)	Normal	1.70 V
Tek TDS 3024 (standard)	Normal	4.05 V

The first part of the case study quantifies the uncertainty of the standard Tektronix TDS 3024 digitizer. The Type A standard uncertainty associated with this digitizer, based on an average of 30 measurements, is estimated (from Table 6.1) by

$$u_{A,\text{std}}(y) = \frac{2.43}{\sqrt{30}} = 0.44\ V. \tag{6.17}$$

The Type B uncertainties for the two digitizers are reported in Table 6.2, obtained from calibration reports for each digitizer.

The Type B standard uncertainty associated with the standard digitizer is thus

$$u_{B,\text{std}}(y) = 4.05\ V.$$

Combining terms, the combined standard uncertainty associated with the standard digitizer is

$$u_{c,\text{std}}(y) = \sqrt{u_{A,\text{std}}^2(y) + u_{B,\text{std}}^2(y)} = \sqrt{(0.44)^2 + (4.05)^2} = 4.07\ V. \tag{6.18}$$

The effective degrees of freedom, calculated via the W-S approximation, with degrees of freedom $+\infty$ for the Type B component, is

$$v_{\text{eff}} = \frac{u_{c,\text{std}}^4(y)}{\frac{u_A^4(y)}{v_{A,\text{std}}} + \frac{u_B^4}{v_{B,\text{std}}}} = \frac{(4.07)^4}{\frac{(0.44)^4}{29} + \frac{(4.05)^4}{+\infty}} \gg 30 \tag{6.19}$$

The coverage factor k is computed from the t-distribution using the effective degrees of freedom and chosen confidence level. In this case, for a 95% confidence interval with degrees of freedom much greater than 30,

$$k = t_{0.975}(30) = 2.04$$

is a conservative upper bound on the value of k. The expanded uncertainty is thus

$$\begin{aligned} U &= k u_c(y) \\ &= (2.04)(4.07) \\ &\cong 8.3 V \end{aligned} \tag{6.20}$$

and the resulting 95% ($k = 2$) uncertainty interval for the measurand, based on the average of 30 measurements using the standard digitizer, is

$$\bar{y}_{std} \pm U$$
$$= 353.5 \pm 8.3 V \tag{6.21}$$

The resulting interval (345V, 362V) provides a range of values that could reasonably be expected from the standard digitizer under study as the outcome of the average of 30 measurements with the trigger voltage source set to 350 V. This uncertainty interval would be used if the standard digitizer was used in the calibration of a secondary standard digitizer.

The second part of the voltage study quantifies the uncertainty and bias in the PXI-5124 digitizer (the tester), using the standard to estimate the bias component of the tester uncertainty.

The Type A standard uncertainty associated with the tester, for a *single measurement*, is the sample standard deviation of the 30 voltage measurements (Table 6.1),

$$u_A(y) = 0.91 \ V,$$

and the Type B standard uncertainty associated with the tester (Table 6.2) is

$$u_B(y) = 1.70 \ V.$$

The bias (δ) is estimated as the difference between the standard and the tester, averaged over the 30 tests:

$$\hat{\delta} = \frac{\sum_{i=1}^{30}(\text{Standard}_i - \text{Tester}_i)}{30} = \bar{y}_{\text{Standard}} - \bar{y}_{\text{Tester}} \tag{6.22}$$
$$= 353.5 - 337.6 = 15.9 \ V.$$

The standard uncertainty $u(\hat{\delta})$ of the estimated bias should also be included in the total uncertainty calculation. It is estimated by the standard deviation of the average difference. From Table 6.2:

$$u(\hat{\delta}) = \frac{2.90}{\sqrt{30}} = 0.53. \tag{6.23}$$

Note that the numerator of this quantity is the standard deviation of the 30 differences, used in this case because the tester and standard values are paired by sample number. Combining terms, the combined standard uncertainty associated with the tester digitizer is

6.3 Direct Measurements and the Basic Measurement Model

$$u_c(y) = \sqrt{u_A^2(y) + u_B^2(y) + u^2\left(\hat{\delta}\right)}$$
$$= \sqrt{(0.91)^2 + (1.70)^2 + (0.53)^2} = 1.99 \ V.$$
(6.24)

The effective degrees of freedom, calculated via the W-S approximation, with degrees of freedom $+\infty$ for the Type B component, is

$$v_{\text{eff}} = \frac{u_c^4(y)}{\frac{u_A^4(y)}{v_A} + \frac{u_B^4(y)}{v_B} + \frac{u^4(\hat{\delta})}{v_\delta}}$$
$$= \frac{(1.99)^4}{\frac{(0.91)^4}{29} + \frac{(1.70)^4}{+\infty} + \frac{(0.53)^4}{29}} \gg 30$$
(6.25)

The coverage factor k is computed from the t-distribution using the effective degrees of freedom and chosen confidence level. In this case, for a 95% confidence interval with degrees of freedom much greater than 30,

$$k = t_{0.975}(30) = 2.04$$

is a conservative upper bound on the value of k. The expanded uncertainty is thus

$$U = ku_c(y)$$
$$= (2.04)(1.99)$$
$$\cong 4.1$$
(6.26)

and the resulting 95% ($k = 2$) uncertainty interval for the actual value, adjusted for bias, is

$$\left(\bar{y}_{\text{Tester}} + \hat{\delta}\right) \pm U$$
$$= (337.6 + 15.9) \pm 4.1$$
$$= 353.5 \pm 4.1.$$
(6.27)

The GUM (7.2.6) recommends that the standard uncertainties and expanded uncertainties not be reported with an "excessive number of digits." In practice usually two or three significant digits will suffice, depending on the particular measurand. In this case, the final uncertainty interval for the tester digitizer (adjusted for bias) would be reported as $(350V, 358V)$.

The resulting interval Eq. (6.27) provides a range of values that could reasonably be expected from the tester under study as the outcome of a *single measurement* with the voltage source set to 350 V. The amount of uncertainty associated with testers such as this would potentially be used to inform decisions regarding product acceptance and product state of health over time.

6.3.2 Discussion

The case study above illustrates the recommended approach for determining the expanded uncertainty associated with direct measurements. A tester digitizer was compared to a standard digitizer for 30 shots of an output voltage source set to 350 V. The standard digitizer uncertainty, tester digitizer uncertainty, and bias uncertainty were determined. The resulting expanded uncertainties were used to derive the tester digitizer's bias-adjusted 95% ($k = 2$) uncertainty interval.

Note that the uncertainty interval above Eq. (6.27) is the uncertainty associated with a single measurement of voltage using the tester digitizer. The uncertainty associated with measuring an average of multiple measurements would use a Type A uncertainty adjusted by the sample size. That is, the Type A uncertainty of the tester digitizer based on sample size n would be

$$u_A(\bar{y}) = \frac{0.90}{\sqrt{n}}. \tag{6.28}$$

The standard digitizer added uncertainty to the tester digitizer uncertainty only through the bias calculation. This is because the Type B uncertainty for the tester digitizer was derived from its tabled calibration result which did not involve the standard digitizer. The standard's Type A uncertainty was larger than the tester's Type A uncertainty because the standard available at the time-of-test was an old standard that had yet to be updated. Despite this limitation, the standard digitizer did provide useful information regarding the bias without significantly inflating the calculated uncertainty of the tester digitizer.

As recommended in both the GUM (3.2.4) and in Phillips and Eberhardt (1997), the uncertainty in the estimate of bias should be included in the expanded uncertainty calculation. Because the estimate of bias was based on 30 paired differences, however, it did not contribute much to the total uncertainty.

6.4 Indirect Measurements and the Indirect Measurement Model

In most measurement studies, the measurand is not measured directly, but is determined through a functional relationship. An *indirect measurement* is a measurement for which the value of the measurand is obtained by measuring other quantities functionally related to the measurand. In the NSE, examples include the

6.4 Indirect Measurements and the Indirect Measurement Model

measurement of neutron yield using a lead probe, film thickness using X-ray fluorescence, and voltage using a Josephson volt.

An indirect measurement model will be represented by the equation

$$Y = f(X_1, X_2, \ldots, X_N). \tag{6.29}$$

The "realization" or "estimate" (GUM 4.1.4) of the random variable X_i is denoted by x_i, and the estimate of the measurand Y is denoted by y. The function f defines the mathematical relationship between the measurand and the input quantities, the $X_i's$, which are the known sources of variation affecting the measurement. The form of the function f may be based on a known physical relationship between Y and the $X_i's$ or may be determined experimentally via a statistical design of experiment (see Chap. 9). The $X_i's$ themselves may also be thought of as measurands whose uncertainty must be understood to evaluate the uncertainty in Y.

The function f should contain each variable, including possible correction factors, that may contribute a significant amount of uncertainty to the measurement. In practice, development of an appropriate function f will often be the most critical step in quantifying the measurand's uncertainty.

A simple example of a measurement equation based on a known relationship is Ohm's Law

$$V = IR, \tag{6.30}$$

relating voltage (V) to current (I) and resistance (R). In this case,

$$V = f(I, R) = IR. \tag{6.31}$$

If the initial function f does not model the measurement adequately, additional terms may be added to the model. For example, in the case of a temperature-dependent resistor, the above equation could be modified to reflect that dependence:

$$V = IR_0[1 + a(T - T_0)], \tag{6.32}$$

where R_0 represents the resistance at temperature T_0, and a is the temperature coefficient of resistance. Then the measurement equation becomes

$$V = f(I, R_0, T, a) = IR_0[1 + a(T - T_0)]. \tag{6.33}$$

In practice, the goodness-of-fit of the measurement equation f should be assessed and the equation adjusted if the lack-of-fit is not acceptable. This type of model is sometimes referred to as a "hybrid" model in that it is developed from a physics-based model ($V=IR$) that is adjusted with the addition of an empirically determined constant (a).

The GUM's approach to quantifying the uncertainty of a measurand derived from an indirect measurement equation is summarized in the following steps:

1. The first step in an uncertainty analysis of a measurand observed indirectly is to formulate a measurement equation. The function $Y = f(X_1, X_2, \ldots, X_N)$ should contain every variable, including corrections and correction factors, that can contribute a significant amount of uncertainty to the result of the measurement. These variables may be identified through experimentation, engineering judgement, or known physical relationships.
2. The probability distributions for each input quantity (the $X_i's$) should be determined and the standard uncertainties evaluated. The distributions are determined in terms of both Type A and Type B uncertainties. In the derivation of the combined standard uncertainty $u_c(y)$ in Step 3, we use $u^2(x_i)$ to represent an estimate of $Var(X_i)$.
3. Given this probabilistic framework, the combined standard uncertainty of the measured value y is determined from the standard uncertainties of the input quantities (the $X_i's$). The combined standard uncertainty, $u_c(y)$, is the positive square root of the combined variance, given by

$$u_c^2(y) = \sum_{i=1}^{N} \left(\frac{\partial f}{\partial x_i}\right)^2 u^2(x_i). \qquad (6.34)$$

Each $u(x_i)$ is a standard uncertainty evaluated as in Sect. 6.2 using both Type A and Type B evaluations. This expression is the result of computing the variance, term by term, of a first-order Taylor series approximation of f, assuming uncorrelated inputs. Equation (6.34) and its parallel expression for correlated inputs are referred to in the GUM as the *law of propagation of uncertainty*. Extensions to correlated inputs and higher-order terms are described in Sects. 7.4 and 7.5.

4. The partial derivatives $\frac{\partial f}{\partial x_i}$ are evaluated at the averages of the $x_i's$. These derivatives are referred to as *sensitivity coefficients*. They describe how the output y varies with changes in the $x_i's$. With the sensitivity coefficients defined as $c_i = \frac{\partial f}{\partial x_i}$, the standard uncertainty in y associated with the input quantity x_i is $|c_i|u(x_i)$.
5. The estimate of the measurand from the measurement equation

$$Y = f(X_1, X_2, \ldots, X_N) \qquad (6.35)$$

is well approximated by

$$\bar{y} = f(\bar{x}_1, \bar{x}_2, \ldots, \bar{x}_N) \qquad (6.36)$$

if the measurement equation is approximately linear in the $X_i's$, or by

6.4 Indirect Measurements and the Indirect Measurement Model

$$\bar{y} = \frac{1}{n}\sum_{k=1}^{n} y_k = \frac{1}{n}\sum_{k=1}^{n} f(x_{1,k}, x_{2,k}, \ldots, x_{N,k}) \qquad (6.37)$$

if the measurement equation is highly nonlinear. That is, $\bar{y}$ is taken as the arithmetic mean of the n independent determinations y_k.

6. The next step is to determine the expanded uncertainty, U, which provides an uncertainty interval for the measurand of the form $(\bar{y} - U, \bar{y} + U)$. This interval is expected to contain a large fraction of the values that could reasonably be attributed to the measurand Y. U is determined by multiplying the combined standard uncertainty by a coverage factor $k = t_p(\nu_{\mathit{eff}})$. The effective degrees of freedom ν_{eff} are determined from the W-S approximation Eq. (6.50) with sensitivity coefficients included and $t_p(\nu_{\mathit{eff}})$ is the t-value discussed in Sect. 6.2.
7. This results in the expanded uncertainty:

$$U = k u_c(y). \qquad (6.38)$$

8. The final step is to report the uncertainty. The interval should be expressed as

$$\bar{y} \pm U. \qquad (6.39)$$

The confidence level p should be reported as well as the value of k used to obtain U. The methodology presented here for indirect measurements is illustrated in Sect. 6.4.1 with a case study involving the lead probe, a secondary standard used to measure neutron yield.

6.4.1 Case Study: Neutron Yield Measurement

Many neutron measurements made within the NSE are referenced to a lead probe measurement via calibration. The lead probe is a secondary standard, maintained by the Primary Standards Lab (PSL), and is itself referenced to a primary standard lead probe, which is referenced to an ion beam neutron source that produces a known number of neutrons. The uncertainty of any neutron measurement made using a neutron detector is therefore linked to the uncertainty of the lead probe used to calibrate that detector. This uncertainty helps to inform decisions regarding characterization and qualification of various neutron sources.

The case study in this section illustrates the uncertainty analysis of a lead probe measurement of neutron yield following the steps outlined in Sect. 6.4 above for indirect measurements. It is adapted from Walsh et al. (2017). The R code used to perform the calculations is included in the text.

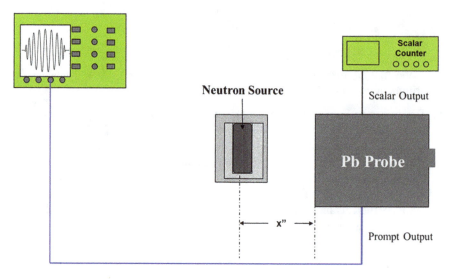

Fig. 6.1 Neutron measurement test setup

The sources of uncertainty in measuring neutron yield via a lead probe include a calibration factor, F, the background radiation counts, B, the total counts from the lead probe, S, and the attenuation factor, A, determined by the geometry of the test setup. The test setup is illustrated in Fig. 6.1.

The lead probe in Fig. 6.1 is set up perpendicular to the neutron source. This orientation and the distance x" between the two affects the calibration factor F. The attenuation factor A is 1.00 in air. The lead probe produces the scalar count (neutron count + background count) used in the measurement equation.

The resulting measurement equation for the total number of neutrons produced, with the measurand denoted by η, is

$$\eta = AfF(S-B). \tag{6.40}$$

```
# Measurement equation
eta <- function(A, f, F, S, B) {A*f*F*(S - B)}
```

The Type A evaluation of uncertainty is based on the scalar counts (S) and the background counts (B). The Type B evaluation is based on calibration reports for the attenuation factor (A) and the calibration factor (F). The constant f is a conversion factor related to the efficiency of neutron detection using a lead probe. It is determined by the PSL, and is treated as essentially known with zero Type A or Type B uncertainty. The experimental test plan consisted of 30 scalar measurements taken at multiple settings of the neutron source, along with 10 background measurements made before beginning the test. The results presented here are based on 30 scalar measurements taken at a single setting of the neutron source. The purpose of the

6.4 Indirect Measurements and the Indirect Measurement Model

uncertainty analysis was to quantify the uncertainty of a *single* measurement of neutrons from a neutron source.

For the indirect measurement case with multiple factors, it is helpful to construct a summary table for the uncertainty analysis. The table should include the distributional assumptions, as well as the means and standard uncertainties used in the analysis (Table 6.3).

```
# Mean values for input quantities and standard uncertainties
f <- 4353
A <- c(1, 0.020)
F <- c(1.27, 0.032)
S <- c(9700, 98.5)
B <- c(80, 2.83)
```

The scalar count (S) had mean 9700 based on the setting of the neutron source used for the 30 test shots. Assuming a Poisson distribution (Sect. 4.4.1.2) for the counting process, the Type A standard uncertainty for a *single* scalar neutron measurement is $u(S) = \sqrt{9700} = 98.5$. Concern is with the uncertainty associated with a single measurement of neutron yield because it can be measured only once due to the destructive nature of the test (see Sect. 11.4). The background radiation count (B) had mean 80 based on ten measurements prior to the 30 test shots. A single background count B is assumed to have a Poisson distribution, while the mean of ten background counts is approximately Normal, based on the central limit theorem. The Type A standard uncertainty for the average of ten background counts is thus approximated by

$$u(B) = \frac{\sqrt{80}}{\sqrt{10}} = 2.83. \tag{6.41}$$

In this case the numerator is divided by $\sqrt{10}$ because the average of the ten measurements is the best estimate of the background count at the time-of-test. This assumes a fairly constant background rate over the short time interval making the ten measurements.

The Type A standard uncertainties in neutron count η, associated with S and B, are based on the following sensitivity coefficients determined from Eq. (6.34):

$$\frac{\partial \eta}{\partial S} = AfF, \tag{6.42}$$

and

$$\frac{\partial \eta}{\partial B} = -AfF. \tag{6.43}$$

Table 6.3 Distributions for input quantities and standard uncertainties

Input quantity	Mean (μ)	Standard Uncertainty type A	PDF type A	Standard Uncertainty type B	PDF type B
f	4353				
A	1.00			0.020	Uniform
F	1.27			0.032	Uniform
S scalar count	9700	98.5	Poisson		
B background count	80	2.83	Normal		

```
# Partial derivative with respect to S
dNdS <- Deriv(eta, "S")
dNdS
## function (A, f, F, S, B)
## A * f * F
# Partial derivative with respect to B
dNdB <- Deriv(eta, "B")
dNdB
## function (A, f, F, S, B)
## - (A * f * F)
```

The Type A combined standard uncertainty, using Table 6.3 and the standard uncertainties and sensitivity coefficients, is

$$u_A(\eta) = \sqrt{(AfF)^2 u^2(S) + (-AfF)^2 u^2(B)}$$
$$= \sqrt{(544,539)^2 + (15,645)^2} \qquad (6.44)$$
$$= 544,800.$$

```
# Type A combined standard uncertainty
uS <- dNdS(A = A[1], f = f, F = F[1])*S[2]
uB <- dNdB(A = A[1], f = f, F = F[1])*B[2]
uTypeA <- sqrt(uS^2 + uB^2)
uTypeA
## [1] 544763.2
```

The Type B combined standard uncertainties for A and F are based on the following sensitivity coefficients from Eq. (6.34):

$$\frac{\partial \eta}{\partial A} = fF(S - B), \qquad (6.45)$$

and

6.4 Indirect Measurements and the Indirect Measurement Model

$$\frac{\partial \eta}{\partial F} = Af(S - B). \tag{6.46}$$

```
# Partial derivative with respect to A
dNdA <- Deriv(eta, "A")
dNdA
## function (A, f, F, S, B)
## f * F * (S - B)
# Parital derivative with respect to F
dNdF <- Deriv(eta, "F")
dNdF
## function (A, f, F, S, B)
## A * f * (S - B)
```

The Type B combined standard uncertainty, using Table 6.3 and the standard uncertainties and sensitivity coefficients, is

$$\begin{aligned} u_B(\eta) &= \sqrt{(fF(S-B))^2 u^2(A) + (Af(S-B))^2 u^2(F)} \\ &= \sqrt{(1,063,647)^2 + (1,340,028)^2} \\ &= 1,710,853. \end{aligned} \tag{6.47}$$

```
# Type B combined standard uncertainty
uA <- dNdA(f = f, F = F[1], S = S[1], B = B[1])*A[2]
uF <- dNdF(A = A[1], f = f, S = S[1], B = B[1])*F[2]
uTypeB<- sqrt(uA^2 + uF^2)
uTypeB
## [1] 1710853
```

The combined standard uncertainty is calculated as the RSS of the Type A and Type B combined standard uncertainties:

$$\begin{aligned} u_c(\eta) &= \sqrt{u_A^2(\eta) + u_B^2(\eta)} \\ &= \sqrt{(544,800)^2 + (1,710,853)^2} \\ &= 1.796 \times 10^6. \end{aligned} \tag{6.48}$$

```
# Combined Uncertainty
uC <- sqrt(uTypeA^2 + uTypeB^2)
uC
## [1] 1795501
```

The estimate of the total neutron yield comes from Eq. (6.40), substituting means,

Table 6.4 Type A and Type B degrees of freedom for lead probe example

Input quantity	Type A degrees of freedom	Type B degrees of freedom
f		
A		12.5
F		12.5
S	29	
B	9	

$$\eta = AfF(S - B)$$
$$= (1.00)(4353)(1.27)(9700 - 80) \qquad (6.49)$$
$$= 5.32 \times 10^7 \text{ neutrons.}$$

```
# Total neutron count using means
n <- eta(A = A[1], f = f, F = F[1], S = S[1], B = B[1])
n
## [1] 53182340
```

A second table should be constructed to display the Type A and Type B degrees of freedom for each factor (Table 6.4).

In this experiment, the degrees of freedom associated with each Type A standard uncertainty are equal to the number of independent observations of the related measurement. The mean scalar count (S) was based on 30 measurements of the neutron source, and the background count (B) was based on ten measurements, resulting in $\nu_S = 29$ and $\nu_B = 9$. The uncertainty intervals for both A and F were based on calibration reports, and the relative error in their respective uncertainties was thought to be no more than 20%. Applying Eq. (6.16), the degrees of freedom associated with each Type B source of uncertainty were set equal to 12.5. Note that the degrees of freedom associated with the t-distribution need not be an integer.

The effective degrees of freedom were calculated using the W-S formula

$$\nu_{\mathit{eff}} = \frac{u_c^4(\eta)}{\sum_{i=1}^{4} \frac{(c_i u(x_i))^4}{\nu_i}}$$

$$= \frac{(1.796 \times 10^6)^4}{\left(\frac{(544,539)^4}{29} + \frac{(15,645)^4}{9} + \frac{(1,063,647)^4}{12.5} + \frac{(1,340,028)^4}{12.5}\right)} \cong 29$$

$$(6.50)$$

```
# Effective degrees of freedom
dof <- data.frame(A = 12.5, F = 12.5, S = 29, B = 9)
veff <- uC^4/ ((uS^4/dof$S) + (uB^4/dof$B) + (uA^4/dof$A) + (uF^4/dof
```

6.4 Indirect Measurements and the Indirect Measurement Model

```
$F))
veff
## [1] 28.63258
```

It should be noted that the W-S approximation for the indirect measurement case appears slightly different than for the direct measurement equation Eq. (6.19), as the sensitivity coefficients are included here. The expanded uncertainty is calculated using the coverage factor determined from the t-distribution for a 95% level of confidence (two-sided interval):

$$\begin{aligned}U_\eta^{95} &= t_{0.975}(29) \times u_c(\eta) \\ &= (2.05) \times (1.796 \times 10^6) \\ &= 3.672 \times 10^6\end{aligned} \qquad (6.51)$$

```
# Uncertainty interval
uEta <- qt(p = .975, df = 29) * uC
uEta
## [1] 3672213
```

The uncertainty interval (95% confidence) for a single measurement of total neutron yield is thus:

$$\begin{aligned}\eta \pm U_\eta^{95} &= 5.32 \times 10^7 \pm 3.67 \times 10^6 \\ &= 5.32 \times 10^7 \pm 6.9\% \text{ neutrons.}\end{aligned} \qquad (6.52)$$

```
interval <- c(n - uEta, n + uEta)
interval
## [1] 49510130 56854555
```

The resulting interval Eq. (6.52) provides a range of values that could reasonably be expected from a single measurement of neutron yield from a laboratory neutron source using the lead probe secondary standard. The uncertainties, not reported here, were evaluated over a wide range of neutron source settings. These results were used, via a traceability chain, to quantify the uncertainty associated with neutron detectors measuring total output of neutron sources. The uncertainties are used both in product acceptance and characterization of neutron source performance.

6.4.2 Discussion

The case study in Sect. 6.3.1 illustrates the recommended approach for determining the expanded uncertainty associated with indirect measurements. Neutron yield was measured using a lead probe secondary standard that is used to calibrate neutron detectors. Thirty measurements from the neutron source and ten background measurements were used to quantify the Type A standard uncertainty. Calibration reports were used to quantify the Type B standard uncertainties associated with the conversion factors. The standard uncertainty and expanded uncertainty terms were used, along with the k value determined from the degrees of freedom and t-distribution, to quantify the expanded uncertainty and resulting uncertainty interval for the measurand.

Given the measurement equation $\eta = AfF(S - B)$ it is of interest to quantify the effect that each variable in the equation has on the total uncertainty of the neutron measurement. This information can then be used to determine which variables may need more attention in terms of reducing the overall uncertainty.

For each variable X_i in a measurement equation $Y = f(X_1, X_2, \ldots, X_N)$, a simple estimate of the relative contribution of that variable to total uncertainty is % of total uncertainty associated with variable X_i

$$= \frac{(c_i u(x_i))^2}{u_c^2(y)} \times 100\%. \qquad (6.53)$$

This ratio is the standard uncertainty associated with the input quantity X_i (squared) as a percent of the combined standard uncertainty (squared). The resulting percent of total uncertainty for each variable in the expression for neutron yield was determined using Eq. (6.53) and appear in Table 6.5 below.

```
# Vary A only
sample <- runif(m, A[1]-A[2]/(sqrt(1/12)*2), A[1]+A[2]/(sqrt(1/12)
*2))
ABoot <- eta(A = sample, f = f, F = F[1], S = S[1], B = B[1])
varA <- var(ABoot)

# Vary F only
sample <- runif(m, F[1]-F[2]/(sqrt(1/12)*2), F[1]+F[2]/(sqrt(1/12)
*2))
FBoot <- eta(A = A[1], f = f, F = sample, S = S[1], B = B[1])
varF <- var(FBoot)

# Vary S only
sample <- rpois(m, S[1])
```

Table 6.5 Percent of total uncertainty by variable with $(S, B) = (9700, 80)$

Variable	% of total uncertainty
f	0.0
A	35.1
F	55.7
S	9.2
B	0.0

6.5 Related Reading

```
SBoot <- eta(A = A[1], f = f, F = F[1], S = sample, B = B[1])
varS <- var(SBoot)

# Vary B only
sample <- rnorm(m, B[1], B[2])
BBoot <- eta(A = A[1], f = f, F = F[1], S = S[1], B = sample)
varB <- var(BBoot)
# Calculate variance
totalVar <- sum(varA, varF, varS, varB)
varResults <- t(data.frame(f = 0,
 A = varA/totalVar*100,
 F = varF/totalVar*100,
 S = varS/totalVar*100,
 B = varB/totalVar*100))
colnames(varResults) <- c('Percent of Standard Uncertainty')
varResults
## Percent of Standard Uncertainty
## f 0.000
## A 35.09
## F 55.70
## S 9.20
## B 0.012
```

Knowledge of the relative "components of variation" can suggest ways to improve the measurement. Key results from this analysis were that the measurement uncertainty could be reduced by reducing the uncertainty in the attenuation factor A and the calibration factor F. Examination of Eqs. (6.44) and (6.47), however, shows that significant reduction in uncertainty can be realized by increasing the net scalar counts per source neutron (i.e., reducing f). However, this would require a new detector technology with improved efficiency, requiring substantial design effort.

6.5 Related Reading

The direct measurement model (Eq. 6.8) presented here is typically not covered in metrology guidelines such as the JCGM:100 (GUM). This may be because direct measurements are less common in practice, and their uncertainty analyses often result in intervals of the form

$$\bar{y} \pm k u_c(y),$$

where $\bar{y}$ and $u_c(y)$ are based on simple summary statistics. The model was included in Chap. 6 primarily to introduce the idea of an additive statistical model that could be extended to include multiple fixed and random terms in a more complicated ANOVA (Sect. 9.3). A metrology guideline that does cover the direct measurement model is ISO 5725-1: 1994, which provides guidelines in designing, implementing, and analyzing interlaboratory repeatability and reproducibility studies. The basic model of ISO 5725 is similar to Eq. (6.8), with a bias term that represents the systematic deviations of each laboratory from an overall mean.

Certain measurands, such as length or weight, can be measured either directly or indirectly. A comparison of factors such as cost, complexity, accuracy, and repeatability will usually favor one technique over the other. Agamloh (2009) compared direct and indirect measurements of induction motor efficiency. The direct measurement was simpler and faster than the more complicated, costlier indirect measurement. His uncertainty analysis, however, indicated that the direct measurement was not accurate enough to be used as a substitute for the indirect measurement. Comparisons such as these are often necessary as technology advances both the accuracy and complexity of measurement techniques.

Uncertainty analyses associated with indirect measurements are the focus of numerous metrology guidelines. The JCGM 100 (GUM) includes the most detail on using "propagation of uncertainties" to analyze indirect measurements. NIST Technical Note 1297 (Taylor and Kuyatt 1994) was written as a companion guide to the GUM, with emphasis on NIST measurement results. JCGM 102 generalizes the GUM to the case of multiple output quantities, and JCGM 103 (in development) provides guidance on how to define the measurand and develop a measurement model. Numerous case studies illustrating the approaches outlined in this chapter can be found in the archives of journals such as *Metrologia*, *Measurement Science and Technology*, and *Measure* and in the volumes of the NIST Research Library. These case studies tend to be very discipline-specific, typically including the instrumentation, test setup, and procedures tailored to the measurement of interest.

Practitioners are encouraged to consider the assumptions and limitations associated with applying the basic GUM methodology. An overview of these limitations is given by Akdogan (2018). The GUM recommendation of using a first-order Taylor series expansion for propagating uncertainties usually works well. However, if the nonlinearities in the measurement equation are significant or if the input quantities are highly non-normal or correlated, the GUM methodology may not work well. Becerra and Hernandez (2006) investigate the effect of correlated input variables on air density uncertainty. They conclude that if the Type A contribution to uncertainty is smaller than the Type B contribution, the effect of correlation is negligible, but if the Type A contribution is larger than the Type B contribution, the effect of correlation can be significant. Crowder and Moyer (2006) investigate the performance of the GUM with nonlinear measurement equations and small sample size. They conclude that the GUM prescription may not work well when the inputs to the measurement equation (the $X_i's$) have low signal to noise ratios, that is, when the μ_{X_i}/σ_{X_i} values are less than one. In this case, higher-order terms will be needed in the Taylor series approximation. Wang and Iyer (2005) propose a generalized inference approach as an alternative method to the GUM for cases like these when the Taylor series expansion does not lead to satisfactory approximations.

The use of the Welch–Satterthwaite approximation Eq. (6.5) to estimate effective degrees of freedom has also been investigated. Ballico (2000) shows that this approximation may not work well when there is a significant difference in the magnitude of uncertainty terms. In particular, adding small uncertainty terms with large degrees of freedom to a much larger term with small degrees of freedom results in a distribution not well approximated by the W-S approximation.

The Monte Carlo method outlined in the GUM Supplement 1 will often be an appropriate alternative in these special cases. That is the subject of Chap. 8.

6.6 Exercises

6.1. Scientists in Hawaii want to assess the uncertainty of a direct thermocouple measurement of lava temperature when it first flows from a volcano. A sample of ten measurements appears in Table 6.6. Compute the Type A standard uncertainty for this measurement.

6.2. A calibration report for the thermocouple under investigation reports a tolerance of $\pm 75\,^\circ\mathrm{C}$ over the range 1000–1300 $^\circ\mathrm{C}$. Compute the Type B standard uncertainty associated with the calibration of this thermocouple. Assume a uniform distribution (4.4.1.1) over the stated tolerance interval.

6.3. Compute the combined standard uncertainty and the effective degrees of freedom using the Welch–Satterthwaite approximation. What assumption would you make regarding Type B degrees of freedom?

6.4. Compute the coverage factor for a confidence level of 95.45%. Report the result of the measurement along with the expanded uncertainty.

6.5. How could you improve the measurement of volcanic lava temperature?

6.6. Suppose an indirect measurement is defined by the equation
$Y = f(X_1, X_2) = X_1^2 + X_2^2$, where X_1 and X_2 are independent with $N(0, 1)$ distributions. Using a first-order approximation, estimate the expanded uncertainty of the measurand Y. Is this a good approximation? Why or why not?

Table 6.6 Lava temperature data

Reading number	Temperature ($^\circ$C)
1	1130.60
2	1210.64
3	1211.25
4	1047.01
5	1135.67
6	1189.19
7	1237.59
8	1164.33
9	1225.51
10	1165.79
$\bar{y}$	1171.76

Table 6.7 Mean input quantities and standard uncertainties for XRF measurement

Input quantity	Mean (μ)	Standard uncertainty Type A	Type A degrees of freedom	Type A PDF
$Y_{uncorrected}$	0.4409 μm	0.00360	19	Normal
X_1	0.1820 μm	0.00093	9	Normal
X_2	0.1823 μm	0.00058	19	Normal

6.7. Now assume X_1 and X_2 are independent with $N(1000, 1)$ distributions. Using a first-order approximation, estimate the expanded uncertainty of the measurand Y. Is this a good approximation? Why or why not?

6.8. What steps should be followed in developing an indirect measurement equation? What is a "hybrid" measurement equation? When might you use one?

6.9. A corrected X-ray fluorescence (XRF) thickness measurement, $Y_{corrected}$, is estimated indirectly from an uncorrected thickness measurement, $Y_{uncorrected}$, via the relationship

$$Y_{corrected} = \left(\frac{X_1}{X_2}\right) \cdot Y_{uncorrected},$$

where X_1 and X_2 are independent correction factors. Table 6.7 above provides the information needed to complete an uncertainty analysis. The Type B uncertainties were negligible and were not part of the analysis.

Using the GUM methodology (Sect. 6.4), perform an uncertainty analysis for the measurand $Y_{corrected}$.

6.10. How is a "components of variation" analysis used as part of an uncertainty analysis? Perform this analysis for the measurand $Y_{corrected}$.

References

Agamloh, E. A.: Comparison of direct and indirect measurement of induction motor efficiency. https://doi.org/10.1109/IEMDC.2009.5075180 (2009)

Akdogan, A.: Metrology. Intechopen Limited, London (2018)

Ballico, M.: Limitations of the Welch-Satterthwaite approximation for measurement uncertainty calculations. Metrologia. **37**, 61–64 (2000)

Becerra, L., Hernandez, I.: Evaluation of the air density uncertainty: the effect of the correlation of input quantities and higher order terms in the Taylor series expansion. Meas. Sci. Technol. **17**, 2545–2550 (2006)

Crowder, S.V., Moyer, R.D.: A two-stage Monte Carlo approach to the expression of uncertainty with non-linear measurement equation and small sample size. Metrologia. **43**(1), 34–41 (2006)

ISO 5725-2: Accuracy (trueness and precision) of measurement methods and results - Part 2: Basic method for the determination of repeatability and reproducibility of a standard measurement method. (1994)

References

JCGM 100: Evaluation of Measurement Data - Guide to the Expression of Uncertainty in Measurement. (2008)

JCGM 102: Evaluation of Measurement Data - Supplement 2 to the -Guide to the Expression of Uncertainty in Measurement - Extension to any Number of Output Quantities. (2011)

Montgomery, D.: Introduction to Statistical Quality Control. Wiley, New York (2013)

Phillips, S.D., Eberhardt, K.R.: Guidelines for expressing the uncertainty of measurement results containing uncorrected bias. J. Res. NIST. **102**, 577–585 (1997)

Satterthwaite, F.: An approximate distribution of estimates of variance components. Biom. Bull. **2**(6), 110–114 (1946)

Taylor, B.N., Kuyatt, C.E.: Guidelines for Evaluating and Expressing the Uncertainty of NIST Measurement Results. NIST Technical Note 1297. National Institute of Standards and Technology, Gaithersburg, MD (1994)

Walsh, D., Crowder, S., Burns, E., Thacher, P.: Estimating uncertainty in laboratory neutron measurements made with a lead probe scaler detector. SAND2017-6863J. Sandia National Laboratories. (2017)

Wang, C.M., Iyer, H.K.: On higher-order corrections for propagating uncertainties. Metrologia. **42**, 406–410 (2005)

Chapter 7
Analytical Methods for the Propagation of Uncertainties

7.1 Introduction

This chapter presents analytical methods for propagating uncertainties through an indirect measurement model. The method was first introduced by Carl Gauss in the 1820s (Gauss 1821), and was further developed by Kline and McClintock (1953). In 1995, the Joint Committee for Guides in Metrology standardized the method using it as the basis for its GUM (JCGM 100 2008), so the technique is often referred to as the GUM method for uncertainty propagation.

Given an indirect measurement model, the measurand is a function of multiple measured values:

$$Y = f(X_1, X_2, \ldots, X_N), \qquad (7.1)$$

where the inputs $X_1, X_2, \ldots, X_N$ are the physical quantities to be measured and the function f relates the measurands to the output quantity Y. As an example, the output quantity Y could represent the density of a cylinder, while X_1, X_2, and X_3 are quantities of mass, diameter, and height, and f is the functional relationship between the quantities (density = mass/volume).

An estimate of Y, denoted y, is determined by making measurements—or estimates—of each quantity. These estimates are denoted $x_1, x_2, \ldots, x_N$ (see GUM Sect. 4.1.4). Then y is computed using $y = f(x_1, x_2, \ldots x_n)$. The input estimates are often the average of multiple independent measurements on the same quantity. Each input estimate has an associated standard uncertainty estimate $u(x_i)$, computed from Type A and Type B uncertainty components as described in Chap. 6. Continuing the density example, an estimate of cylinder density is $y = \rho = f(m, d, h) = \frac{4m}{\pi h d^2}$, where m, d, and h are the average measured values of mass, diameter, and height, respectively.

The remainder of this section considers how to determine $u_c(y)$, the combined standard uncertainty in the output estimate y. For a small subset of problems, $u_c(y)$

can be calculated exactly, but in most cases, the method makes an approximation to linearize the measurement function. A derivation of the uncertainty expression introduced in Chap. 6 (Eq. 6.34) assumes the measurement model can be linearly approximated and the inputs are uncorrelated. The correlated terms are then retained in the uncertainty derivation. Finally, the case where nonlinear terms in the measurement model become significant is examined.

7.2 Mathematical Basis

This section reviews several important properties of the sums of random variables. These results will be used, along with a Taylor series approximation of Eq. (7.1), to estimate the combined uncertainty of any arbitrary measurement equation.

First, define $E(\cdot)$ as the expectation operator, $\text{Var}(\cdot)$ as the variance operator, and $\text{Cov}(\cdot, \cdot)$ as the covariance operator for random variables (see Casella and Berger 2002). The expectation is simply the weighted average of a probability distribution, the variance $\text{Var}(X) = E(X - E(X))^2$ is a measure of spread of the distribution, equal to the square of the standard deviation, and the covariance is a measure of the strength of the relationship between two random variables. If $X_1, X_2, \ldots, X_N$ are random variables (see GUM C.2.2) and $a_1, a_2, \ldots, a_N$ are constants, then some useful properties of expectation and variance are:

$$E(X + a) = E(X) + a$$
$$E(aX) = aE(X)$$
$$E(X_1 + X_2) = E(X_1) + E(X_2)$$
$$\text{Var}(X + a) = \text{Var}(X)$$
$$\text{Var}(aX) = a^2 \text{Var}(X)$$
$$\text{Var}(X_1 + X_2) = \text{Var}(X_1) + \text{Var}(X_2) + 2\text{Cov}(X_1, X_2)$$

The above properties can be combined and generalized to multiple random variables:

$$E\left(\sum_{i=1}^{N} a_i X_i\right) = \sum_{i=1}^{N} a_i E(X_i),$$

and

7.3 The Simple Case: First-Order Terms with Uncorrelated Inputs

$$\mathrm{Var}\left(\sum_{i=1}^{N} a_i X_i\right) = \sum_{i=1}^{N} a_i^2 \mathrm{Var}(X_i) + 2\sum_{i=1}^{N-1}\sum_{j=i+1}^{N} a_i a_j \mathrm{Cov}(X_i, X_j) \quad (7.2)$$

If the X_i's are uncorrelated, the covariance terms are 0, and Eq. (7.2) reduces to

$$\mathrm{Var}\left(\sum_{i=1}^{N} a_i X_i\right) = \sum_{i=1}^{N} a_i^2 \mathrm{Var}(X_i) \quad (7.3)$$

To determine the expectation and variance of an indirect measurand Y, one must find the expected value of Y, $E(Y) = E[f(X_1, X_2, \ldots, X_N)]$, and the variance of Y, $\mathrm{Var}(Y) = \mathrm{Var}[f(X_1, X_2, \ldots, X_N)]$. For example, consider a model that converts measurements units such as meters to centimeters: $f(X) = aX$. Using 7.3 the variance of $f(X)$ is $a^2\mathrm{Var}(X) = a^2 u^2(x)$. Therefore, the uncertainty of the converted value is the uncertainty of the measurement of x in original units multiplied by the units conversion factor a.

Equations (7.2) and (7.3) allow for finding the expectation and variance of a measurement model consisting of a linear combination of variables, such as simple shifting ($f(X) = X + a$), scaling ($f(X) = aX$), or adding multiple scaled variables ($f(X_1, X_2) = a_1 X_1 + a_2 X_2$). In these cases, an exact variance is computed. Of course, most measurement models do not fall into one of these simple linear combinations, so approximation of more complex measurement equations must be considered. The simplest approximate case uses a linear approximation of the nonlinear model with uncorrelated variables, considered in Sect. 7.3.

7.3 The Simple Case: First-Order Terms with Uncorrelated Inputs

The previous section considered propagating uncertainty for linear combinations of multiple measurements. Most measurement models, however, will be nonlinear. The approach for an arbitrary (nonlinear) measurement equation $Y = f(X_1, X_2)$ is to approximate the equation with a first-order (linear) Taylor series expansion. The derivation of the estimated combined standard uncertainty $u_c(y)$ uses $u^2(x_i)$ to represent an estimate of $\mathrm{Var}(X_i)$ and $u(x_i, x_j)$ to represent an estimate of $\mathrm{Cov}(X_i, X_j)$. The "realization" or "estimate" of the random variable X_i is denoted by x_i, and the estimate of the measurand Y is denoted by y.

Expanding about the input variables' respective means, μ_{X_1} and μ_{X_2}, the first-order approximation takes the form:

$$Y = f(X_1, X_2) \approx f(\mu_{X_1}, \mu_{X_2}) + \frac{\partial f}{\partial X_1}(X_1 - \mu_{X_1}) + \frac{\partial f}{\partial X_2}(X_2 - \mu_{X_2}). \quad (7.4)$$

The arbitrary function f, approximated this way, becomes a linear combination of multiple variables, and therefore may be addressed using the variance properties of Sect. 7.2. Note that the partial derivatives $\left(\frac{\partial f}{\partial X_1}\text{ and }\frac{\partial f}{\partial X_2}\right)$ are evaluated at the average measured values x_1 and x_2, so they reduce to constant multipliers.

The variance of the approximation is

$$\begin{aligned}\text{Var}(Y) &= \text{Var}(f(X_1, X_2)) \\ &\approx \text{Var}\left(f(\mu_{X_1}, \mu_{X_2}) + \frac{\partial f}{\partial X_1}(X_1 - \mu_{X_1}) + \frac{\partial f}{\partial X_2}(X_2 - \mu_{X_2})\right)\end{aligned} \quad (7.5)$$

Applying Eq. 7.3, the first term $f(\mu_{X_1}, \mu_{X_2})$ is a constant (having 0 variance) that affects the mean of the approximating function but does not affect its variance. The second and third terms apply constant multipliers ($\frac{\partial f}{\partial X_1}$ and $\frac{\partial f}{\partial X_2}$), so the estimated variances of those terms are $\left(\frac{\partial f}{\partial x_1}\right)^2 u^2(x_1)$ and $\left(\frac{\partial f}{\partial x_2}\right)^2 u^2(x_2)$, respectively. The straightforward addition of these two components gives the estimated combined variance:

$$u_c^2(y) = \left(\frac{\partial f}{\partial x_1}\right)^2 u^2(x_1) + \left(\frac{\partial f}{\partial x_2}\right)^2 u^2(x_2). \quad (7.6)$$

This result can then be generalized to any number of input variables:

$$u_c^2(y) = \sum_{i=1}^{N} \left(\frac{\partial f}{\partial x_i}\right)^2 u^2(x_i), \quad (7.7)$$

which is the standard expression for uncertainty from the GUM and from Chap. 6 (Eq. 6.34).

This equation assumes that all the input measurements are uncorrelated. It also assumes that the measurement model is approximately linear in the region of interest, so that using a first-order Taylor series approximation introduces only minimal error in the result.

Finally, the standard uncertainty $u_c(y)$ can be expanded to the desired confidence level using an appropriate coverage factor k. The expanded uncertainty is

$$U = k \cdot u_c(y), \quad (7.8)$$

7.3 The Simple Case: First-Order Terms with Uncorrelated Inputs

where $k = t_p(\nu_{\text{eff}})$ is computed from the t-distribution with confidence level p and effective degrees of freedom ν_{eff}. The Welch–Satterthwaite approximation can be used to calculate ν_{eff}:

$$\nu_{\text{eff}} = \frac{u_c^4(y)}{\sum_{i=1}^{N} \frac{(c_i u(x_i))^4}{\nu_i}} \tag{7.9}$$

Here, the c_i variables are the "sensitivity coefficients," which are simply shorthand for the partial derivative terms computed previously: $c_i = \frac{\partial f}{\partial x_i}$.

7.3.1 Measurement Examples

Consider measuring the density of a cylinder. The mass (m), diameter (d), and height (h) are measured, and the density found from the measurement equation:

$$\rho = \frac{m}{\pi \left(\frac{d}{2}\right)^2 h} \tag{7.10}$$

```
# Measurement equation
rho <- function(m, h, d) {m/(pi*(d/2)^2*h)}
```

The measured m, h, and d values, and their standard uncertainties $u(m)$, $u(h)$, and $u(d)$ are all known from the measurement results and setup (Table 7.1).

```
# Mean values and uncertainties for variables
m <- c(5.0, 0.001)
h <- c(2.0, 0.005)
d <- c(0.5, 0.005)
```

Applying Eq. (7.7), the expression for the combined variance in density is

$$u_c^2(\rho) = \left(\frac{\partial \rho}{\partial m}\right)^2 u^2(m) + \left(\frac{\partial \rho}{\partial h}\right)^2 u^2(h) + \left(\frac{\partial \rho}{\partial d}\right)^2 u^2(d) \tag{7.11}$$

The sensitivity coefficients (denoted as c_x) are:

Table 7.1 Measured values and uncertainties of cylinder density determination

Variable	Mean	Uncertainty
m	5.0 g	0.001 g
h	2.0 cm	0.005 cm
d	0.5 cm	0.005 cm

$$c_m = \frac{\partial \rho}{\partial m} = \frac{4}{\pi d^2 h}, c_h = \frac{\partial \rho}{\partial h} = \frac{4m}{\pi d^2 h^2}, c_d = \frac{\partial \rho}{\partial d} = -\frac{8m}{\pi d^3 h} \qquad (7.12)$$

```
# Partial derivative with respect to m
dpdm <- Deriv(rho, "m")
dpdm
## function (m, h, d)
## 1/(h * pi * (d/2)^2)

# Partial derivative with respect to h
dpdh <- Deriv(rho, "h")
dpdh
## function (m, h, d)
## {
##     .e1 <- (d/2)^2
##     -(m * pi * .e1/(h * pi * .e1)^2)
## }

# Partial derivative with respect to d
dpdd <- Deriv(rho, "d")
dpdd
## function (m, h, d)
## -(0.5 * (d * h * m * pi/(h * pi * (d/2)^2)^2))
```

Substituting into Eq. (7.11) and simplifying, the combined variance is

$$u_c^2(\rho) = \frac{16}{\pi^2 d^4 h^2} u^2(m) + \frac{16 m^2}{\pi^2 d^4 h^4} u^2(h) + \frac{64 m^2}{\pi^2 d^6 h^2} u^2(d) \qquad (7.13)$$

Substituting in the measured numbers gives a standard ($k = 1$) uncertainty of

$$u_c(\rho) = 0.26 \qquad (7.14)$$

```
# Value of rho
p <- rho(m = m[1], h = h[1], d = d[1])
p
## [1] 12.7324

# Combined standard uncertainty
um <- dpdm(m = m[1], h = h[1], d = d[1])*m[2]
uh <- dpdh(m = m[1], h = h[1], d = d[1])*h[2]
ud <- dpdd(m = m[1], h = h[1], d = d[1])*d[2]
uComb <- sqrt(um^2 + uh^2 + ud^2)
uComb
## [1] 0.2566423
```

7.4 First-Order Terms with Correlated Inputs

If all the measurement uncertainties have infinite degrees of freedom, $k = 1.96$ gives a 95% confidence interval, and the density is

$$\rho = 12.73 \text{g/cm}^3 \pm 0.52 \text{g/cm}^3 \ (95\%, k = 1.96). \tag{7.15}$$

```
#95% Confidence interval
k <- qnorm((1 - (.05/2)))
confInt <- k * uComb
confInt
## [1] 0.5030096
```

To identify the variable that is the biggest contributor to total uncertainty, we compute the proportion of the total variance that comes from each variable via the formula: $\frac{c_x^2 u^2(x)}{u_c^2(y)} \times 100\%$.

For this example, $u(m)$ contributes 0.01%, $u(h)$ contributes 1.54%, and $u(d)$ contributes 98.45% of the total uncertainty. If the goal is to reduce the total uncertainty, the focus should be on improving the uncertainty in the diameter measurement.

```
# Proportion of Variance
umProp <- um^2/uComb^2
umProp
## [1] 9.845184e-05

uhProp <- uh^2/uComb^2
uhProp
## [1] 0.0153831

udProp <- ud^2/uComb^2
udProp
## [1] 0.9845184
```

7.4 First-Order Terms with Correlated Inputs

In the previous derivation, the covariance term was set to 0 because the input variables were assumed to be uncorrelated. This is often not the case, so a complete derivation beginning with the Taylor series and retaining the covariance term is given here. Taking the same approach as the simple models in Sect. 7.3, f is approximated by the first-order Taylor series:

$$Y = f(X_1, X_2) \approx f(\mu_{X_1}, \mu_{X_2}) + \frac{\partial f}{\partial X_1}(X_1 - \mu_{X_1}) + \frac{\partial f}{\partial X_2}(X_2 - \mu_{X_2}). \quad (7.16)$$

Applying Eq. (7.2) term by term (including the covariance term), the combined variance is

$$\begin{aligned}
\text{Var}(Y) &\approx \text{Var}\left(f(\mu_{X_1}, \mu_{X_2}) + \frac{\partial f}{\partial X_1}(X_1 - \mu_{X_1}) + \frac{\partial f}{\partial X_2}(X_2 - \mu_{X_2}) \right) \\
&= \text{Var}\left(\frac{\partial f}{\partial X_1}(X_1 - \mu_{X_1}) \right) + \text{Var}\left(\frac{\partial f}{\partial X_2}(X_2 - \mu_{X_2}) \right) \\
&\quad + 2\text{Cov}\left(\frac{\partial f}{\partial X_1}(X_1 - \mu_{X_1}), \frac{\partial f}{\partial X_2}(X_2 - \mu_{X_2}) \right),
\end{aligned} \quad (7.17)$$

with all other variance and covariance terms equal to 0. This expression yields the following estimated combined variance, recalling that $u(x_1, x_2)$ represents an estimate of the covariance of X_1 and X_2:

$$u_c^2(y) = \left(\frac{\partial f}{\partial x_1} \right)^2 u^2(x_1) + \left(\frac{\partial f}{\partial x_2} \right)^2 u^2(x_2) + 2 \frac{\partial f}{\partial x_1} \frac{\partial f}{\partial x_2} u(x_1, x_2). \quad (7.18)$$

This result can also be generalized to any number of input variables as

$$u_c^2(y) = \sum_{i=1}^{N} \left(\frac{\partial f}{\partial x_i} \right)^2 u^2(x_i) + 2 \sum_{i=1}^{N-1} \sum_{j=i+1}^{N} \frac{\partial f}{\partial x_i} \frac{\partial f}{\partial x_j} u(x_i, x_j). \quad (7.19)$$

Note that when the covariance for all combinations of inputs is zero, this reduces to the expression found in Sect. 7.2 (Eq. 7.7).

7.4.1 Covariance, Correlation, and Effect on Uncertainty

It is often easier to work in terms of a correlation coefficient (Eq. 4.4) than a covariance. The correlation coefficient always ranges from -1 to $+1$, with uncorrelated values having a coefficient of 0. To convert covariance into correlation $r(x_i, x_j)$, use

$$r(x_i, x_j) = \frac{u(x_i, x_j)}{u(x_i) u(x_j)}. \quad (7.20)$$

Some common cases of correlated inputs occur when two quantities in the model are measured using the same piece of equipment, when temperature affects two or

7.4 First-Order Terms with Correlated Inputs

more measurements assuming the measurements are made at the same temperature, or when multiple measurements are made simultaneously (as in the resistance example in Sect. 7.4.2). In many cases, the correlation is insignificant, but it can also be difficult to quantify. One technique is to take multiple measurements of the same sample and estimate the correlation between each input using the sampled data (Eq. 4.4). In some cases, for example, with temperature correlations, it may be easier to include temperature as an independent variable in the model itself rather than trying to determine a correlation coefficient between the other variables.

Correlated variables do not necessarily make the total uncertainty higher or lower. The direction of change due to correlation depends entirely on the sign of the sensitivity coefficient and the sign of the correlation coefficient. Consider measurements of electrical resistance using the model $R = V/I$. Applying Eq. (7.18) to this expression and computing the partial derivatives, it is easy to show that the constant multiplier in the covariance term of the uncertainty expression is $-(2V/I^3)$. Voltage and current always have the same sign in this situation, so a positive covariance always *decreases* the total uncertainty. However, if the same measurements were used to determine electrical power using the model $P = IV$, the multiplier in the covariance term is $2IV$, and a positive covariance *increases* the total uncertainty.

7.4.2 Measurement Examples

Semiconductor nanowires are grown by a Vapor–Liquid–Solid method in a furnace under tightly controlled conditions of temperature, pressure, and atmospheric composition. To use a very simplified model of the complex chemistry of nanowire growth, material properties of the nanowires, such as electrical resistivity, depend on the furnace conditions, while the length and diameter dimensions depend on the amount of time spent in the furnace. Because both length and diameter depend on time, these variables are correlated.

The expected resistance of a batch of nanowires can be calculated using the resistivity (ρ), length (L), and diameter (d):

$$R = \frac{4\rho L}{d^2} \qquad (7.21)$$

```
# Measurement equation
R <- function(rho, L, d) { (4*rho*L)/d^2 }
```

The furnace has been well characterized, so for a given process the resistivity is known to be 9.6×10^{-5} Ω-m with uncertainty of 2.0×10^{-6} Ω-m. After growing a batch of wires using this process, 20 wires were sampled and the dimensions measured to be $d = 180$ nm $\pm$ 20 nm and $L = 5.2$ μm $\pm$ 0.20 μm (all $k = 1$ standard uncertainties). Using Eq. (4.4) to calculate the correlation coefficients for the set of 20 measurement samples, the correlation coefficient was determined to be +0.64.

To determine the expected value and uncertainty of the nanowire resistance, calculate the sensitivity coefficients (first converting all length dimensions to meters):

$$c_\rho = \frac{\partial R}{\partial \rho} = \frac{4L}{d^2} = 6.4 \times 10^8$$

$$c_L = \frac{\partial R}{\partial L} = \frac{4\rho}{d^2} = 1.2 \times 10^{10}$$

$$c_d = \frac{\partial R}{\partial d} = -\frac{8L\rho}{d^3} = -6.9 \times 10^{11} \qquad (7.22)$$

```
# Mean values and uncertainties for variables
rho <- c(9.6e-5, 2.0e-6)
L <- c(5.2e-6, 0.20e-6)
d <- c(180e-9, 20e-9)
# Partial derivative with respect to rho
dRdrho <- Deriv(R, "rho")
dRdrho
## function (rho, L, d)
## 4 * (L/d^2)
# Partial derivative with respect to L
dRdL <- Deriv(R, "L")
dRdL
## function (rho, L, d)
## 4 * (rho/d^2)
# Partial derivative with respect to d
dRdd <- Deriv(R, "d")
dRdd
## function (rho, L, d)
## - (8 * (L * rho/d^3))
```

The combined standard (uncorrelated) uncertainty is

$$u_c(R) = \sqrt{\left(\frac{\partial R}{\partial \rho}\right)^2 u^2(\rho) + \left(\frac{\partial R}{\partial L}\right)^2 u^2(L) + \left(\frac{\partial R}{\partial d}\right)^2 u^2(d)}$$

$$= \sqrt{\begin{array}{c}(6.4 \times 10^8)^2 (2.0 \times 10^{-6})^2 + (1.2 \times 10^{10})^2 (0.2 \times 10^{-6})^2 \\ + (-6.9 \times 10^{11})^2 (20 \times 10^{-9})^2\end{array}}$$

$$= 14066\ \Omega. \qquad (7.23)$$

7.4 First-Order Terms with Correlated Inputs

```
# Uncorrelated combined standard uncertainty
urho <- dRdrho(rho = rho[1], L = L[1], d = d[1])*rho[2]
uL <- dRdL(rho = rho[1], L = L[1], d = d[1])*L[2]
ud <- dRdd(rho = rho[1], L = L[1], d = d[1])*d[2]
uComb <- sqrt(urho^2 + uL^2 + ud^2)
uComb
## [1] 13958.27
```

Adding in the terms due to the correlated variables, expressed in terms of the correlation coefficient r (instead of covariance):

$$u_{c,\text{corr}}(R) = \sqrt{u_c^2(R) + 2\frac{\partial R}{\partial L}\frac{\partial R}{\partial d}u(L)u(d)r(L,d)}$$

$$= \sqrt{14066 + 2(1.2\times 10^{10})(-6.9\times 10^{11})(0.2\times 10^{-6})(20\times 10^{-9})(0.64)}$$

$$= 12468\ \Omega. \tag{7.24}$$

```
# Correlated combined standard uncertainty
uLd <- uL*ud
uCombCorr <- sqrt(uComb^2 + 2*uLd*0.64)
uCombCorr
## [1] 12380.63
```

Because the sign of the sensitivity coefficients for the correlated term is negative, the total uncertainty is reduced when correlation is accounted for. Note that the correlation between resistivity and the dimension variables is assumed to be zero, so they do not contribute.

The W-S formula (Eq. 7.9) is used to expand the standard uncertainty to a 95% confidence level. Assuming infinite degrees of freedom for resistivity, and $(20 - 1) = 19$ degrees of freedom for length and diameter, the W–S formula gives

$$v_{\text{eff}} = \frac{u_{c,\text{corr}}^4(R)}{\frac{c_d^4 u^4(d)}{v_d} + \frac{c_L^4 u^4(L)}{v_L}} = \frac{12468^4}{\frac{(-6.9\times 10^{11})^4(20\times 10^{-9})^4}{19} + \frac{(1.2\times 10^{10})^4(0.2\times 10^{-6})^4}{19}}$$

$$= 12.67 \tag{7.25}$$

```
# Effective degrees of freedom
dof <- data.frame(L = 19, d = 19)
veff <- uCombCorr^4/((ud^4/dof$d) + (uL^4/dof$L))
veff
## [1] 12.67732
```

which gives a 95% confidence k-factor of 2.16. Therefore, the expected resistance for this batch of nanowires, to 95% confidence, reported to the typical 2 significant figures in uncertainty, is

$$R = 62 \text{ k}\Omega \pm 27 \text{ k}\Omega \ (95\%, k = 2.16). \tag{7.26}$$

```
# Value of R
Rval <- R(rho = rho[1], L = L[1], d = d[1])
Rval
## [1] 61629.63
#95% Confidence interval
k <- qt((1 - (.05/2)), 13)
confInt <- k * uCombCorr
confInt
## [1] 26746.73
```

7.5 Higher-Order Terms with Uncorrelated Inputs

The analytical method discussed so far uses a first-order Taylor series approximation of the measurement function. This assumes that the function can be modelled linearly in the region of interest (near the mean values of the measurands).

What does it mean to be nonlinear in this sense? Consider the model $f(x) = e^x$ shown in Fig. 7.1. On the left side, a small x uncertainty is given. Over a small range like this, the exponential can be treated as linear without introducing significant error. On the right side, x has a much larger uncertainty, and a linear approximation is no longer sufficient to cover the entire range of possible $f(x)$ values. Any function can be approximated as linear if the uncertainties are small enough. Unfortunately, it is difficult to know beforehand whether this is true for a particular case without calculating the nonlinear effects and comparing to the required number of significant digits.

To include nonlinear effects in the uncertainty expression, a higher-order Taylor series can be used. Here the second-order terms of the approximation are included for two input variables:

$$\begin{aligned} Y = f(X_1, X_2) \approx & f\left(\mu_{X_1}, \mu_{X_2}\right) + \frac{\partial f}{\partial X_1}\left(X_1 - \mu_{X_1}\right) + \frac{\partial f}{\partial X_2}\left(X_2 - \mu_{X_2}\right) \\ & + \frac{1}{2}\frac{\partial^2 f}{\partial X_1^2}\left(X_1 - \mu_{X_1}\right)^2 + \frac{\partial f}{\partial X_1}\frac{\partial f}{\partial X_2}\left(X_1 - \mu_{X_1}\right)\left(X_2 - \mu_{X_2}\right) \\ & + \frac{1}{2}\frac{\partial^2 f}{\partial X_2^2}\left(X_2 - \mu_{X_2}\right)^2 \end{aligned} \tag{7.27}$$

7.5 Higher-Order Terms with Uncorrelated Inputs

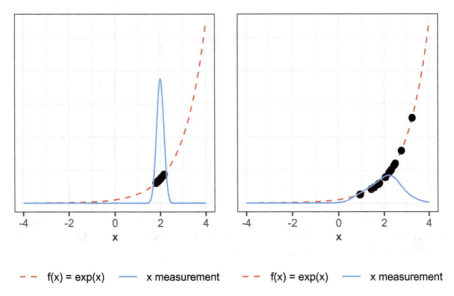

Fig. 7.1 The model $f(x) = e^x$. The dashed line is the exponential function while the solid line is the x measurement with small uncertainty (left) and large uncertainty (right)

While the same process in Sect. 7.3 can be followed to derive an analytical expression including these extra terms, the algebra quickly becomes overwhelming. Practical solutions typically justify ignoring the least significant terms and the correlated terms when this series is expanded. The GUM provides a generalized solution for Eq. (7.27) that includes the "most important" terms in the estimated combined variance:

$$u_c^2(y) = \sum_{i=1}^{N} \left(\frac{\partial f}{\partial x_i}\right)^2 u^2(x_i)$$
$$+ \sum_{i=1}^{N} \sum_{j=1}^{N} \left[\frac{1}{2}\left(\frac{\partial^2 f}{\partial x_i \partial x_j}\right)^2 + \frac{\partial f}{\partial x_i}\frac{\partial^3 f}{\partial x_i \partial x_j^2}\right] u^2(x_i) u^2(x_j). \tag{7.28}$$

A more complicated, but more accurate (retaining more of the higher-order terms), expression for the estimated combined variance is fully derived in Wang and Iyer (2005):

$$u_c^2(y) = \sum_{i=1}^{N} \left(\frac{\partial f}{\partial x_i}\right)^2 u^2(x_i) + \frac{1}{N} \sum_{i=1}^{N} \sum_{j=1}^{N} \left(\frac{\partial^2 f}{\partial x_i \partial x_j}\right)^2 u^2(x_i) u^2(x_j)$$

$$+ \sum_{i=1}^{N} \frac{\partial f}{\partial x_i} \frac{\partial^3 f}{\partial x_i^3} u^4(x_i) +$$

$$\sum_{i=1}^{N} \sum_{\substack{j=1 \\ j \neq i}}^{N} \frac{\partial^2 f}{\partial x_i^2} \left(\frac{\partial f}{\partial x_j}\right)^2 u^2(x_i) u^2(x_j) + \frac{1}{4} \left(\sum_{i=1}^{N} \frac{\partial^2 f}{\partial x_i^2} u^2(x_i)\right)^2 \qquad (7.29)$$

Because of the complexity of these formulas, the need for higher-order partial derivatives, and the ignoring of correlated variables, it is often easier to use numerical approaches such as Monte Carlo methods described in Chap. 8 to account for nonlinearities in the measurement model.

7.5.1 Measurement Examples

Returning to the density measurement example, the same cylinder was measured using different equipment. Because of the larger uncertainties in this equipment, it is suspected that the linearity assumption may no longer be valid (Table 7.2).

$$\rho = \frac{m}{\pi \left(\frac{d}{2}\right)^2 h} \qquad (7.30)$$

Following the same calculation as in Sect. 7.3.1, using the first-order uncertainty expression, and assuming all variables are independent, the density ($k = 1$) is

$$\rho = 12.7 \text{g/cm}^3 \pm 2.57 \text{g/cm}^3 \qquad (7.31)$$

Applying the GUM's expression of nonlinear uncertainty (Eq. 7.28) adds in the higher-order terms:

Table 7.2 Cylinder density measurement values using different equipment

Variable	Mean	Uncertainty
m	5.0 g	0.01 g
h	2.0 cm	0.05 cm
d	0.5 cm	0.05 cm

$$u_{c,2}^2(\rho) = \frac{16}{\pi^2 d^4 h^2} u^2(m) + \frac{16m^2}{\pi^2 d^4 h^4} u^2(h) + \frac{64m^2}{\pi^2 d^6 h^2} u^2(d)$$
$$+ \frac{16}{\pi^2 d^4 h^4} u^2(h) u^2(m)$$
$$+ \frac{32m^2}{\pi^2 d^4 h^6} u^4(h) + \frac{64}{\pi^2 d^6 h^2} u^2(d) u^2(m) \quad (7.32)$$
$$+ \frac{64m^2}{\pi^2 d^6 h^4} u^2(d) u^2(h) + \frac{288m^2}{\pi^2 d^8 h^2} u^4(d)$$
$$+ \frac{96m}{\pi d^2 h^5} u^4(h) + \frac{96m}{\pi d^4 h^3} u^2(d) u^2(h) + \frac{480m}{\pi d^6 h} u^4(d)$$

which simplifies to a density of

$$\rho = 12.7 \text{g/cm}^3 \pm 2.65 \text{g/cm}^3 \quad (7.33)$$

Note that this result is only a slightly different uncertainty than using the first-order approximation, but it is a slightly better approximation to the true uncertainty. Using the even more complete expression in Wang and Iyer (2005) (Eq. 7.29) results in a density of

$$\rho = 12.7 \text{g/cm}^3 \pm 2.72 \text{g/cm}^3. \quad (7.34)$$

In theory, one could keep adding terms to the Taylor expansion until the resulting combined uncertainty did not change within the reported significant digits. If the above results were rounded and reported to the typical two significant digits, adding the additional terms of the Wang and Iyer method would not change the result compared to the GUM method.

7.6 Multiple Output Quantities

Occasionally, a measurement model will result in multiple output quantities based on the same set of input measurements. One common example is encountered in microwave measurements, where the real and imaginary components of a complex-valued microwave signal are input to a measurement model that results in magnitude and phase as outputs. Other examples include measurement of voltage and current used to calculate both resistance and power, or a set of Coordinate Measuring Machine (CMM) measurements used to compute both the lengths and angle of intersection of two components.

Cases with multiple output quantities, dealt with extensively in JCGM 102 (2011), result in an "uncertainty region" in multiple dimensions, rather than the simple one-dimensional "plus or minus" uncertainty region evaluated previously in this chapter. Part of a full solution of a multiple output uncertainty evaluation is determining the joint probability between the multiple outputs in order to fully define the region of uncertainty.

In the bivariate case, the joint PDF of the resulting uncertainty calculation can be described by a bivariate normal distribution that defines an ellipse. A bivariate normal PDF requires the mean values of the two output quantities y_1 and y_2 and their covariance matrix U_y. Defining U_x as the covariance matrix of the input measurements, the sensitivity matrix

$$C_x = \begin{bmatrix} \dfrac{\partial f_1}{\partial X_1} & \dfrac{\partial f_1}{\partial X_2} \\ \dfrac{\partial f_2}{\partial X_1} & \dfrac{\partial f_2}{\partial X_2} \end{bmatrix} \tag{7.35}$$

with the partial derivatives evaluated at the measured values $X_i = x_i$, the covariance matrix of the two output quantities is given by

$$U_y = C_x U_x C_x^T = \begin{bmatrix} u^2(y_1) & u(y_1, y_2) \\ u(y_1, y_2) & u^2(y_2) \end{bmatrix} \tag{7.36}$$

These matrices can be expanded to any number of output quantities.

7.7 Limitations of the Analytical Approach

The analytical approach to uncertainty propagation has its limitations. Except for a few very simple cases (those described in Sect. 7.2), the solution is an approximation based on a Taylor series expansion. It is not obvious when nonlinearities in the model will become significant, requiring use of higher-order terms. The solutions all require finding partial derivatives of the measurement model, which in some cases can be difficult or impossible to solve analytically.

Another limitation is the normality assumption. All input variables are defined by their mean and standard deviation, regardless of their actual distribution. Any information about a distribution (e.g., uniform or triangular) is lost after converting to a standard deviation as described in Sect. 6.3. As such, the distribution of the function's output will always be treated as a normal distribution, defined by a mean and standard deviation.

Finally, using the W-S formula to determine effective degrees of freedom and an expanded uncertainty is only justified when the model is first-order with uncorrelated inputs. For these reasons, the Monte Carlo method described in Chap. 8 is often used in parallel with this section's analytical methods to help confirm the results.

7.8 Related Reading

The paper by Kline and McClintock (1953), and the follow-up paper by Kline (1985), on which the uncertainty propagation methods described in this chapter are based, is recommended reading simply for historical reasons. NIST published Technical Note 1297 (Taylor and Kuyatt 1994) which provides additional guidance on using the GUM method and expressing the results. NIST Technical Note 1900 (Possolo 2015) provides many practical examples demonstrating use of the GUM uncertainty methods.

The GUM (JCGM 2008), also released as ISO/IEC Guide 98-3:2008, is the industry standard for guidance on uncertainty propagation, covering many additional aspects of uncertainty analysis. ISO 21748:2017 provides guidance for evaluating uncertainty due to repeatability and reproducibility in experimental measurements. NCSLI's recommended practice RP-12 National Conference of Calibration and Standards Laboratories International (1995) provides additional recommendations on how to practically implement uncertainty calculations. Suggestions include methods of identifying sources of uncertainty in a measurement, selecting appropriate probability distributions, and choosing a coverage factor.

Hall and Willink (2001) investigate the Welch–Satterthwaite approximation and when it does or does not apply. Work on adapting the W-S approximation for correlated uncertainties is ongoing in Castrup (2010), Wang and Iyer (2005), Ballico (2000), Willink (2007), and Kuster (2013). Some other shortcomings of the GUM uncertainty method, such as additional correlation effects and nonlinearities, are explored in detail in Rabinovich (2005).

7.9 Exercises

1. Use the GUM uncertainty formula (Eq. 7.7) to find an expression for the standard uncertainty in:

   (a) $f = x$
   (b) $f = x + y$
   (c) $f = x + y + z$

   Given uncertainties $u(x)$, $u(y)$, and $u(z)$ and assuming x, y, and z are independent (uncorrelated). Write a general expression for uncertainty in $f = \sum x_i$ with uncertainties $u(x_i)$.

2. Use the GUM uncertainty formula (Eq. 7.7) to compute the standard uncertainty in $f(x) = cx$, where c is a constant with no uncertainty. Apply the result to find the uncertainty in kilograms of a mass is given as 355 g with uncertainty 5.2 g.

3. Find an expression for the uncertainty in $f(x, y) = x^2 + y$. Then solve for:

   (a) $x = 4$, $y = 3$, $u(x) = 0.4$, $u(y) = 0.15$,
   (b) $x = 4$, $y = 3$, $u(x) = 10\%$, $u(y) = 5\%$.

In part b, keep the uncertainties in percent to obtain an output as a percent. Then convert to an absolute uncertainty and compare with part a.

4. For the following measurement models, determine whether the GUM uncertainty method provides an exact solution or approximation. Hint: consider the Taylor expansion (Eq. 7.4) used to approximate the measurement function, and whether it gives an exact representation of the model. Terms a, b, and c are constants (zero uncertainty).

   (a) $f(x, y) = x + y$
   (b) $f(x, y) = ax + by$
   (c) $f(x) = \exp(ax)$
   (d) $f(x, y) = a + bx + cy^2$
   (e) $f(x, y) = axy$

5. Resistance varies with temperature following the first-order equation $R = R_0(1 + \alpha(T - T_0))$, where R_0 is the nominal resistance at temperature T_0, T is the operating temperature, and α the temperature coefficient of resistance. Write an expression for $u(R)$, the uncertainty in corrected resistance, in terms of the variables and their uncertainties $u(R_0)$, $u(T)$, $u(T_0)$, and $u(\alpha)$.

6. The original uncertainty propagation paper by Kline and McClintock (1953) uses the example of a pitot-static tube—a pressure sensor used to measure the airs peed of an aircraft in flight. The measurement equation is

$$v = \sqrt{\frac{2\Delta p R_s T_a}{p_a}},$$

where v is velocity (air speed), Δp is the pressure differential measured by the pitot-static tube, R_s the specific gas constant for dry air, T_a the ambient temperature, and p_a the ambient atmospheric pressure. The values and $k = 2$ uncertainties given in the paper, converted to SI units, are $\Delta p = 1.993 \pm 0.025$ kPa; $R_s = 287.06$ J/K/kg (assume no uncertainty); $T_a = 292.8 \pm 0.11$ K; and $P_a = 101.4 \pm 2.1$ kPa. Calculate the air speed and its expanded ($k = 2$) uncertainty in meters per second.

7. The ancient Greek mathematician Eratosthenes was the first to estimate the circumference of the earth. Unfortunately for Eratosthenes, the concepts of uncertainty propagation would not be fully developed for millennia, but we can estimate the uncertainty in his calculation. His measurement model was

$$\text{circumference} = \frac{\text{distance}}{\text{shadow angle}/360},$$

where the distance is the measured distance between the cities of Alexandria and Syene and the shadow angle was the angle in degrees made by a vertical pole in Alexandria on the equinox (Walkup 2005). The distance was determined to be 5000 stades (a Greek measurement of length, estimated to be around 185 modern

7.9 Exercises

meters) and the angle measured (using what is basically a sundial) to be 7.2°. If we assume the distance measurement had $k = 2$ uncertainty of 5% (250 stades) and the angular measurement uncertainty of 0.5°, what is the $k = 2$ uncertainty, in stades and in percent, of the circumference calculation? (As a side note, the choice of measurement model itself adds additional uncertainty, unaccounted for in this exercise, due to its assumptions—such as the Earth being a perfect sphere, the vertical poles being perpendicular to the center of the Earth, and the two cities falling on the same meridian for this equation to apply.)

8. Consider the conversion of real and imaginary components into magnitude and phase.

$$\text{mag} = \sqrt{\text{Re}^2 + \text{Im}^2}$$

$$\text{phase} = \tan^{-1} \frac{\text{Im}}{\text{Re}}$$

For both equations, use the correlated GUM uncertainty calculation (Eq. 7.19) to find an expression for uncertainty in magnitude and phase. Given Re and Im both equal to 1.0 with standard uncertainty of 0.05, find the magnitude and phase uncertainties when there is (a) no correlation, (b) −0.8 correlation between Re and Im, (c) +0.8 correlation between Re and Im. How does adding the correlation coefficient affect the uncertainty in magnitude and phase?

9. Using Eqs. (7.35) and (7.36), find the covariance matrix of the magnitude and phase outputs in exercise 8. Assume the real and imaginary input variables are uncorrelated.

10. The fundamental audio frequency of a vibrating column of air, such as an open-ended organ pipe, is given by

$$f_1 = \frac{v}{2L},$$

where v is the velocity of sound through the air and L the length of the pipe. Neglecting minor effects of pressure and humidity, the velocity of sound can be calculated using the semi-empirical formula $v = (331.3 + 0.606T)$ m/s, where T is the temperature in Celsius (Zuckerwar 2002). Calculate the length of an organ pipe that produces a middle A tone, frequency of 440 Hz, at a nominal temperature of 23 °C.

The minimum human-perceptible difference at this frequency is approximately 1.3 Hz (Time and Frequency from A to Z 2010). Determine the largest acceptable uncertainty in length that would keep the frequency within ±1.3 Hz (normal, $k = 2$) of 440 Hz, assuming the temperature is tightly controlled to ±1.5 °C (normal, $k = 2$). Hint: write an expression for $u(f_1)$ and algebraically solve it for $u(L)$.

Table 7.3 SEM calibration data

Reference standard, x_2 (μm)	Device under test, y_m (μm)
0.998	0.656
1.001	0.635
1.005	0.642
0.999	0.652
0.999	0.649
1.010	0.650
1.007	0.647
1.001	0.646
1.009	0.652
1.012	0.645

11. The total resistance of resistors connected in series is the sum of n individual resistances: $R = R_1 + R_2 + \ldots + R_n$. Assume all resistors have 5% uncertainty ($k = 2$) and equal value. While keeping a constant total resistance of $R = 100\ \Omega$, compute the uncertainty in total resistance when $n = 1$ ($R_1 = 100\ \Omega$), $n = 2$ ($R_1 = R_2 = 50\ \Omega$), etc. up to $n = 10$. Plot uncertainty as a function of the number of series resistors.

12. The voltage gain of an inverting operational amplifier circuit is given by $A = -R_f/R_i$, where R_f is the feedback resistor and R_i the input resistor. Determine the uncertainty in gain of an amplifier built with $R_f = 10{,}000\ \Omega$ and $R_i = 1000\ \Omega$, both with manufacturer's tolerances of 5%. Assume a uniform distribution for the tolerances.

13. To calibrate dimensional measurements made using a scanning electron microscope (SEM), a ratio method can be used where a known reference standard is measured on the SEM and this value compared against the reference's certified value. Then the ratio of measured to certified value is used to adjust subsequent length measurements made on the SEM. The measurement equation becomes

$$y_c = \frac{x_1}{x_2} y_m,$$

where x_1 and x_2 are the certified and SEM-measured values of the reference standard, respectively, y_m the uncorrected SEM measurement, and y_c the corrected measurement. An SEM is used to take 10 measurements of a reference standard consisting of a set of 1.000 μm pitch lines certified to 30 nm uncertainty ($k = 2$, 95% confidence). The 10 repeated measurements provide a Type A uncertainty of the SEM calibration measurement. The pixel-resolution of the SEM adds a Type B component of uncertainty on all length measurements is estimated to be a rectangular distribution of full width 5 nm. Once calibrated, the SEM is used to measure an experimental device, again taking 10 measurements. Using the SEM measurement values in Table 7.3, determine the length and uncertainty of the device. Report the uncertainty expanded to 95% confidence.

References

Ballico, M.: Limitations of the Welch-Satterthwaite approximation for measurement uncertainty calculations. Metrologia. **37**(1), 61–64 (2000)

Casella, G., Berger, R.L.: Statistical Inference, 2nd edn. Duxbury, Pacific Grove, CA (2002)

Castrup, H.: A Welch-Satterthwaite relation for correlated errors. In: Measurement Science Conference, Pasadena, CA (2010)

Gauss, C.F.: Theoria combinationis observationum erroribus minimis obnoxiae. Carl Friedrich Gauss Werke. **4**, 95–100 (1821)

Hall, B.D., Willink, R.: Does Welch-Satterthwaite make a good uncertainty estimate? Metrologia. **38**, 9–15 (2001)

ISO 21748: Guidance for Use of Repeatability, Reproducibility, and Trueness Estimates in Measurement Uncertainty and Evaluation. (2017)

JCGM 100: Evaluation of Measurement Data - Guide to the Expression of Uncertainty in Measurement. (2008)

JCGM 102: Evaluation of Measurement Data - Supplement 2 to the - Guide to the Expression of Uncertainty in Measurement - Extension to any Number of Output Quantities. (2011)

Kline, S. J., McClintock, F. A.: Describing uncertainties in single sample experiments. Mech. Eng., 75 3–8 (1953)

Kline, S.J.: The purposes of uncertainty analysis. J. Fluids Eng. **107**, 153–160 (1985)

Kuster, M.: Applying the Welch-Satterthwaite formula to correlated errors. NCSLI Meas. **8**(1), 42–55 (2013)

National Conference of Calibration and Standards Laboratories International: Recommended Practice 12 - Determining and Reporting Measurement Uncertainties. (1995)

Possolo, A.: Simple guide for evaluating and expressing the uncertainty of NIST measurement results. https://doi.org/10.6028/NIST.TN.1900 (2015). Accessed 29 Apr 2020

Rabinovich, S.G.: Measurement Errors and Uncertainties, 3rd edn. Springer, New York, NY (2005)

Taylor, B.N., Kuyatt, C.E.: Guidelines for evaluating and expressing the uncertainty of NIST measurement results. NIST Technical Note 1297. National Institute of Standards and Technology, Gaithersburg, MD (1994)

Time and Frequency from A to Z. https://www.nist.gov/pml/time-and-frequency-division/popular-links/time-frequency-z (2010). Accessed 29 Apr 2020

Walkup, N.: Eratosthenes and the mystery of the Stades. Mathematical Association of America. https://www.maa.org/press/periodicals/convergence/eratosthenes-and-the-mystery-of-the-stades-introduction (2005). Accessed 29 Apr 2020

Wang, C.M., Iyer, H.K.: On higher-order corrections for propagating uncertainties. Metrologia. **42**, 406–410 (2005)

Willink, R.: A generalization of the Welch-Satterthwaite formula for use with correlated uncertainty components. Metrologia. **44**, 340–349 (2007)

Zuckerwar, A.J.: Handbook of the Speed of Sound in Real Gases. Academic Press, San Diego, CA (2002)

Chapter 8
Monte Carlo Methods for the Propagation of Uncertainties

8.1 Introduction to Monte Carlo Methods

Random sampling to investigate physical and mathematical problems has been in use for several centuries (De Comte Buffon 1774; Kelvin 1901). However, the first practical use, and coining of the term "Monte Carlo" (after the casino in Monaco) for the method, date to Los Alamos National Laboratory during the Manhattan Project. The Monte Carlo method was first proposed for studying diffusion of neutrons into radiation shielding in 1947, and the first computerized Monte Carlo calculations were carried out on the Electronic Numerical Integrator and Computer (ENIAC) in 1948 (Metropolis 1987; Richtmyer and Von Neumann 1947). It was not until computers became more powerful and more available in the 1960s and 1970s that Monte Carlo methods started gaining widespread traction as a way to simulate various engineering problems.

The reliance on computers is due to the very nature of Monte Carlo methods: drawing conclusions from numerous trial calculations using large quantities of random numbers. Applying the method requires each input variable in an equation to be randomly sampled from some probability distribution. The sets of sampled input variables are combined to determine one possible output value. Repeated enough times, statistical conclusions can be drawn from the set of calculated possible outputs. According to The Law of Large Numbers (Graham and Talay 2013), as the number of samples approaches infinity, the expected output value approaches its true value. Because of the large quantity of calculations and random numbers required, early Monte Carlo methods computed by hand were impractical.

Example—Calculating Pi A classic use of Monte Carlo is calculating the value of π. Knowing that the ratio of a square's area to the area of an inscribed quarter-circle is $\pi/4$, points can be sampled at random from within the square to see how many fall inside the circle. By counting the dots where $\sqrt{x^2 + y^2} \leq 1$, the ratio of points inside the circle to the total number of points can be computed. Figure 8.1 shows three cases, with increasing numbers of random sampled points. When multiplied by 4, the

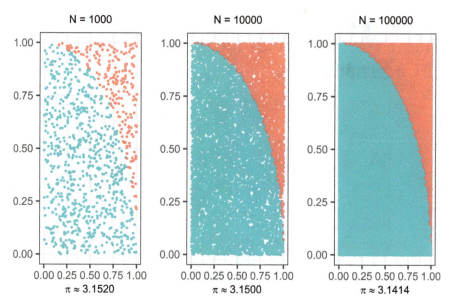

Fig. 8.1 Calculation of π using Monte Carlo method with 1,000 (left), 10,000 (middle), and 100,000 (right) samples

result for this trial with $N = 100,000$ was 3.1414, or approximately π. As seen in the figure, the more points sampled, the closer this approximation will come to the true value of π.

8.1.1 Random Sampling Techniques and Random Number Generation

Because the Monte Carlo method relies on random sampling from a distribution, a good random number generator is an essential, but often overlooked, requirement. While precursors to Monte Carlo analysis created randomness by physical means, such as rolling dice, computer algorithms are required to efficiently obtain a sufficient quantity of random values. Most modern computer programming languages or statistical packages already implement algorithms for random number generation and sampling from distributions, but the basic concepts are described below.

Computerized random number generators fall into two categories: true-random number generators and pseudo-random number generators (PRNGs). True random number generators produce completely random, independent, unpredictable samples. These generators are essential in applications such as encryption, where predictability in the random sequence could mean finding a way to break the encryption. True random generators are almost impossible to implement completely within software. PRNGs, on the other hand, appear random, yet follow a prescribed

8.1 Introduction to Monte Carlo Methods

algorithm. They are predictable if the initial state of the generator and details of the algorithm are known. In engineering applications such as Monte Carlo analysis, the repeatability of a PRNG is desirable for debugging and testing the algorithms. PRNGs are also much simpler and can be implemented completely within software. The remainder of this section discusses PRNGs and their use in Monte Carlo analysis.

Several properties of a PRNG should be evaluated for its suitability to Monte Carlo analysis:

1. *Sequence Length* PRNGs are essentially state machines where the next random number depends on the previous output and the internal state of the generator, meaning a PRNG can only generate a specific number of samples before repeating itself. Obviously, this sequence length must be longer than the number of samples required. Due to possible long-term hidden correlations between the sections of the output, it is recommended that the sequence length must be at least the square of the required number of samples (Hellekalek 1998). For typical uncertainty propagation problems, 1 to 10 million samples for each variable are required to converge on a solution. By this metric, for a simple problem with 3 variables needing 10 million samples each, the PRNG must be able to generate at least $\sim 2^{50}$ ($\sim 10^{15}$) samples before repeating (sequence lengths are frequently reported as a power of 2). For problems with many input variables, this number should be even higher.
2. *Uncorrelated Output* The generated values must be independent and uncorrelated. Many algorithms appear independent over short sequences of numbers yet contain some amount of long- or short-term order to the values they produce. Reducing the number of samples used from the sequence as described above can mitigate the longer-term correlation effects.
3. *Uniform Generation* A PRNG must generate samples uniformly over its range to be useful for simulation. In other words, it should be unbiased in producing its output values.
4. *Reproducibility and Portability* For Monte Carlo and other engineering applications, the same sequence of numbers should be generated at different times and on different platforms. This is to aid in debugging and testing the simulation and model.

For some recommended algorithms that meet these requirements, and an assessment of the PRNGs in popular programming languages, see Sect. 8.1.3.

8.1.1.1 Sampling from Normal and Non-Normal Distributions

Most PRNG algorithms generate a uniform distribution of numbers, usually in the range of 0–1. However, in Monte Carlo analysis, normal or other distribution shapes are required. Transformations are necessary to convert the uniform random numbers into the required distribution. The simplest transformation takes the uniform distribution on the interval [0, 1] and applies a shift and scale to move the uniform distribution to the desired range.

The Box–Muller transformation (Box and Muller 1958), for example, transforms two uniformly distributed independent random samples U_1 and U_2 into two normally distributed independent samples Z_1 and Z_2:

$$\begin{aligned} Z_1 &= \sqrt{-2 \ln U_1} \cos(2\pi U_2) \\ Z_2 &= \sqrt{-2 \ln U_1} \sin(2\pi U_2) \end{aligned} \tag{8.1}$$

Similar transformations have been developed to produce other distribution types; see Robert and Casella (2004) for examples.

8.1.1.2 Generating Correlated Random Samples (Normal Distribution)

Correlated random variables are generated using similar transformations. These transformations take two or more uniform random samples and a covariance matrix and convert them into correlated random variables of another distribution. The multivariate normal distribution is the most common of these transformations. For implementation details, see Gentle (2009) and Cox and Harris (2006).

When generating correlated random samples, it is often useful to generate pairwise scatter plots of the samples. If there is slope or other order to the scatter points, the two variables are correlated (Fig. 8.2).

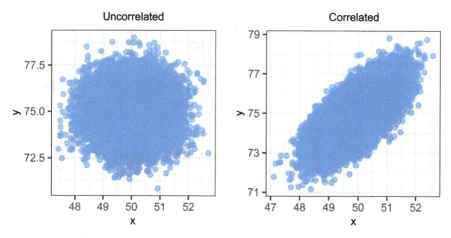

Fig. 8.2 Example scatter plots showing correlated and uncorrelated random samples

8.1.2 Generation of Probability Density Functions Using Random Data

Once the N output values of a Monte Carlo analysis are computed, those samples must be compiled into meaningful statistics. While the output samples are not necessarily normally distributed, it is common to compute the mean and standard deviation of all N samples using the formulas in Chap. 4.

When the output distribution deviates from normal, a more precise description of the distribution is sometimes desired. Plotting a histogram of the samples provides insight into the true distribution. The probability density function of a distribution can be approximated by taking a histogram and connecting the top of each bar to obtain a discrete representation of the data. Changing the binning of the histogram will change the accuracy and smoothness of the result. The KDE method described in Chap. 4 is a more robust method of approximating a PDF from sampled data.

Figure 8.3 shows the histogram of a non-normal distribution with the line showing a KDE approximation of the PDF. While the mean of this distribution is 0.29, the most probable value is about 0.20.

8.1.3 Computational Approaches

This section describes and evaluates some PRNG algorithms and the PRNGs implemented in common programming languages.

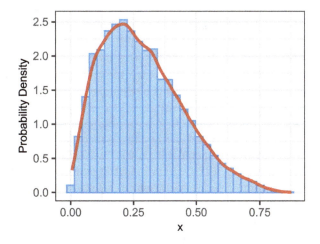

Fig. 8.3 KDE approximation of probability density function

8.1.3.1 Linear Congruential Generator

One of the oldest (if not the best) PRNGs is an algorithm called the Linear Congruential Generator (LCG). While generally not suitable for Monte Carlo analysis, it is described here for its simplicity and as a starting point for understanding other PRNG algorithms. In particular, it provides the basis for the recommended Enhanced Wichmann–Hill PRNG described in the next section. The LCG operates on integers and is implemented by a single iterative equation:

$$X_{n+1} = (aX_n + c) \bmod m, \quad (8.2)$$

where mod is the modulo (remainder) operator.

Three constant parameters are used: a, c, and m. The selection of these constants, and a starting value X_0, called a "seed," determines the sequence of random numbers. The choice of a and c is important for the quality of the random sequence. For example, choosing $a = 4$, $c = 1$, $m = 9$, and an initial value of $X_0 = 0$ leads to a sequence of 0, 1, 5, 3, 4, 8, 6, 7, 2, 0, 1, 5, 3, 4, ... Notice that after nine numbers, the sequence begins to repeat itself. This LCG algorithm with a sequence length of nine (the value of m) is not suitable for generating the few million samples required for typical Monte Carlo methods. In reality the parameters are carefully chosen to be much larger numbers with a large sequence length. For example, using $a = 1664525$, $c = 1013904223$, and $m = 2^{32}$ is recommended by Press et al. (2002). This results in just over 4 billion samples before repeating. Using the sequence length criteria above, this setup is good for about 65,000 samples. In addition, the LCG does not produce uncorrelated values over long sequences (Marsaglia 1968). For these reasons, LCG is generally not suitable for Monte Carlo analysis.

8.1.3.2 Better PRNG Algorithms

Two PRNGs that satisfy the requirements are the Mersenne Twister (Matsumoto and Nishimura 1998) and the Enhanced Wichmann–Hill Generator (Wichmann and Hill 2006), not to be confused with their original 1982 version (Wichmann and Hill 1982). These generators have a large sequence length and pass various randomness tests. While they operate on similar principles as the LCG, they overcome its limitations. The Wichmann–Hill algorithm is simply a combination of four parallel independent LCGs. The Mersenne Twister, instead of using a single seed (X_0 in the LCG) and three constants to record the internal state of the generator, uses 2500 bytes of state. It is also slower and requires much more code to implement.

Typically, one should rely on the PRNG implementation built into a computing package such as R, Python, or Matlab after evaluating it for the above requirements rather than attempting to implement it independently. Most modern programming languages use the Mersenne Twister, but be aware that some older languages, or ones that are slow to update, still use PRNGs unsuitable for the task. If you must

Table 8.1 PRNG implementations in common programming languages

Algorithm	Languages	Sequence length	Maximum samples	Suitable for MC?
Mersenne twister	R, Python, Matlab, Mathematica, C/C++11, Excel >=2010[a]	$2^{19937}-1$	$\sim 10^{3000}$	Yes
Enhanced Wichmann and Hill (2006)	Available as optional packages in R and Python. Recommended by National Physical Laboratory and GUM.	2^{120}	$\sim 10^{18}$	Yes
Original Wichmann and Hill (1982)	LabVIEW, Excel <=2007[b], Excel VBA (all versions to date)	2^{42}	$\sim 2 \times 10^6$	No
LCG	Visual Basic <= 6	2^{24}	~ 4096	No
LCG	Java	2^{48}	$\sim 16 \times 10^6$	No
LCG	Older C/C++	2^{31}	$\sim 46,000$	No

[a]No seeding is available in Excel. Numbers will be recomputed every update with a different seed, so it fails criteria #4. Excel VBA still uses Original Wichmann–Hill. Only uniform distributions are available directly

[b]Excel 2007 and earlier versions also had errors in the PRNG implementation (McCullough 2008)
References: CRAN (2018), Python Software Foundation (2018), Mathworks Inc. (2018), Wolfram (2018), Brown (2018), Melard (2014), National Instruments (2018), McCullough (2008), Javamex (2018), Barreto & Howland (2006)

implement your own PRNG from scratch, the Enhanced 2006 Wichmann–Hill algorithm as concisely described by Cox and Harris (2006) is recommended for its relative ease of implementation.

See Table 8.1 for a list of some common programming languages, their built-in PRNG algorithm (as of versions released in 2020), and whether they are suitable for Monte Carlo uncertainty propagation based on the above requirements.

Similarly, transformations from uniform to other distributions are typically implemented by each programming language and should be used when available.

8.2 Standard Monte Carlo for Uncertainty Propagation

This section introduces the use of Monte Carlo techniques for quantifying the uncertainty associated with an indirect measurement using a measurement equation, known as the "propagation of distributions" (JCGM 101: 2008).

8.2.1 Monte Carlo Techniques

When applied to uncertainty propagation problems, Monte Carlo methods are straightforward to implement. Figure 8.4 shows a graphical illustration of the Monte Carlo approach to uncertainty propagation. Each input variable is randomly

Fig. 8.4 Illustration of Monte Carlo uncertainty propagation

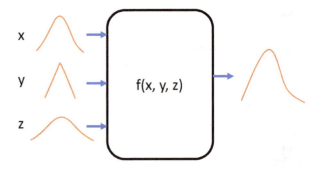

Table 8.2 Measured values for cylinder density calculation

Measurement	Mean	Uncertainty ($k = 1$)
Mass	5 g	0.01 g
Height	2 cm	0.05 cm
Diameter	0.5 cm	0.05 cm

sampled N times from its distribution, and N possible output values are computed using the measurement model. The mean and standard deviation of the output distribution are calculated to provide the expected value and standard uncertainty of the result.

The method is best illustrated with an example.

8.2.1.1 Case Study: Calculating Density

To measure the density of a cylinder, measurements are made on the height h, diameter d, and mass m. The density is calculated using

$$\rho = \frac{m}{\pi \left(\frac{d}{2}\right)^2 h}. \tag{8.3}$$

```
# Measurement equation
rho <- function(m, h, d) { (m/(pi*(d/2)^2*h)) }
```

Measured values and their uncertainties (all assumed normal) are given in Table 8.2.

```
# Assign mean values and uncertainties for variables
m <- c(5.00, 0.01)
h <- c(2.00, 0.05)
d <- c(0.50, 0.05)
```

To apply Monte Carlo uncertainty propagation, mass, height, and diameter were each sampled 1,000,000 times from their respective probability distributions.

8.2 Standard Monte Carlo for Uncertainty Propagation

Table 8.3 First five Monte Carlo samples

d (cm)	h (cm)	m (g)	ρ (g/cm^3)
0.482	1.976	4.997	13.854
0.551	2.051	4.987	10.204
0.500	2.067	5.002	12.318
0.444	1.977	4.987	16.292
0.399	1.992	4.998	19.976

Combining each of the samples through the density equation results in 1,000,000 possible density values. The first few sampled input values and the resulting output are listed in Table 8.3.

```
# Generate Monte Carlo input samples
n <- 1e6
mSamp <- rnorm(n, m[1], m[2])
hSamp <- rnorm(n, h[1], h[2])
dSamp <- rnorm(n, d[1], d[2])

# Plot input histograms
p1 <- ggplot(mapping = aes(x = mSamp, y = ..density..)) +
  geom_histogram(fill = "steelblue3", alpha = .8,
  col = "steelblue3", bins = 100) +
  xlab("m") +
  thm + theme(axis.text = element_text(size = 10))

p2 <- ggplot(mapping = aes(x = hSamp, y = ..density..)) +
  geom_histogram(fill = "steelblue3", alpha = .8,
  col = "steelblue3", bins = 100) +
  xlab("h") +
  thm + theme(axis.text = element_text(size = 10))

p3 <- ggplot(mapping = aes(x = dSamp, y = ..density..)) +
  geom_histogram(fill = "steelblue3", alpha = .8,
  col = "steelblue3", bins = 100) +
  xlab("d") +
  thm + theme(axis.text = element_text(size = 10))

g <- grid.arrange(p1, p2, p3, ncol = 3)

# Generate Monte Carlo output samples
out <- rho(m = mSamp, h = hSamp, d = dSamp)

# Plot output histogram
ggplot(mapping = aes(x = out, y = ..density..)) +
  geom_histogram(fill = "steelblue3", alpha = .8,
  col = "steelblue3", bins = 100) +
  xlab(expression(rho)) +
  thm
```

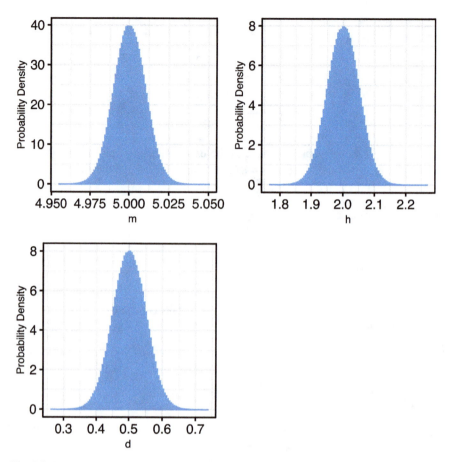

Fig. 8.5 Histograms of Monte Carlo sampled input values

Histograms of the entire set of input samples are shown in Fig. 8.5, and the output histogram is shown in Fig. 8.6. Taking the mean value and standard deviation of the output samples gives a value and ($k = 1$) uncertainty:

$$13.1 \pm 2.8 \text{g}/\text{cm}^3 \tag{8.4}$$

```
# Calculate mean and standard deviation
mean(out)
## [1] 13.14602
sd(out)
## [1] 2.804819
```

Unlike the GUM approximation, no assumptions are made. The actual measurement function is used to propagate each input sample through the model, so there are no requirements on linearity. As long as inputs can be sampled over the appropriate

Fig. 8.6 Histogram of Monte Carlo cylinder density output

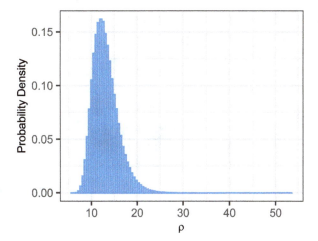

distribution, the inputs do not need to be normally distributed. Correlated variables are handled simply by drawing correlated samples from a joint distribution.

Note that shape of the density output distribution is not normal. There are two recommended ways to compute a coverage interval containing 95% of the distribution. The simplest is to calculate a symmetric distribution by taking the 2.5% and 97.5% percentiles. However, this may not be the shortest interval covering 95% of the samples. The shortest confidence interval is found by locating the endpoints of all possible 95% ranges (for example, the 1% and 96% percentiles) and taking the one with the smallest distance between the endpoints. When the distribution is symmetric about its mean, these two methods produce the same results. Figure 8.7 shows the two intervals for the density calculation.

```
# Symmetric coverage interval
symInterval <- quantile(out, probs = c(0.025, 0.975))
symInterval
##    2.5%    97.5%
## 8.872158 19.773273

# Shortest coverage interval
y <- sort(out)
N <- length(y)
q <- round(.95*N)
ridx <- 0
rmin <- Inf
for(r in 1:(N-q)){
  if((y[r+q] - y[r]) < rmin){
    rmin = y[r+q] - y[r]
    ridx = r
  }
  shortest <- c(y[ridx], y[ridx+q])
}
shortest
## [1] 8.322284 18.752868
```

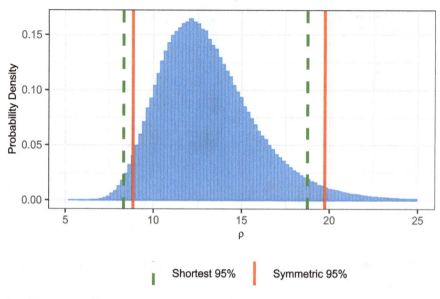

Fig. 8.7 Shortest and symmetric confidence intervals

8.2.1.2 Sensitivity Coefficients

The GUM method calculates sensitivity coefficients from the partial derivatives of the function with respect to each input, which provides information on the relative contribution to total uncertainty of each variable. The Monte Carlo method does not directly provide sensitivity coefficients, although they can still be estimated using the nonlinear model.

The technique for estimating these coefficients is to first hold all variables constant except one. Sample the selected variable N times and determine the uncertainty of the output due to the change in that variable alone. To find the sensitivity coefficient for the input being varied, divide the standard deviation in output due to the single variable by the standard deviation of the variable itself (JCGM 101:2008). This results in a "nonlinear sensitivity coefficient" for that variable (Cox and Harris 2006). If the model is linear, this value will be close to the sensitivity coefficient calculated by the GUM. To find the proportion of total variance, divide the variance in the output due to the single variable by the total variance in the output when all inputs are randomly sampled. The proportion results may not exactly add to 100% due to the random numerical nature of the calculation.

In the cylinder density example, sensitivity coefficients and proportion of total variance are given in Table 8.4. Here, the diameter measurement is by far the biggest contributor to uncertainty in the density.

8.2 Standard Monte Carlo for Uncertainty Propagation

Table 8.4 Nonlinear sensitivity coefficients

Variable	Sensitivity	Proportion (%)
m	2.55	0.01
h	6.37	1.29
d	55.6	98.41

```
# Calculate sensitivity coefficients
n <- 1e6 # Number of bootstrap samples

# Vary m only
sample <- rnorm(n, m[1], m[2])
mBoot <- rho(m = sample, h = h[1], d = d[1])
sdM <- sd(mBoot)

# Vary h only
sample <- rnorm(n, h[1], h[2])
hBoot <- rho(m = m[1], h = sample, d = d[1])
sdH <- sd(hBoot)

# Vary d only
sample <- rnorm(n, d[1], d[2])
dBoot <- rho(m = m[1], h = h[1], d = sample)
sdD <- sd(dBoot)

sens <- t(data.frame(m = sdM/m[2], h = sdH/h[2], d = sdD/d[2]))

# Calculate proportion
totalVar <- sd(out)
prop <- t(data.frame(m = (sdM/totalVar)^2*100, h = (sdH/totalVar)
^2*100, d = (sdD/totalVar)^2*100))
varResults <- cbind(sens, prop)
colnames(varResults) <- c('Sensitivity', 'Proportion')
varResults
##   Sensitivity Proportion
## m  2.545723  0.008230318
## h  6.375892  1.290673314
## d 55.599243 98.145899963
```

8.2.1.3 Convergence Plots and Adaptive Sampling

One problem with Monte Carlo methods is the inherent numerical nature of the calculation. For example, if only 10 samples were taken, the output distribution cannot possibly provide enough information to predict the uncertainty range. The Law of Large Numbers says that with infinite samples, the solution will converge

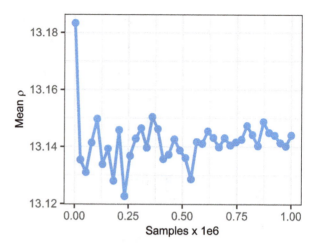

Fig. 8.8 Convergence plot

exactly. Unfortunately, no one has infinite time to solve the problem, so how many samples is enough to be reasonably sure you have a solution?

A convergence plot can help determine if the solution has settled on a "final" value. This plot is generated by gradually increasing the number of samples N, and estimating the unknown parameter for each value. The value of N should be increased until the plot levels off, with little additional change observed by further increasing N.

In the cylinder density example, the solution has reasonably converged after about 500,000 samples (see Fig. 8.8). Similar problems generally require 500,000 to 1,000,000 samples for a solution but occasionally need even more.

Algorithms such as this can be made more formal by continually adding samples until some pre-determined convergence criteria is met. Typically, a tolerance is set and a subset of samples in the neighborhood of 10,000 is computed. A new set of 10,000 samples is computed, and if the two solutions differ by more than the tolerance, the calculation continues by increasing N. The calculation stops once adding a new set of samples does not significantly change the result. Unless the model is quite complicated, modern computers are fast enough that several million samples can be computed in seconds and the additional complexity of adaptive sampling is not worth the effort.

8.3 Comparison to the GUM

The preceding discussion of Monte Carlo uncertainty propagation should make its advantages over the GUM approach apparent. Monte Carlo does not require the Taylor series expansion and makes no assumptions about the linearity of the measurement model. Each input variable can maintain its probability distribution

8.3 Comparison to the GUM

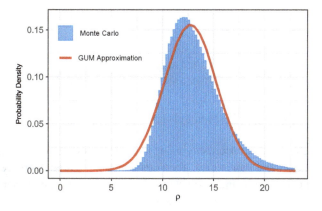

Fig. 8.9 Comparison of Monte Carlo and GUM for uncertainty propagation example

shape through the entire calculation, whereas the GUM requires converting all inputs into normal distributions before computing.

The downsides to Monte Carlo are the computation time required to generate many random samples, the need for quality random number generator algorithms, and the potential variability of the results from one calculation to the next due to the random nature of the algorithm. In addition, the lack of a single analytical expression makes the solution less portable to other similar uncertainty calculation problems based on the same measurement model.

Both GUM and Monte Carlo approaches only approximate the true uncertainty of a measurement model. However, comparing the two results can provide insight into the validity of the approximations. A qualitative comparison can be made by superimposing plots of the GUM resulting probability density function over the Monte Carlo output histogram. Close agreement provides validation of the GUM approximation to the problem.

Figure 8.9 compares the GUM and Monte Carlo methods for the cylinder density uncertainty example. The metrologist will often rely on experience and judgement to decide if the difference is enough to discount the GUM approach.

8.3.1 Quantitative GUM Validity Test

A more quantitative comparison of the GUM and Monte Carlo approaches can be used to validate the GUM approximation (Cox and Harris 2006; JCGM 2008). The approach is:

1. Compute 95% confidence intervals using both the GUM method and the Monte Carlo method.

Table 8.5 GUM validity test

95.00% coverage	Lower limit	Upper limit
GUM	7.70	17.76
Monte Carlo	8.34	18.75
abs(GUM - MC)	0.64	0.99
abs(GUM - MC) < δ	FAIL	FAIL

2. Determine the number of meaningful significant digits n_{dig}. Typically use $n_{dig} = 1$ or $n_{dig} = 2$.
3. Express the computed uncertainty value $u(y)$ in the form $a \times 10^r$ where a is an n_{dig} digit integer and r an integer. Then compute the required tolerance $\delta = \frac{1}{2} 10^r$.
4. Compare the 95% endpoints computed in step 1, both upper and lower bounds. If the magnitude difference in endpoints is less than δ, the GUM method is a valid approximation.

Applying this test to the density example (to one significant digit):

$$u(y) = 2.8 \approx 3 \times 10^0$$

$$r = 0$$

$$\delta = 0.5 \times 10^0 = 0.5 \tag{8.5}$$

```
# Table 8.5 Gum Validity Test
k <- 1.96 # Use 95% for infinite deg. freedom
gumLimits <- c(gumMean - gumSD*k, gumMean + gumSD*k)
mcLimits <- shortest # Use shortest interval calculated above (step 1)

ndig <- 1 # Number of significant digits determined by user (Step 2)
r <- floor(log(gumSD)) - (ndig-1) # Determine the r exponent (step 3)
delta <- 0.5 * 10.0^r # Determine delta tolerance (step 3)
diff <- abs(gumLimits - mcLimits)
testResult <- ifelse(diff < delta, 'PASS', 'FAIL')
validity <- t(data.frame(GUM = gumLimits, MC = mcLimits, Difference = diff, Result = testResult))
colnames(validity) <- c("Lower Limit", "Upper Limit")
validity
##            Lower Limit Upper Limit
## GUM        " 7.6928"   "17.7672"
## MC         " 8.322284" "18.752868"
## Difference "0.6294838" "0.9856681"
## Result     "FAIL"      "FAIL"
```

Table 8.5 compares the endpoints on each side of the distribution between GUM and Monte Carlo methods. Both endpoints differ by more than δ, so by this metric, the Monte Carlo method would be preferred to the GUM method for this problem.

8.4 Monte Carlo Case Studies

This section provides two practical examples of using Monte Carlo for uncertainty propagation: the neutron yield measurement first introduced in Chapter 6, and the time constant of a resistor-capacitor (RC) circuit.

8.4.1 Case Study: Neutron Yield Measurement

The neutron yield measurement using a lead probe detector was introduced in Sect. 6.4.1 and calculated using the GUM approximation. Here, it is calculated again using the Monte Carlo approach. Recall that the number of neutrons, η, is calculated from

$$\eta = AfF(S - B), \tag{8.6}$$

where A is the calibration factor, f is a conversion factor, and F is the attenuation factor. Variables S and B are the scalar and background counts from the detector. The mean and standard uncertainties for each input variable are repeated in Table 8.6.

```
# Measurement equation
eta <- function(A, f, F, S, B) {A*f*F*(S - B)}
```

```
# Mean values for input quantities and standard uncertainties
f <- 4353
A <- c(1, 0.020)
F <- c(1.27, 0.032)
S <- c(9700, 98.5)
B <- c(80, 2.83)
```

To use the Monte Carlo approach for uncertainty propagation, random samples for each of these variables must be pulled from their respective probability distributions. Variable f is treated as known without uncertainty, so it may be considered a

Table 8.6 Distributions for input quantities and standard uncertainties

Input quantity	Mean (μ)	Standard uncertainty	PDF type A	Standard uncertainty type B	PDF type B
f	4353				
A	1.00			0.020	Uniform
F	1.270			0.032	Uniform
S scalar counts	9700	98.5	Poisson		
B background counts	80	2.83	Normal		

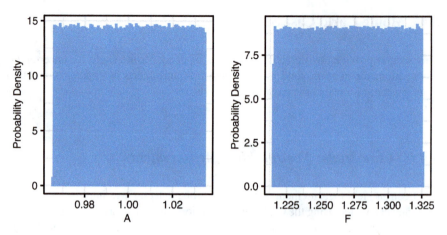

Fig. 8.10 Monte Carlo samples drawn from uniform distributions

constant. Variables A and F are both uniform distributions. The Type B uncertainties for these variables are already provided as standard uncertainties, but typical random number generator software requires the half-width of the uniform distribution, not the standard uncertainty. Equation 4.6 can be used to compute the half-width of the uniform distribution from its standard deviation:

$$U_A = u_B(A) \times \sqrt{3} = 0.020 \times \sqrt{3} = 0.035$$
$$U_F = u_B(F) \times \sqrt{3} = 0.032 \times \sqrt{3} = 0.055$$

```
# Compute half-width of the uniform distribution
A[3] <- A[2]*sqrt(3)
F[3] <- F[2]*sqrt(3)
```

Using the half-widths U_A and U_F, with their mean values, one million random samples can be generated from these uniform distributions as shown in Fig. 8.10. The distribution for variable A is centered about its mean of 1.0, and is constant in the range 1 ± 0.035. Similarly, the distribution for variable F is centered about its mean of 1.270 and is constant in the range 1.270 ± 0.055.

```
# Generate Monte Carlo Input Samples
n <- 1e6
ASamp <- runif(n, A[1]-A[3], A[1]+A[3])
FSamp <- runif(n, F[1]-F[3], F[1]+F[3])
SSamp <- rpois(n, 9700)
BSamp <- rnorm(n, 80, 2.83)
```

Variable B is approximated by a normal distribution, which is straightforward to sample using its mean of 80 and standard deviation of 2.83, as shown in Fig. 8.11.

Fig. 8.11 Monte Carlo samples drawn from a normal distribution

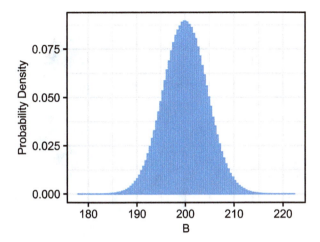

Fig. 8.12 Monte Carlo samples drawn from a Poisson distribution with $\lambda = 9700$

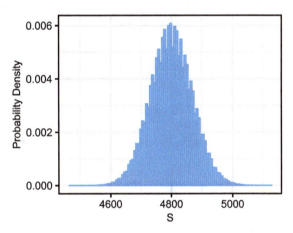

The Poisson distribution (Eq. 4.10) of variable S requires more care in implementation. In this case, drawing random samples from a Poisson distribution with $\lambda = 9700$ produces the correct result. Recall that a Poisson distribution has both mean and variance equal to λ. Although the Poisson distribution is discrete, for large values of λ it is well approximated by a normal distribution (Fig. 8.12).

Combining all the sampled values using Eq. (8.6) results in the combined Monte Carlo distribution. For this set of samples, the mean and standard deviation can be estimated using standard techniques described in Chapter 4. This results in a mean value of 5.32×10^7 neutrons and standard deviation of 1.79×10^6 neutrons. Because the output of a Monte Carlo will not necessarily be normally distributed, the 95% coverage range is determined from the percentiles of the sampled distribution. Calculating the 2.5% and 97.5% percentiles results in the 95% coverage range from 4.98×10^7 to 5.66×10^7 neutrons.

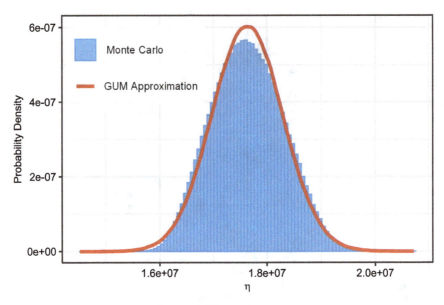

Fig. 8.13 Comparison of Monte Carlo and GUM results

The GUM method calculated in Sect. 6 resulted in a 95% coverage range from 4.95×10^7 to 5.69×10^7 neutrons. Results from the two methods are compared in Fig. 8.13. Although slightly different, the difference is not significant.

```
# Generate Monte Carlo output samples and compare to the GUM
out <- eta(A = ASamp, f = f, F = FSamp, S = SSamp, B = BSamp)

# Symmetric coverage interval
symInterval <- quantile(out, probs = c(0.025, 0.975))
symInterval
##    2.5%    97.5%
## 49832000 56647167
```

The difference in the two methods can be quantified to validate the GUM approximation using the method described in Sect. 8.3.1.

Step 1: The 95% coverage range for the GUM approach is $(4.95 \times 10^7, 5.69 \times 10^7)$ and for the Monte Carlo approach is $(4.98 \times 10^7, 5.66 \times 10^7)$.

Step 2: Use $n_{\text{dig}} = 1$.

Step 3: The standard uncertainty is $1.8 \times 10^6 \approx 2 \times 10^6$, so $r = 6$. Then the tolerance $\delta = \frac{1}{2} 10^6 = 5 \times 10^5$.

Step 4: The difference in upper bounds between the GUM and Monte Carlo is $5.69 \times 10^7 - 5.66 \times 10^7 = 3.0 \times 10^5$ which is less than δ. The upper limit passes the test. The lower limit difference is $4.98 \times 10^7 - 4.95 \times 10^7 = 3.0 \times 10^5$ which is also less than δ.

8.4.2 Case Study: RC Circuit

An RC circuit is constructed using commercial off-the-shelf (COTS) components (Fig. 8.14). The components are a 5 kΩ resistor with 1% tolerance, a 0.22 μF capacitor with 5% tolerance, and a parallel 0.1 μF capacitor with 1% tolerance. Because the component uncertainties are given as tolerances, uniform distributions are used.

The time constant of the circuit is found using the measurement model:

$$\tau = R(C_1 + C_2). \tag{8.7}$$

The distribution of 1 million Monte Carlo samples, along with the results calculated by the GUM method, are shown in Fig. 8.15. Note that the GUM method results in a normal distribution whereas the Monte Carlo method produces a more trapezoidal shape, and the 95% limits are not the same. In this case, the difference is not from nonlinearity of the measurement model, but in the GUM's conversion of all inputs into normal distributions.

```
# Measurement equation
tau <- function(R, C1, C2) {R*(C1+C2)}

# Mean values and uncertainties for variables
R <- c(5.00, 0.01)
C1 <- c(0.22, 0.05)
C2 <- c(0.10, 0.01)

# Convert uncertainties to percent and normalize uniform distribution
R[3] <- (R[1]*R[2])/sqrt(3)
C1[3] <- (C1[1]*C1[2])/sqrt(3)
C2[3] <- (C2[1]*C2[2])/sqrt(3)
```

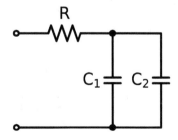

Fig. 8.14 Resistor-capacitor circuit

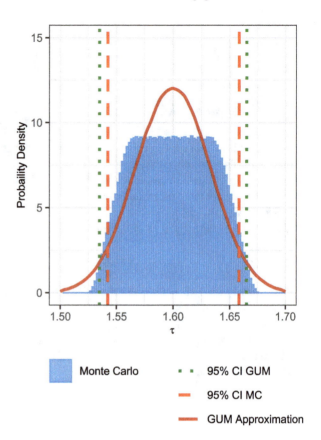

Fig. 8.15 Results of RC circuit uncertainty propagation

```
# Calculate GUM
# Partial derivative with respect to R
dtdR <- Deriv(tau, "R")

# Partial derivative with respect to C1
dtdC1 <- Deriv(tau, "C1")

# Partial derivative with respect to d
dtdC2 <- Deriv(tau, "C2")

# Mean value
gumMean <- tau(R = R[1], C1 = C1[1], C2 = C2[1])
gumMean
## [1] 1.6

# Combined standard uncertainty
uR <- dtdR(R = R[1], C1 = C1[1], C2 = C2[1])*R[3]
uC1 <- dtdC1(R = R[1], C1 = C1[1], C2 = C2[1])*C1[3]
uC2 <- dtdC2(R = R[1], C1 = C1[1], C2 = C2[1])*C2[3]
uComb <- sqrt(uR^2 + uC1^2 + uC2^2)
```

```
#95% Confidence interval
gum <- rnorm(n, gumMean, uComb)
k <- qnorm((1 - (.05/2)))
gumConfInt <- c(gumMean - k * uComb, gumMean + k* uComb)
gumConfInt
## 1.534936 1.665064

# Generate Monte Carlo Input Samples
n <- 1e6
RSamp <- runif(n, R[1]-R[1]*R[2], R[1]+R[1]*R[2])
C1Samp <- runif(n, C1[1]-C1[1]*C1[2], C1[1]+C1[1]*C1[2])
C2Samp <- runif(n, C2[1]-C2[1]*C2[2], C2[1]+C2[1]*C2[2])

# Generate Monte Carlo Output Samples
out <- tau(R = RSamp, C1 = C1Samp, C2 = C2Samp)

# Symmetric coverage interval
symInterval <- quantile(out, probs = c(0.025, 0.975))
symInterval
##    2.5%    97.5%
## 1.542273 1.658352
```

8.5 Summary

This chapter introduced the Monte Carlo method of uncertainty propagation. The Monte Carlo method uses large sets of random samples to approximate the distribution of possible output values from the measurement model. Because of the numerical nature, it does not require a measurement model that can be written down in closed form. Unlike the GUM, it works in situations where the measurement model cannot be differentiated symbolically, for example, in functions that must be solved iteratively. It also accounts for non-normal input probability distributions and nonlinearities in the measurement model, resulting in an output probability distribution that is not necessarily normal.

It is typically recommended to perform a Monte Carlo uncertainty calculation in parallel with the GUM approximation in order to fully understand the results of both methods.

8.6 Related Reading

Donald Knuth's classic book series *The Art of Computer Programming* is essential reading for any programmer, but Volume 2 (Knuth 1998) is most relevant to the Monte Carlo methods of this chapter. It discusses random number generation in detail, including justification for choice of particular constants in LCGs and various empirical and theoretical statistical tests to ensure randomness in the output. Because Monte Carlo is by nature a parallel computation, it lends itself to parallel implementations such as on a GPU. However, care must be taken to ensure the random numbers are distributed between processes and recompiled correctly. Kindratenko (2014) gives a thorough overview of GPU MC implementation.

The GUM's Supplement 1 (JCGM 101, 2008), also released as ISO/IEC Guide 98-3, details use of Monte Carlo methods for uncertainty propagation. ISO/TR 10017 mentions Monte Carlo as an applicable statistical method but that it should not be used when exact analytical methods are available.

Cox and Harris (2006) and Cox and Siebert (2006) provide guidance on Monte Carlo specifically for uncertainty propagation. They explore convergence criteria, sampling from joint probability distributions for correlated inputs, adaptive sampling, and the multivariate output case where the result is an "uncertainty region" rather than a univariate distribution. Crowder and Moyer (2006) detail an approach to Monte Carlo that accounts for small sample sizes by using a two-stage bootstrap sampling technique. Willink (2007) analyzes the two approaches to extracting coverage intervals from Monte Carlo results (briefly mentioned in Sect. 8.2.1.2). Another example of using Monte Carlo methods is evaluation of uncertainty of weighing in air, including a thorough evaluation of the PRNG used in the evaluation, given by Flicker and Tran (2016).

8.7 Exercises

1. Use your favorite programming language to write an algorithm that estimates π using a Monte Carlo method. Make a convergence plot showing the estimated π value as a function of number of Monte Carlo trials up to 1 million trials.

2. Monte Carlo Methods can be used to approximate definite integrals, and may have advantages over other numeric integration approximations when the dimensionality becomes high. For the function $f(x)$, continuous in the range from a to b, the integral is approximated by:

$$\int_a^b f(x)\, dx \cong \frac{(b-a)}{N} \sum_{i=0}^{N} f(x_i),$$

8.7 Exercises

where x_i is a random sample drawn from a uniform distribution in the range from a to b. Use this Monte Carlo integration technique, with N = 10000 samples, to find the value of the integral

$$\int_0^{2\pi} e^{-x/2} \sin\left(\frac{\pi x}{2}\right) dx.$$

Compare the result to a trapezoidal approximation of the integral.

3. Implement a function to generate random numbers using an LCG as given by Eq. (8.2). The function should take a, c, m, and N as arguments, and return an array of N random numbers. Using $m = 256$, select integer values for a and c and use the LCG to generate 10000 random samples. Plot a histogram of the samples with 256 bins. Does it appear uniformly distributed, with each bin having approximately the same count? Plot a scatter plot of random number vs. sample number. Does there appear to be order or a pattern to the scatter plot? Experiment with several values of a and c.

4. Make a histogram and scatter plot of 10,000 random samples from your programming language's built-in uniform random number generator. Does the histogram appear uniformly distributed? Does there appear to be any order or pattern to the scatter plot?

5. For the following distributions x_1 and x_2, generate 100,000 random samples from each (using your programming language's built-in random generator) and plot a histogram of the function $f = x_1 + x_2$. The uniform distributions' a parameter refers to the half-width of the rectangle.

    (a) $x_1 = \text{Normal}(\mu = 0; \sigma = 1)$ $x_2 = \text{Normal}(\mu = 5; \sigma = 0.5)$
    (b) $x_1 = \text{Normal}(\mu = 0; \sigma = 0.5)$ $x_2 = \text{Uniform}(\mu = 5, a = 2)$
    (c) $x_1 = \text{Uniform}(\mu = 2; a = 1)$ $x_2 = \text{Uniform}(\mu = 5, a = 2)$

6. Repeat exercise 5 with the function $f = x_1^2 + x_2^2$.

7. Use the Monte Carlo method to calculate the uncertainty in the volume of a box. The measured lengths of the three sides are 3 cm, 4 cm, and 5 cm, all with 0.1 cm standard uncertainty. From the resulting Monte Carlo samples, plot a histogram of possible output values and determine a 95% coverage range.

8. After measuring the voltage and resistance, the power dissipated in a resistor is calculated from the equation $P = V^2/R$. Assume the measured voltage is 10 V ± 0.1 V ($k = 1$, normal distribution), and the resistor uses the manufactured specifications of 1000 Ω ± 5% (assume uniform distribution with half-width 5% of 1000). Use Monte Carlo uncertainty propagation with 10_6 samples to plot a histogram of the distribution of possible values of power. Determine the mean, standard uncertainty, and a 95% coverage interval for this distribution.

9. For the power measurement in exercise 8, make a convergence plot. Plot the standard uncertainty as a function of samples for the first 100,000 samples, the first 200,000 samples, and so on up to all 10^6 samples. Were enough samples generated to converge on a reasonable solution?

10. Recalculate the uncertainty in the pitot-static tube exercise from Chap. 7 (exercise 7.6) using the Monte Carlo method, assuming all inputs are normally distributed. Do the results agree with the GUM method? Quantify the difference in the two methods by applying the Quantitative GUM Validity Test described in Sect. 8.3.1. Use $n_{\text{dig}} = 1$, determine the tolerance δ, and check if the 95% interval endpoints from the GUM and Monte Carlo methods agree within δ. Try the test again using $n_{\text{dig}} = 2$.
11. Recalculate the uncertainty in the gain ($A = -Rf/Ri$) of the inverting amplifier exercise from Chap. 7 (exercise 7.12) using the Monte Carlo method. Recall Rf = 100,00 Ω and Ri = 1000 Ω with 5% tolerance (uniform distribution). Compare the Monte Carlo histogram with the GUM solution.
12. Recalculate the uncertainty in the Scanning Electron Microscope exercise from Chap. 7 (exercise 7.13) using the Monte Carlo method. Because x_2 and y_m have both Type A and Type B uncertainties, a random sample must be drawn from both the Type A and Type B distributions and added together for each of these variables to obtain the variable's sampled value for each trial. Take care that the mean of the distribution of Monte Carlo samples for x_2 and y_m represents their measured values, i.e. draw the Type A sample from $N(\mu_A, \sigma_A)$ but draw the Type B sample from Uniform(0, a_B), centered at zero rather than the measured value μ_A, to avoid shifting the distribution twice. Does the Monte Carlo agree with the GUM calculation?
13. The arcsine distribution (PDF $f(x) = \frac{1}{\pi\sqrt{x(1-x)}}$) is sometimes used to model the distribution of a sinusoidally fluctuating value, such as the temperature of an environmentally controlled lab (JCGM 101:2008). For a lab controlled to 23 ± 3 °C, generate random samples from the arcsine distribution centered at 23 with half-width 3 and plot a histogram of the samples. If your software does not have a built-in method for sampling from an arcsine distribution, a transformation can be used: generate uniform samples in the range [0, 1], apply the transform $x = \sin^2\left(\frac{\pi F}{2}\right)$ where F are the uniform samples, and then scale and shift to the appropriate range.

Next, perform a Monte Carlo uncertainty propagation to find the uncertainty of a resistor used in this lab using the formula $R = R_0(1 + \alpha(T - T_0))$. Here, R_0 is the calibrated resistance of $1000 \pm 5\ \Omega$ at temperature $T_0 = 20 \pm 0.025$ °C. The temperature coefficient of resistance, α, is given as 0.0039 °C^{-1} $\pm$ 1.9×10^{-5} °C^{-1} from the resistor's calibration certificate. Plot a histogram of values for the temperature-adjusted resistance R, then determine the mean and standard deviation of the histogram.
14. Perform a Monte Carlo uncertainty propagation for the measurement model $f(x, y, z) = x^2 + y^2 + z^2$. Given x defined by a normal distribution with mean 0 and standard deviation 1, y defined by a uniform distribution with mean 0 and half-width parameter $a = 0.5$, and z given by a symmetric triangular distribution with mean 0 and half-width 2, plot a histogram of the output samples and determine the mean value of the distribution. All the input distributions have mean value of

0, suggesting the "true" value of the function $f(x, y, z)$ is 0. Does the PDF given by the Monte Carlo histogram suggest a true value of 0?
15. In addition to airspeed indication, pitot-static tubes can provide an aircraft with altitude information. The formulas for calculating altitude, h, and velocity, v, are

$$v = \sqrt{\frac{2(p_a - p_d)R_s T_a}{p_a}}$$

$$h = -\frac{R_s}{g_0} T_a \cdot \ln\left(\frac{p_a}{p_0}\right)$$

where R_s is the specific gas constant for dry air (287.06 J/K/kg), g_0 the gravitational acceleration (9.80664 m/s^2), and p_0 the atmospheric pressure measured at sea level. The aircraft measures the ambient temperature T_a, ambient static pressure, p_a, and dynamic pressure p_d (expanding the difference Δp from the previous exercise). The value for p_0 is provided by air traffic control as 101.4 ± 2.1 kPa. Given readings of $T_a = 236 \pm 2.5$ K, $p_a = 21 \pm 0.5$ kPa, and $p_d = 15 \pm 0.5$ kPa, perform a Monte Carlo uncertainty evaluation for v and h, plotting histograms for both. Assume R_s and g_0 have no uncertainty, and all other uncertainties are given at $k = 2$. (With the units given, results will be in meters/second and meters, respectively. Convert to preferred units of the aviation industry, such as knots—nautical miles per hour—and feet, if desired.) Then plot a scatter plot of h versus v. Do the output values appear correlated? Use your software to calculate the correlation coefficient between h and v.

References

Barreto, H., Howland, F.M.: Introductory econometrics using Monte Carlo simulation with microsoft excel. Cambridge University Press, Cambridge (2006)

Box, G.E., Muller, M.E.: A note on the generation of random normal deviates. Ann. Math. Stat. **29**(2), 610–611 (1958)

Brown, W. E.: Random number generation in C++11 (WG21 N3551). Standard C++ Foundation. https://isocpp.org/files/papers/n3551.pdf (2018). Accessed April 28, 2020

Cox, M.G., Harris, P.M.: Software specifications for uncertainty evaluation, DEM-ES-010. Middlesex, UK, National Physical Laboratory (2006)

Cox, M.G., Siebert, B.R.L.: The use of a Monte Carlo method for evaluating uncertainty and expanded uncertainty. Metrologia. **43**(4), S178–S188 (2006)

CRAN Task view: probability distributions. (2018). https://cran.r-project.org/web/views/Distributions.html. Accessed June10, 2018

Crowder, S.V., Moyer, R.D.: A two-stage Monte Carlo approach to the expression of uncertainty with non-linear measurement equation and small sample size. Metrologia. **43**(1), 34–41 (2006)

De Comte Buffon, G.: Essai d'arithmetique morale. In: Supplement a l'histoire naturelle. Nabu Press, Charleston, SC (1774)

Flicker, C., Tran, H.: Calculating measurement uncertainty of the "conventional value of the result of weighing air". NCSLI Measure. **11**(2), 22–37 (2016)

Gentle, J.E.: Computational statistics. Springer, New York (2009)

Graham, C., Talay, D.: Strong law of large numbers and Monte Carlo methods. Springer-Verlag, Berlin (2013)

Hellekalek, P.: Don't trust parallel Monte Carlo! ACM SIGSIM Simul Dig. **28**(1), (1998)

Javamex.: How does java.util.Random work and how good is it? https://www.javamex.com/tutorials/random_numbers/java_util_random_algorithm.shtml (2018). Accessed April 29, 2020.

JCGM 101: Evaluation of Measurement Data - Supplement 1 to the Guide to the Expression of Uncertainty in Measurement - Propagation of Distributions Using a Monte Carlo Method. (2008)

Kelvin, L.: Ninteenth century clouds over the dynamical theory of heat and light. Phil Mag. **6**(2), 1–40 (1901)

Kindratenko, V.: Numerical computations with GPUs. Springer, New York (2014)

Knuth, D.: The art of computer programming, Volume 2: Semi-numerical algorithms, 3rd edn. Addison Wesley Longman, Boston, MA (1998)

Marsaglia, G.: Random numbers fall mainly in the planes. Proc. Natl. Acad. Sci. **61**(1), 25–28 (1968)

Mathworks Inc.: RandStream - random number stream. https://www.mathworks.com/help/matlab/ref/randstream.html (2018). Accessed April 29, 2020

Matsumoto, M., Nishimura, T.: Mersenne twister: A 623-dimensionally equidistributed uniform pseudo-random number generator. ACM Trans Model Comput Simul. **8**(1), 3–30 (1998)

McCullough, B.D.: Microsoft Excel's 'Not the Wichmann-Hill' random number generator. Comput. Stat. Data Anal. **52**, 4587–4593 (2008)

Melard, G.: On the accuracy of statistical procedures in Microsoft Excel 2010. Comput. Stat. **29**, 1095–1128 (2014)

Metropolis, N.: The beginning of the Monte Carlo method. Los Alamos Sci.15,125–130 (1987)

National Instruments: What is the algorithm used by the LabVIEW random number (0-1) function? http://digital.ni.com/public.nsf/allkb/9D0878A2A596A3DE86256C29007A6B4A (2018). Accessed April 29, 2020

Press, W.H., Teukolsky, S.A., Vetterling, W.T., Flannery, B.P.: Numerical recipes in C - the art of scientific computing, 2nd edn. Cambridge University Press, Cambridge (2002)

Python Software Foundation: Python documentation. https://docs.python.org/3/library/random.html (2018). Accessed April 29, 2020

Richtmyer, R.D., Von Neumann, J.: Statistical methods in neutron diffusion. Los Alamos National Laboratory, Los Alamos, NM (1947)

Robert, C.P., Casella, G.: Monte Carlo statistical methods, 2nd edn. Springer, New York (2004)

Wichmann, B.A., Hill, I.D.: Algorithm AS 183: An efficient and portable pseudo-random number generator. J. R. Stat. Soc. Ser. C. **31**(2), 188–190 (1982)

Wichmann, B.A., Hill, I.D.: Generating good pseudo-random numbers. Comput. Stat. Data Anal. **51**, 1614–1622 (2006)

Willink, R.: A generalization of the Welch-Satterthwaite formula for use with correlated uncertainty components. Metrologia. **44**, 340–349 (2007)

Wolfram Research: Wolfram Language & System Documentation Center. http://reference.wolfram.com/language/tutorial/RandomNumberGeneration.html (2018). Accessed April 29, 2020

Chapter 9
Design of Experiments in Metrology

One of the topics that has received relatively little attention in the metrology literature is the statistical design of experiments (DOEx). A premise of statistical DOEx is that efficiency is paramount in terms of the resources (cost, time, personnel, equipment, etc....) used to perform an experiment. In the case of metrology studies, individual runs in an experiment may be very costly, and the experiment must be designed in such a way that information regarding the measurement process is obtained efficiently. In this chapter we present a variety of experimental designs and approaches to efficiently developing, evaluating, and optimizing a measurement process.

9.1 Introduction

Metrology experiments will typically be used in the design and optimization of a new measurement process or in the evaluation of an existing measurement process. To design and optimize a new measurement process, factorial and fractional factorial designs (DOEx) will often be used to identify factors and quantify their effect on the measurement. To evaluate an existing measurement process, an analysis of variance (ANOVA) approach will often be used to estimate sources of variation that affect the measurement. In this chapter the basics of these types of designs are presented along with case studies to illustrate the approach. The strategies for experimental design presented here will rely on historical knowledge of the measurement process, careful determination of the objectives of the experiment, and sample size considerations to guide the choice of experimental design.

9.2 Factorial Experiments in Metrology

Section 6.4 introduced the notion of an indirect measurement as a measurement in which the value of the measurand is functionally related to the values of other measured variables. The measurement equation was represented by (Eq. 6.29):

$$y = f(X_1, X_2, \ldots, X_N).$$

Following the approach for indirect measurements presented in Chap. 6, the measurement equation is approximated via the Taylor series expansion:

$$y = f(X_1, X_2, \ldots, X_N) \cong \beta_0 + \beta_1 X_1 + \beta_2 X_2 + \ldots + \beta_N X_N$$

$$= \beta_0 + \sum_{i=1}^{N} \beta_i X_i. \tag{9.1}$$

In this relationship, the input variables, the $X_i s$, are the sources of variation affecting the measurand. The coefficients, the $\beta_i's$, correspond to the partial derivatives from the Taylor series expansion (see Eq. 6.34). In some cases, the form of f will not be well known and must be determined experimentally. The DOEx approach can be used to efficiently identify input variables and determine the relationship between the measurand and the identified variables. A vast literature exists on the topic of DOEx (see, for example, Box et al. 2005 and Montgomery 2017), but little attention has been given to the use of DOEx in metrology. In metrology, the $X_i's$ may include physical quantities such as voltage, equipment factors, test setup factors, environmental factors, or human factors that may have an impact on the outcome of the measurement.

Factorial DOEx attempts to systematically identify important factors and interrelationships between factors and to quantify their effect on the measurand. The approach first attempts to identify the important $X_i's$ and associated $\beta_i's$ in Eq. (9.1). A second experiment may then be required to improve the estimates of the $\beta_i's$ and to check for the presence of second order terms. This sequential approach to experimentation may often be more efficient than a single large experiment. With results from a well-planned DOEx, the measurement process can be designed or optimized by reducing the effect of any variables that increase measurement uncertainty. The analysis may also point to improvements in equipment, materials, or procedures that reduce measurement uncertainty.

The sections that follow below outline a DOEx planning strategy for factorial experiments that we use in our work. The four basic steps are: (1) Select the appropriate measurand(s) and determine the objective of the experiment, (2) Select factors to be included in the experiment, (3) Select factor levels, and (4) Choose the design pattern. If the purpose of the experiment is to characterize an established measurement process, the methods of ANOVA that follow in Sect. 9.3 should be considered in addition to the factorial design strategy described in this section.

9.2 Factorial Experiments in Metrology 183

9.2.1 Defining the Measurand and Objective of the Experiment

The first step in any metrology experiment is to carefully define the measurand of interest. The definition of the measurand should include, in detail, how the measurement will be made, and any special physical conditions present during the taking of the measurement. The GUM specifies that "the measurand should be defined in sufficient detail that any uncertainty arising from its incomplete definition is negligible in comparison with the required accuracy of the measurement." This first step in the DOEx process should also include a statement of the objective of the study. The objective could be to (1) identify a measurement equation including inputs and coefficients, (2) optimize a measurement process given either a known or unknown measurement equation, or (3) perform an uncertainty analysis based on a known measurement equation.

9.2.2 Selecting Factors to Incorporate in the Experiment

Given a measurand and the objective of the experiment, the second step is to list all factors that might impact the measurement. Expert knowledge, previous experimental results, historical data analysis, literature research, brainstorming, cause and effect diagramming, etc. may be used to develop a list of potential factors. As an input to the measurement equation, each factor may itself be thought of as a measurand.

The next step is to determine which factors can be controlled experimentally. For each of the factors deemed controllable, it must be determined if the factor can be measurable with sufficient accuracy. The possibility of interactions between factors and how that might impact the measurand should also be considered.

Finally, it must be determined how the factors will be handled during the experiment. The factors identified in the steps above could be *controlled* at various levels, *held constant*, *left uncontrolled but monitored*, or *ignored* during the experiment. Logistics, costs, and suspected impact will determine which of the potential factors will be carried forward into the metrology experiment.

9.2.3 Selecting Factor Levels and Design Pattern

Next, the factor levels and design pattern for the experiment should be chosen. Factorial design patterns under consideration should include full factorials, in which the experimental trials consist of all possible combinations of factor settings, and fractional factorials, in which the experimental trials consist of a carefully chosen subset of the corresponding full factorial design.

Table 9.1 Choosing type of design

If the objective is	Use this type of design
Identify the important factors affecting a measurand. This information may be used to design a measurement process and identify the form of the resulting measurement equation	Depending on the number of factors to be investigated, a full factorial or fractional factorial, usually with factors at two levels
Estimate coefficients of a measurement equation and optimize the measurement process	Usually a full factorial with factors at two or three levels. Fewer factors than in the first stage
For an established measurement process, estimate components of variance and quantify measurement uncertainties	Factorial design or an ANOVA-based design with a combination of random and fixed factors in a crossed or nested classification

The choice of factor levels and design pattern will depend on the objective of the experiment. Table 9.1 lists the most common objectives and the experimental plans that correspond to those objectives. After the type of design is determined, the low and high levels for each factor should be selected. Any theoretical relationships between the experimental factors and the measurand should be used to select physically meaningful levels. Levels for each factor could also be determined from previous experience with the same factors. The selected levels should be bold, yet meaningful, within the context of the measurement process being studied. For each qualitative factor, the levels are simply the different categories within that factor. For example, if the qualitative factor is "Operator," the operators used in the experiment define the levels.

Based on the objective, the recommended choice of experimental design will be a full factorial, a fractional factorial, or an ANOVA-based design with random and possibly fixed effects.

For a thorough discussion of the types of designs, models, and analyses associated with full factorial designs and mixed-level factorials, see Montgomery (2017). For two-level fractional factorials and second order designs, see Box et al. (2005), and for random and mixed effects models and estimation of variance components, see Searle and Gruber (2016).

The full factorial and fractional factorial design and analysis approach is illustrated with case studies that follow in Sects. 9.2.4 and 9.2.5. The random and fixed effects design and analysis approach is illustrated with case studies that follow in Sects. 9.3.4 and 9.3.5.

9.2.4 Analysis of CMM Errors via Design of Experiments (2^4 Full Factorial)

The first case study demonstrates the design and analysis of a full factorial design. The experiment investigated the uncertainty of a coordinate measuring machine (CMM) contact probe measurement (Zhang, et al., 2002). The objective of the experiment was to quantify the uncertainty of the measurement with respect to several factors and to optimize use of the CMM.

9.2 Factorial Experiments in Metrology

Table 9.2 Experimental design matrix (2^4 full factorial)

Run	A = sampling speed (mm/s)	B = part size (mm)	C = surface form	D = probe size
1	−1	−1	−1	−1
2	1	−1	−1	−1
3	−1	1	−1	−1
4	1	1	−1	−1
5	−1	−1	1	−1
6	1	−1	1	−1
7	−1	1	1	−1
8	1	1	1	−1
9	−1	−1	−1	1
10	1	−1	−1	1
11	−1	1	−1	1
12	1	1	−1	1
13	−1	−1	1	1
14	1	−1	1	1
15	−1	1	1	1
16	1	1	1	1

Factor levels were: Sampling Speed (−1 = 300 mm/s and 1 = 1000 mm/s), Part Size (−1 = Small Diameter and 1 = Large Diameter), Surface Form (−1 = Rough and 1 = Smooth), and Probe Size (−1 = Small and 1 = Large)

In a full factorial design with N factors, if the number of levels for factor X_i is l_i for $i = 1, 2, \ldots, N$, the total number of possible combinations in the experiment is $l_1 \times l_2 \times \ldots \times l_N$. This number can become prohibitive if many factors or many levels for each factor are under consideration. The full factorial design will thus be used when the number of factors and factor levels are relatively small. This type of design could be used as a second experiment after a fractional factorial has been used to reduce the number of potential factors. The benefit of the full factorial design is that all possible interactions between factors can be thoroughly investigated.

The full factorial design presented here included four factors related to the operation of a CMM: Sampling Speed, Part Size, Surface Form, and Probe Size. The factors in the experiment and the full factorial DOEx appear in Table 9.2. Each factor was investigated at two levels for a total of $2^4 = 16$ combinations.

Setting the factor levels to −1 and +1 is somewhat arbitrary, with −1 representing the "low" level of the factor and +1 representing the "high" level of the factor. These values for factor levels have historically been used to make hand calculations of effects and interactions easy. However, because use of computers for the analyses makes these designations unnecessary, the hand calculations are not presented here.

To provide estimates of repeatability, each combination was run a total of ten times. One small cylinder and one large cylinder were used in the experiment. The response variable was the diameter of the cylinder measured on the CMM. Table 9.3 includes all the data from the experiment, along with summary statistics (mean, standard deviation) across the ten replicates of each combination.

Table 9.3 Measured cylinder diameters (mm)

Run	d1	d2	d3	d4	d5	d6	d7	d8	d9	d10	Mean	StDev
1	10.089	10.098	10.103	10.083	10.092	10.083	10.082	10.103	10.098	10.090	10.092	0.008089
2	10.093	10.099	10.086	10.096	10.093	10.083	10.087	10.085	10.097	10.088	10.091	0.005599
3	36.909	36.909	36.939	36.925	36.924	36.930	36.929	36.916	36.914	36.918	36.921	0.009799
4	36.931	36.927	36.922	36.910	36.934	36.912	36.917	36.915	36.911	36.913	36.919	0.008766
5	7.986	7.985	7.985	7.986	7.985	7.987	7.987	7.987	7.985	7.986	7.986	0.000876
6	7.985	7.984	7.984	7.987	7.988	7.983	7.984	7.987	7.987	7.985	7.985	0.001713
7	31.047	31.049	31.048	31.047	31.048	31.047	31.049	31.049	31.048	31.048	31.048	0.000816
8	31.049	31.048	31.047	31.049	31.048	31.048	31.049	31.049	31.049	31.050	31.049	0.000843
9	10.028	10.029	10.025	10.022	10.026	10.019	10.025	10.018	10.017	10.029	10.024	0.004541
10	10.019	10.026	10.028	10.029	10.031	10.030	10.028	10.027	10.027	10.018	10.026	0.004373
11	36.843	36.844	36.835	36.836	36.841	36.840	36.844	36.838	36.845	36.839	36.841	0.003504
12	36.838	36.843	36.837	36.842	36.844	36.844	36.846	36.840	36.838	36.842	36.841	0.003026
13	7.978	7.977	7.976	7.976	7.978	7.978	7.974	7.978	7.978	7.977	7.977	0.001333
14	7.977	7.979	7.977	7.981	7.978	7.975	7.980	7.978	7.977	7.977	7.978	0.001729
15	31.037	31.035	31.035	31.034	31.036	31.035	31.037	31.035	31.035	31.036	31.036	0.000972
16	31.034	31.034	31.034	31.035	31.034	31.035	31.034	31.035	31.035	31.035	31.035	0.000527

Ten replicates were performed for each experimental combinations

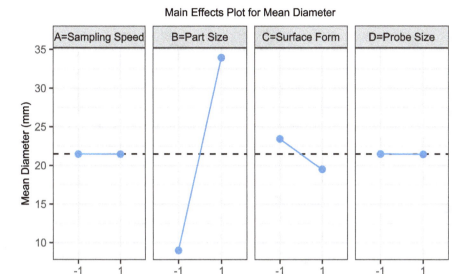

Fig. 9.1 Main effects plot for mean cylinder diameter

Analysis of the data consisted of two stages, analyzing the mean diameter measurements and analyzing the standard deviation of the diameter measurements across the ten replicates. Analysis of the means was important to show what factors most affect the diameter measurement in terms of shifting the mean response and potentially introducing bias. These data can also be used to determine the robustness of the CMM in terms of mean performance across the range of factor levels in the experiment. Analysis of the standard deviations was also important to show what factors most affected the repeatability of the diameter measurement. The standard deviations can also be used to determine the robustness of the CMM in terms of uncertainty across the range of factor levels in the experiment. Both analyses were used to characterize the measurement process, identify features that could or could not be measured, and to optimize use of the CMM.

Analysis of Mean Cylinder Diameter

A main effects plot of cylinder mean diameters was performed first (Fig. 9.1). The main effects plot provides a graphical comparison of the means at the low and the high levels for each factor. A visual comparison of slopes suggests which factors may have a statistically significant effect on the measured response. The main effects plot is also used to quantify the effect of each factor when the levels of the other factors are balanced. A specific factor's main effect is simply the difference in mean response comparing its low and high levels.

From Fig. 9.1 it appears that only Part Size and Surface Form affect the mean diameter. Clearly parts of different sizes will have different mean diameters. Surface Form most likely affects mean diameter because a smooth surface will measure shorter than a rough, wavy surface. It is interesting to note that Sampling Speed and

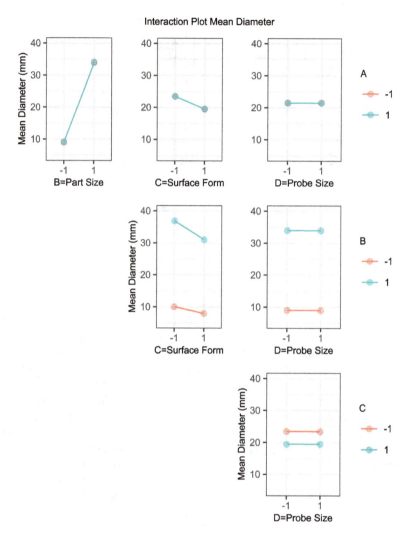

Fig. 9.2 Matrix of two-way interaction plots

Probe Size do not affect mean diameter, suggesting robustness of the CMM with respect to these two factors.

A second important graphical technique to use in the analysis of a factorial experiment is the matrix plot of all two-way interactions (Fig. 9.2). Two factors are said to "interact" if the value of the response at one level of a factor depends on the level of the second factor. An interaction can be thought of as the combined effect of two or more factors. In a two-way interaction plot, lines that are not parallel suggest interaction could be present. Figure 9.2 suggests no major two-way interactions, but a formal statistical test is typically performed to check for interaction. In the top row of the interaction plot, the blue line obscures the nearly identical red line.

9.2 Factorial Experiments in Metrology

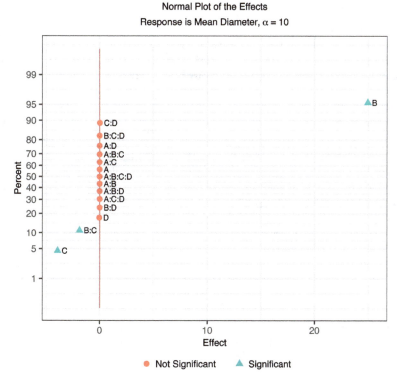

Fig. 9.3 Normal probability plot of main effects and interactions—mean cylinder diameter

A Normal probability plot of main effects and interactions is another graphical tool often used in the analysis of full factorial and fractional factorial designs. This plot (Fig. 9.3) includes the 15 main effects and interactions. If the effects are roughly normal, they will all fall on or near the line. In Fig. 9.3, 12 of the 15 effects fall near the line, but the effects corresponding to factors B, C, and the two-way BC interaction do not. This means that those effects cannot be explained by simple random chance and should be tested formally for statistical significance. The graphical test of significance associated with the Normal effects plot below is known as Lenth's method (see Lenth 1989). A similar test was suggested by Daniel (1959). It is used here only to suggest which variables should be subjected to a formal test of significance and possibly included in the final model.

The factorial regression table (Table 9.4) shows formally via the F-test that Part Size (B) and Surface Form (C) have a statistically significant effect on mean cylinder diameter, while the interaction term Part Size ×Surface Form (BC) has a minimal effect. Note that the "Error" term includes a "Pure Error" term based on replication of each experimental trial. In general, if r is the number of replicates for each of N experimental trials, the number of degrees of freedom for pure error will be

Table 9.4 Factorial regression table for diameter

Analysis of variance

Source	DF	Adjusted SS	Adjusted MS	F-value	P-value
Model	3	25650.7	8550.2	11866751	0.00
Linear	2	25509.2	12754.6	17685491	0.00
B = part size	1	24882.6	24882.6	34502180	0.00
C = surface form	1	626.6	626.6	868802	0.00
2-way interactions	1	141.5	141.5	196269	0.00
B*C	1	141.5	141.5	196269	0.00
Error	156	0.1	0.0		
Lack-of-fit	12	0.1	0.0	426.8	0.00
Pure error	144	0.0	0.0		
Total	159	25650.8			

Model summary

S	R-squared				
0.0046	100%				

Coded coefficients

Term	Effect	Coef.	SE Coef.	T-value	P-value
Constant		21.49	0.0021	10122.0	0.00
B = part size	24.94	12.47	0.0021	5873.9	0.00
C = surface form	−3.96	−1.98	0.0021	−932.1	0.00
B*C	−1.88	−0.94	0.0021	−443.0	0.00

$N \times (r - 1)$. In this case, with 10 replicates for each of $2^4 = 16$ experimental trials, the total degrees of freedom for pure error is $16 \times (10 - 1) = 144$. For a factorial experiment of this size, 3 to 5 replicates for each trial will usually suffice. This will result in more than 30 degrees of freedom for pure error, a recommended goal for measurement experiments. We will discuss sample size determination for an uncertainty analysis in more detail in Sect. 11.5.

The pure error term is used as the denominator in F-tests when the experimental trials have been replicated. In the absence of replication, the "Lack-of-Fit" term is used as the "Error" term. An analysis of the full model of main effects and interaction terms, not shown here, confirmed that only these two factors and their interaction should be considered in the final model. The results in the factorial regression table also include estimates of each parameter.

The goodness of fit statistic $R^2 \simeq 100\%$ indicates a good fit for the two-factor plus interaction model, although this statistic is artificially inflated by using very different size cylinders to be measured by the CMM. Based on this analysis, the hypothesized measurement equation $y = f(X_1, X_2, X_3, X_4)$ reduces to an equation in just two variables, $y = f(X_2, X_3)$. The estimated coefficients (Table 9.4) in the model for Part Size (X_2) and Surface Form (X_3) result in the following equation for the mean cylinder diameter:

9.2 Factorial Experiments in Metrology

$$y = f(X_1, X_2, X_3, X_4) = f(X_2, X_3) \cong \beta_0 + \beta_2 X_2 + \beta_3 X_3 + \beta_{23} X_2 \cdot X_3$$
$$\cong 21.49 + 12.47 X_2 - 1.98 X_3 - 0.94 X_2 X_3 \quad (9.2)$$

The pure error term was calculated from the ten replicates of each trial, resulting in $s = 0.0046$ mm with 144 degrees of freedom. This represents the standard uncertainty (Type A) associated with using the CMM for repeated measurements across the levels of the four experimental factors.

It should be noted that the coefficients in Eq. (9.2) are "coded coefficients," meaning that factor levels -1 and $+1$ were used for the levels of X_2 and X_3 in the regression analysis. These coefficients are sometimes referred to as "half-effects" as they are equal to half the "full-effects," or main effects that are plotted on the Normal probability plot of main effects and interactions.

A final step in the analysis is to test the assumption of normality of the residuals from the fitted model (9.2). Normal probability plotting of the residuals provides a graphical diagnostic check. The normality assumption is necessary to ensure that the F-tests in the factorial regression table are valid, although minor departures from normality are not of great concern. In the Normal probability plot, all the residuals should fall close to the line. Fig. 9.4 indicates that the Normality assumption is reasonable for these data.

Based on this analyses, we can conclude that two factors, Part Size and Surface Form, affect the cylinder diameter measurement. Because their two-way interaction term also appears in the model, all three terms must be considered jointly in the interpretation of the model. Obviously, the part size impacts the diameter measurement. An interesting result is that the surface form also impacts the mean cylinder diameter. The negative coefficient for this model term suggests that a rough surface will produce a larger diameter measurement than will a smooth surface using this CMM technology. It is also of interest to note that Sampling Speed and Probe Size do not impact the mean cylinder diameter measurement over the range of experimentation. This means that either level of these two factors could be used without introducing bias into the measurement. These results provide important information regarding optimal use of this CMM.

Analysis of Standard Deviation of Cylinder Diameter

The second part of the analysis focused on the standard deviations of the ten replicates for each experimental trial. Fig. 9.5 is the corresponding main effects plot for the standard deviation of cylinder diameter. This plot suggests which factors may have a statistically significant effect on the *standard deviation* of the diameter, a measure of repeatability. From this plot, it appears that Surface Form and Probe Size affect the standard deviation of the diameter measurement, but that Sampling Speed and Part Size do not. The plot suggests that a smooth surface ($+1$) will have much smaller standard deviation (uncertainty) than a rough surface (-1) and that a large probe ($+1$) will have smaller standard deviation than a small probe (-1). It is interesting to note that Sampling Speed and Part Size do not appear to affect the standard deviation of the diameter, suggesting robustness of the CMM

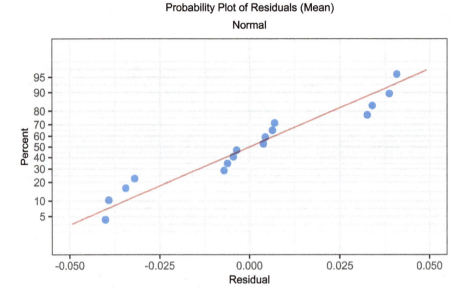

Fig. 9.4 Normal probability plot of residuals – mean cylinder diameter

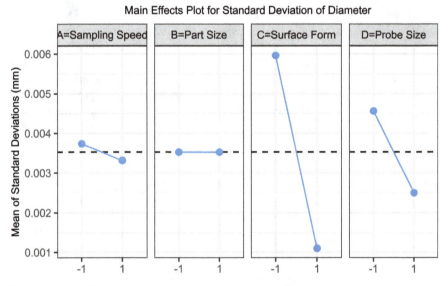

Fig. 9.5 Main Effects plot for standard deviation of cylinder diameter

with respect to these two factors. The interaction plot is used again, but here it is restricted to the two factors, Surface Form (C) and Probe Size (D), that may affect the repeatability measurement. The plot suggests that a rough surface (−1) measured with a small probe (−1) is by far the worst case (largest value) in terms of standard

9.2 Factorial Experiments in Metrology

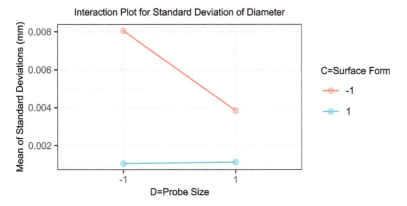

Fig. 9.6 Two-way interaction plot of Surface Form and Probe Size

deviation. A formal test was performed to check for the statistical significance of this two-way interaction (Fig. 9.6). A Normal probability plot of main effects and interactions was again used as a graphical indicator of statistical significance. This plot (Fig. 9.7) includes the 15 main effects and interactions, calculated from the standard deviations. In Fig. 9.7, 12 of the 15 effects plot on or near the line, but the effects corresponding to factors C, D, and the two-way interaction CD, do not. This suggests that those effects cannot be explained by simple random chance and should be tested formally for statistical significance. The factorial regression table (Table 9.5) shows formally via the F-test that Surface Form (C), Probe Size (D), and the interaction term Surface Form $\times$ Probe Size (CD) have a statistically significant impact on the standard deviation of cylinder diameter. Note in this analysis that the "Error" term is based on lack of fit (the 12 non-significant effects) rather than pure error derived from replication of experimental trials.

The goodness of fit statistic $R^2 \cong 91\%$ indicates a good fit for the two-factor plus interaction model. The estimated coefficients in the model for Surface Form (X_3) and Probe Size (X_4) result in the following equation for the standard deviation of cylinder diameter:

$$y = f(X_1, X_2, X_3, X_4) = f(X_3, X_4) \cong \beta_0 + \beta_3 X_3 + \beta_4 X_4 + \beta_{34} X_3 \cdot X_4$$
$$\cong 0.0035 - 0.0024 X_3 - 0.0010 X_4 + 0.0011 X_3 X_4 \quad (9.3)$$

The coefficients in Eq. (9.3) were again "coded coefficients," meaning that the factor levels -1 and $+1$ were used for X_3 and X_4 in the regression analysis.

A final step in the analysis is to assess the assumption of Normality of the residuals from the fitted model (9.3). Figure 9.8 indicates that the normality assumption for the standard deviation data is reasonable for these data. Based on the analyses above, we can conclude that two factors, Surface Form (X_3) and Probe Size

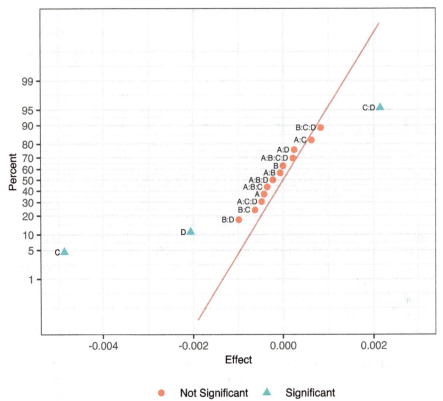

Fig. 9.7 Normal probability plot of main effects and interactions—standard deviation of cylinder diameter

(X_4), affect the standard deviation of the cylinder diameter measurement. Again, because the associated two-way interaction also appears in the model, all three terms must be considered jointly in the interpretation of the model. Based on the interaction plot for these two factors (Fig. 9.6) and the fitted model (Eq. 9.3) for the standard deviation, we can conclude that a smooth surface results in significantly less measurement uncertainty, and for a smooth surface, the probe size does not matter. For a rough surface, however, the larger probe size results in lower standard deviation (less uncertainty). It is also of interest to note that Sampling Speed and Part Size do not affect the standard deviation of the cylinder diameter measurement. This means that either level of these two factors could be used without increasing the uncertainty of the measurement. These results provide additional important information regarding optimal use of this CMM.

9.2 Factorial Experiments in Metrology

Table 9.5 Factorial regression table for standard deviation of diameter

Analysis of variance

Source	DF	Adjusted SS	Adjusted MS	F-value	P-value
Model	3	0.000130	0.000043	41.6	0.00
Linear	2	0.000112	0.000056	53.6	0.00
C = surface form	1	0.000095	0.000095	90.9	0.00
D = probe size	1	0.000017	0.000017	16.4	0.00
2-way interactions	1	0.000018	0.000018	17.6	0.00
C*D	1	0.000018	0.000018	17.6	0.00
Error	12	0.000012	0.000001		
Total	15	0.000142			

Model summary

S	R-squared
0.00102	91.2%

Coded coefficients

Term	Effect	Coef.	SE Coef.	T-value	P-value
Constant		0.0035	0.00026	13.85	0.00
C = surface form	−0.0049	−0.0024	0.00026	−9.53	0.00
D = probe size	−0.0021	−0.0010	0.00026	−4.04	0.00
C*D	0.0021	0.0011	0.00026	4.20	0.00

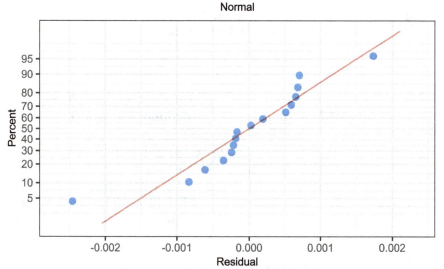

Fig. 9.8 Normal probability plot of residuals—standard deviation of cylinder diameter

Conclusion

Design of Experiments (DOEx) was used to analyze the uncertainty in a CMM's contact probe measurement. In the absence of a pre-defined measurement equation, a four-factor experiment was designed and performed to assess the effect of each factor on both the mean and standard deviation of the cylinder diameter measurement. The four factors included in the experiment were Sampling Speed, Part Size, Surface Form, and Probe Size. The objective of the experiment was to optimize use of the CMM in terms of minimizing uncertainty in the cylinder diameter measurements.

Analysis of the mean data showed that only Part Size and Surface Form impacted the mean cylinder diameter measurement. Both Sampling Speed and Probe Size did not affect the mean response, meaning that these factors did not introduce bias. Analysis of the standard deviations showed that only Surface Form and Probe Size impacted the standard deviation. Sampling Speed and Part Size did not affect the standard deviation, meaning these factors did not influence the uncertainty. The study indicated that the CMM performed better measuring smooth surface parts. When rough surface parts are measured, a larger probe size filters out the effects of the roughness, resulting in a smaller standard deviation. Depending on the intended use of the part, this may not be desired because it could mask the true condition of the part.

This experiment optimized use of the CMM in terms of these four factors, but it did not result in a formal uncertainty analysis. To perform a detailed uncertainty analysis, the CMM would be tested multiple times on multiple dimension using the optimal settings identified in this experiment. Both Type A and Type B uncertainties would be included in such a study. Methods for determining appropriate sample sizes for such an uncertainty study are detailed in Sect. 11.5.

9.2.5 Finite Element Method (FEM) Uncertainty Analysis via Design of Experiments (2^{7-3} Fractional Factorial)

The second case study demonstrates the design and analysis of a fractional factorial experiment. This experiment, taken from Fong et al. (2014), was part of an "uncertainty analysis of the finite element method-based solution of a proposed base design of a magnetic resonance imaging (MRI) RF coil." Figures in the Fong et al. (2014) reference show the usage and dimensions of the birdcage coil investigated in this experiment.

The authors presented a "metrological approach" to the verification and uncertainty quantification of the FEM-based simulations by treating each simulation as a virtual experimental run. The response variable YC (the "measurand") was a dimensionless quantity corresponding to the non-uniformity of the magnetic field profile produced by a run of the FEM code. The objective of the experiment was to identify the key factors affecting this non-uniformity measure. A second experiment,

9.2 Factorial Experiments in Metrology

not presented here, was performed to minimize the non-uniformity by optimizing the design of the RF coil.

Subject matter experts chose seven factors from more than 70 base coil design factors to include in the experiment. The seven factors included were all inputs to the FEM model: Electrical Conductivity of Water (σ), Relative Permittivity of Water (ε), Capacitance (C), Voltage (V), Ring Width (W), Strip Gap Length (L), and Ring Gap Angle (β). It was anticipated that the initial experiment would identify a few critical factors from the seven factors chosen and that additional experimentation would be used to build a more detailed model based on those factors. In particular, a second order model for YC would be used to minimize non-uniformity through the choice of the RF coil design parameters.

Because of cost considerations and the large number of proposed factors, it was determined that the initial experiment would be a fractional factorial design. The design chosen was a $2^{(7-3)}$ fractional factorial. This expression means that seven factors were investigated at two levels each, using a total of $2^{(7-3)} = 16$ combinations. This design is said to be a "$\frac{1}{8}$ of a 2^7" fractional factorial design. Design strategy and reasons for this choice of design are discussed below. The factors in the experiment and the fractional factorial DOEx appear in Table 9.6. Each factor's low and high levels were taken to be within $\pm 10\%$ of the base design of the RF coil. These levels can be thought of as enveloping the uncertainty in the inputs to the FEM.

Note that columns X1–X4 make up a standard 2^4 full factorial design for four factors each at two levels. Columns X5, X6, and X7 were generated by the following relationships, called design "generators":

$$X5 = X2 \times X3 \times X4$$
$$X6 = X1 \times X3 \times X4$$
$$X7 = X1 \times X2 \times X3.$$

This particular choice of design generators results in what is called a "Resolution IV" design (see Box et al. 2005). The resolution number of an experimental design is used to describe the degree of confounding of factors and their interactions. Two factors or interactions are said to be "confounded" with each other if their respective effects are indistinguishable. Fractional factorial designs most useful in practice have Resolution III, IV, or V. These design resolutions are described as:

1. A design of Resolution III has no main effects confounded with each other, but main effects are confounded with two-way interactions.
2. A design of Resolution IV has no main effects confounded with two-way interactions, but two-way interactions are confounded with other two-way interactions, and main effects are confounded with three-way interactions.
3. A design of Resolution V has no main effects or two-way interactions confounded with each other, but two-way interactions are confounded with three-way interactions.

Table 9.6 The experimental design matrix is a $2^{(7-3)}$ fractional factorial

Run	X1	X2	X3	X4	X5	X6	X7	YC
1	−1	−1	−1	−1	−1	−1	−1	50.63
2	1	−1	−1	−1	1	1	−1	42.92
3	−1	1	−1	−1	1	1	1	46.34
4	1	1	−1	−1	−1	−1	1	43.01
5	−1	−1	1	−1	1	−1	1	52.17
6	1	−1	1	−1	−1	1	1	60.62
7	−1	1	1	−1	−1	1	−1	50.58
8	1	1	1	−1	1	−1	−1	51.86
9	−1	−1	−1	1	−1	1	1	49.28
10	1	−1	−1	1	1	−1	1	46.50
11	−1	1	−1	1	1	−1	−1	44.73
12	1	1	−1	1	−1	1	−1	45.52
13	−1	−1	1	1	1	1	−1	47.85
14	1	−1	1	1	−1	−1	−1	57.99
15	−1	1	1	1	−1	−1	1	48.95
16	1	1	1	1	1	1	1	54.25

Factors are: X1 = Electrical Conductivity of Water (σ), X2 = Relative Permittivity of Water (ε), X3 = Capacitance (C), X4 = Voltage (V), X5 = Ring Width (W), X6 = Strip Gap Length (L), X7 = Ring Gap Angle (β)

Using the design generators above results in the factor X5 being confounded with the three-way interaction X2 × X3 × X4. The effects associated with factor X5 and with the three-way interaction X2 × X3 × X4 can therefore not be distinguished from each other. Higher resolution is better, as it implies less confounding, but it comes at the cost of more experimental runs. The full factorial experiment has no confounding, but the number of experimental runs required is often prohibitive. In this experiment, a full factorial design would have required $2^7 = 128$ total combinations. In the design of complex measurement systems, exploratory experiments will often be fractional factorial designs.

The strategy for designing a fractional factorial experiment depends largely on the number of factors, desired resolution, and desired number of replicates per trial. The number of chosen factors will depend on the complexity of the measurement system and the resources available for running a large, multi-factor experiment. The desired resolution of the design will depend on the suspected degree of interaction between factors. The experimenter must chose the resolution and design pattern so that all main effects and interactions deemed important can be estimated without serious confounding. The chosen number of replicates will largely depend on the cost of experimentation and the desired precision in estimating repeatability.

Statistical design of experiments software such as Minitab, JMP, and Design Expert is available to generate acceptable designs given user-specified inputs such as number of factors, desired resolution, and allowable number of runs. The software also typically provides the confounding, or "alias" structure associated with

9.2 Factorial Experiments in Metrology

the resulting fractional factorial design. The experimenter must balance desired resolution and precision with the cost of experimental runs when choosing the design. Table 9.7 shows the confounding structure associated with this $2^{(7-3)}$ fractional factorial.

Lines two through eight of the list of "Aliases" in Table 9.7 show that main effects are confounded with three-way interactions (listed) and higher order interactions (not listed). Interactions higher than three-way are rarely important and are not usually included in these lists. Based on the definitions of resolution above, this particular design is Resolution IV, meaning that main effects are confounded with three-way interactions and two-way interactions are confounded with other two-way interactions. We can thus refer to this design as a $2_{IV}^{(7-3)}$ design. Lines nine through 15 of the list of "Aliases" show how the two-way interactions are confounded with other two-way interactions.

Figure 9.9 is the main effects plot for the non-uniformity metric YC. A visual comparison of slopes suggests which factors may have a statistically significant effect on YC. From this plot it appears that factors B (Relative Permittivity of Water), C (Capacitance), D (Voltage), and E (Ring Width) may have an important effect on YC. Lower values of B and D, along with higher values of C and E are associated with lower non-uniformity in YC, meaning better performance. The factors (A) EC of Water, (F) SG Length, and (G) RG Angle do not appear to affect

Table 9.7 Confounding or "Alias" structure for a $2^{(7-3)}$ fractional factorial

Factor	Name
A	X1
B	X2
C	X3
D	X4
E	X5
F	X6
G	X7
Aliases	
I	
A + BCG + BEF + CDF + DEG	
B + ACG + AEF + CDE + DFG	
C + ABG + ADF + BDE + EFG	
D + ACF + AEG + BCE + BFG	
E + ABF + ADG + BCD + CFG	
F + ABE + ACD + BDG + CEG	
G + ABC + ADE + BDF + CEF	
AB + CG + EF	
AC + BG + DF	
AD + CF + EG	
AE + BF + DG	
AF + BE + CD	
AG + BC + DE	
BD + CE + FG	

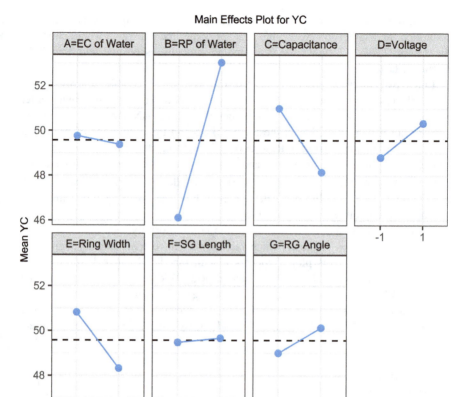

Fig. 9.9 Main effects plot for non-uniformity coefficient YC

YC, suggesting robustness of the RF Coil design with respect to these three factors. Figure 9.10 displays all two-way interactions from the RF Coil experiment. Recall that in a two-way interaction plot, lines that are not parallel suggest interaction could be present. Figure 9.10 shows a dramatic lack of parallelism for two-way interactions BD, CE, and FG. Note from Table 9.7 that these three two-way interactions are confounded with each other, meaning that the estimates of these three interactions would be identical. Thus, only one of the three interaction terms can appear in the final fitted model. The remainder of the matrix suggests no additional major two-way interaction terms. A Normal probability plot was used to graphically assess the relative importance of main effects and interactions from the $2_{IV}^{(7-3)}$ fractional factorial design. With a total of 16 unique experimental runs, a total of 15 main effects and interactions (plus the overall mean) can be estimated from these data. In Fig. 9.11, 11 of the 15 effects plot on or near the line, with the effects corresponding to factors B, C, E, and the two-way interaction CE, falling well off the line. This means that those effects cannot be explained by simple random chance and should be

9.2 Factorial Experiments in Metrology

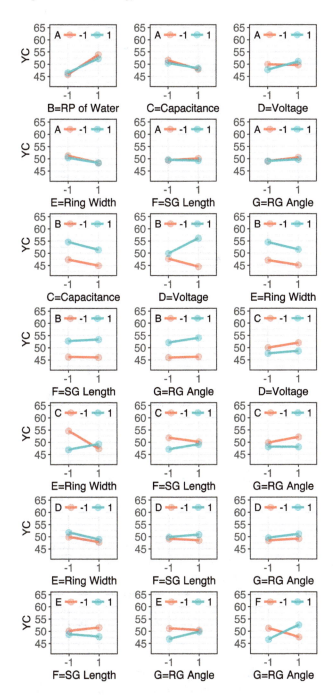

Fig. 9.10 All two-way interaction plots

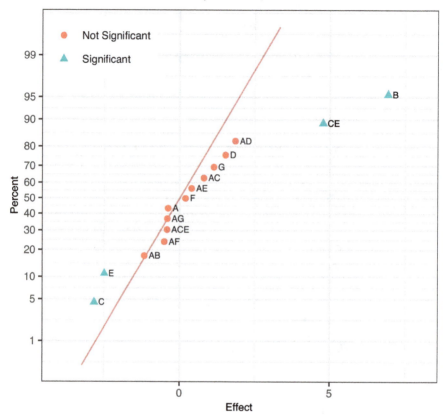

Fig. 9.11 Normal probability plot of main effects and interactions for YC

tested formally for statistical significance. It should be pointed out from Table 9.7 that main effects B, C, and E are confounded with three-way interactions and the CE interaction is confounded with the two-way interactions BD and FG. Because the CE interaction can be thought of as the synergistic effect of main effects C and E, it was investigated further via the factorial regression model. The factorial regression table (Table 9.8) shows formally via the F-test that the Relative Permittivity of Water (B), Capacitance (C), Ring Width (E), and CE interaction terms have a statistically significant effect on non-uniformity YC. An analysis of the model with 15 main effect and interaction terms, not shown here, indicated that only these three main effects and one interaction term should appear in the final model. The results from the factorial regression table include the estimates of each coefficient in the final model.

9.2 Factorial Experiments in Metrology

Table 9.8 Factorial regression table for YC

Analysis of variance					
Source	DF	Adjusted SS	Adjusted MS	F-value	P-value
Model	4	339.77	84.49	23.6	0.00
Linear	3	248.57	82.86	23.0	0.00
B= RP of water	1	191.41	191.41	53.1	0.00
C= Capacitance	1	32.26	32.26	9.0	0.01
E= Ring Width	1	24.90	24.90	6.9	0.02
2-way interactions	1	91.20	91.20	25.3	0.00
C*E	1	91.20	91.20	25.3	0.00
Error	11	39.65	3.60		
Total	15	379.42			

Model summary					
S		R-squared			
1.898		89.6%			

Coded coefficients					
Term	Effect	Coef.	SE Coef.	T-value	P-value
Constant		49.58	0.475	104.45	0.00
B = RP of water	6.92	3.46	0.475	7.29	0.00
C = capacitance	−2.84	−1.42	0.475	−2.99	0.00
E = ring width	−2.50	−1.25	0.475	−2.63	0.02
C*E	4.78	2.39	0.475	5.03	0.00

Based on this analysis, the hypothesized measurement equation

$$y = f(X_1, X_2, X_3, X_4, X_5, X_6, X_7)$$

reduces to an equation in just three variables, $y = f(X_2, X_3, X_5)$. The estimated coefficients (using "coded coefficients") for Relative Permittivity of Water (X_2), Capacitance (X_3), and Ring Width (X_5) result in the following equation for the non-uniformity metric YC:

$$\begin{aligned} YC &= f(X_1, X_2, X_3, X_4, X_5, X_6, X_7) = f(X_2, X_3, X_5) \\ &\cong \beta_0 + \beta_2 X_2 + \beta_3 X_3 + \beta_5 X_5 + \beta_{35} X_3 \cdot X_5 \\ &\cong 49.6 + 3.46 X_2 - 1.42 X_3 - 1.25 X_5 + 2.39 X_3 \cdot X_5 \end{aligned} \quad (9.4)$$

The goodness of fit statistic $R^2 \cong 89\%$ indicates a good fit for the three-factor plus interaction model. The assumption of Normality of the residuals from the fitted model (9.4) is assessed in Fig. 9.12 and indicates that the normality assumption is reasonable for these data.

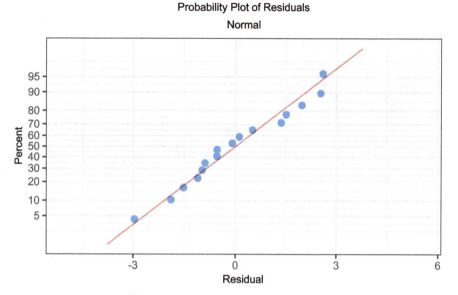

Fig. 9.12 Normal probability plot of residuals

Conclusion

Design of Experiments (DOEx) was used to identify factors that affect a non-uniformity measure resulting from an FEM analysis of a particular RF Coil design. The experiment was not a metrology study in the usual sense, but a "metrological approach" was used as a first step in minimizing the non-uniformity coefficient YC. In the absence of a pre-defined measurement equation, a seven-factor experiment was designed and performed to assess the impact of each factor on YC. The seven factors included in the experiment were all inputs to an FEM model: Electrical Conductivity of Water (σ), Relative Permittivity of Water (ε), Capacitance (C), Voltage (V), Ring Width (W), Strip Gap Length (L), and Ring Gap Angle (β).

The objective of the experiment was to optimize the design of the RF Coil used as part of an MRI imaging setup. The large number of factors necessitated the use of a fractional factorial design, resulting in a $2_{IV}^{(7-3)}$ Resolution IV fractional factorial design. The experiment identified three of the seven factors, Relative Permittivity of Water (ε), Capacitance (C), and Ring Width (W) as significant with respect to the non-uniformity coefficient YC. To improve the design of the RF Coil, an additional experiment would be performed to investigate these three factors in greater detail. A second order response surface design (see Montgomery 2017) could be performed to look for curvature in the YC response variable as a function of these three factors.

A three-variable model to estimate curvature in the response variable YC would be of the form:

9.2 Factorial Experiments in Metrology

$$\text{YC} = f(X_1, X_2, X_3) \cong \beta_0 + \sum_{i=1}^{3} \beta_i X_i + \sum_{i=1}^{3} \sum_{j \geq i}^{3} \beta_{ij} X_i X_j. \tag{9.5}$$

An analysis of this second order model would be used to optimize and finalize the RF Coil design. Once the design was finalized, a formal measurement uncertainty analysis could be performed using actual runs of the MRI imaging setup. Both Type A and Type B uncertainties would be included in such a study. Methods for determining appropriate sample sizes in an uncertainty study are detailed in Sect. 11.5.

The approach to fractional factorial designs presented here assumed all experimental factors were initially investigated at just two levels. In practice, fractional factorial designs may have factors with more than two levels, and factors with different numbers of levels. These more difficult cases are discussed by Montgomery (2017), and Addelman and Kempthorne (1961), who have catalogued orthogonal fractional factorial designs that allow estimation of all main effects with a minimum number of experimental trials.

9.2.6 Summary of Factorial DOEx Method

Full factorial and fractional factorial designs can be used in the design and optimization of a new measurement process or in the evaluation of an existing measurement process. These designs will often be used to identify the most important factors and to quantify their effect on the measurement. The results of these experiments can also be used to optimize a measurement process. The premise of statistical DOEx is that the resulting experimental design will be efficient in terms of the resources used to perform the study. Fractional factorial designs in particular provide an efficient framework for the investigation of many factors in a minimum number of experimental trials.

The design pattern and sample sizes for factorial and fractional factorial designs will depend on the number of factors, resolution of the design, and the desired precision in estimating repeatability. The degrees of freedom for repeatability will depend on the number of replicates of each experimental trial. The resolution must be chosen so that the important main effects and interactions can be estimated without serious confounding. Once the resolution of the design and the design pattern has been chosen, the number of replicates per trial can be chosen. It is recommended that the design have 20–30 degrees of freedom for repeatability, with 15 as minimally acceptable. More detailed recommendations regarding degrees of freedom and sample size are given in Sect. 11.5.

9.3 ANOVA Models in Metrology

Analysis of variance (ANOVA) is a statistical technique used to "identify and quantify individual *random* effects in a measurement so that they may be properly taken into account when the uncertainty of the result of the measurement is evaluated" (GUM, p. 98). It is used to estimate the variance components associated with random effects included in a measurement model. The objective is to gain an understanding of the sources of variance so that the measurement process may be improved. ANOVA is also used to estimate fixed effects that may contribute to the bias in a measurement. It differs from factorial design of experiments in that it is typically not used to identify factors or construct an indirect measurement equation.

ANOVA methods are often used in metrology when certifying reference standards by interlaboratory testing (GUM, p. 98). See ISO Guide 13528 (2015) for a detailed discussion of the statistical methods used in these comparisons. This testing typically consists of a reference standard being tested multiple times on multiple days at each of several laboratories. In this case, differences from day-to-day and differences between repeated measurements are treated as random. Differences between laboratories are treated as fixed. Another common use of ANOVA techniques is to assess the performance of a single measurement device via a Gauge Repeatability and Reproducibility (Gauge R&R) study, commonly used in manufacturing environments. These techniques can be used as key components of tester qualification and tester comparison studies.

ANOVA techniques include the use of many different statistical models (Searle and Gruber 2016). The sections that follow provide an introduction to some of the most commonly used ANOVA models in metrology, including simple random effects models and mixed effects models. In this context, the term "mixed effects model" means an ANOVA model containing both random and fixed effects. These models are particularly useful in measurement studies because they accommodate repeated measurements made on the same test unit under either fixed or varying conditions. The case studies presented below include analyses of both random effects and mixed effects model.

9.3.1 Random Effects Models

Factors with random effects are factors whose levels can be thought of as a random sample from some population (Kutner et al. 2005). As an example, consider a factory that manufactures steel plates. In this factory there are many operators responsible for measuring the thickness of the plates. Selecting a random sample of parts to be measured by a random sample of operators leads to a random effects model for the analysis. Each operator performs the thickness measurement on each plate n times. In this case both Part and Operator would be treated as having random

9.3 ANOVA Models in Metrology

effects. Rather than drawing inferences about the specific parts and operators that happened to be chosen for the experiment, inferences are drawn about the larger population of parts and operators. Since each operator measures each part, the factors Part and Operator are said to be "crossed" factors. If each operator measures different parts, the factor Part is said to be "nested" within the factor Operator. A helpful discussion of the difference between crossed and nested factors is given in Searle and Gruber (2016).

In this example, the associated random effects model can be expressed as

$$y_{ijk} = \text{Ave} + \text{Part}_i + \text{Operator}_j + (\text{Part} \times \text{Operator})_{ij} + \text{Error}_{ijk}, \qquad (9.6)$$

where y_{ijk} is the kth measurement of the ith part by the jth operator, with $i = 1, \ldots, p$ parts, $j = 1, \ldots, o$ operators and $k = 1, \ldots, n$ measurements. In this expression, Ave represents the overall average across all operators and all parts. Part_i is modelled as a Normal random variable with mean zero and variance σ^2_{Part}, called the part-to-part variance. Operator_j is also modelled as a Normal random variable with mean zero and variance σ^2_{Oper}, called the operator-to-operator variance. The $(\text{Part} \times \text{Operator})_{ij}$ term is modelled as a Normal random variable with mean zero and variance $\sigma^2_{\text{Part} \times \text{Oper}}$. This term represents the interaction between the factors Operator and Part. Finally, Error_{ijk} is modelled as a Normal random variable with mean zero and variance σ^2_{Error}. This term is often called the repeatability or error variance. Repeated measurements of each part are used to estimate this term.

The results of an analysis of variance are displayed in a tabular form known as an ANOVA table. The ANOVA table displays the sums of squares, mean squares, F-ratios, and degrees of freedom that are used to test hypotheses about the population means and variances (see Searle and Gruber 2016). These tables are readily produced by standard statistical software, and the mechanics of deriving an ANOVA will not be presented here. Illustrations of how to use the results from an ANOVA table are given in the case studies below.

The expected values of the mean squares that appear in an ANOVA table are used to estimate the variance terms associated with the particular ANOVA model. Table 9.9 provides the expected values of the mean squares from an ANOVA for the above random effects model. Given these expected values, the variance components can be estimated by equating the mean squares with their expected values and solving for the variance terms (Table 9.10).

The random effects model can be generalized to include more factors, depending on the experiment. Guidance on analyzing different-sized random effects models can be found in Kutner, et al. (2005). In this example, the variance terms σ^2_{Error}, σ^2_{Oper}, and $\sigma^2_{\text{Part} \times \text{Oper}}$ would each be part of the Type A combined standard uncertainty in a detailed measurement uncertainty analysis. The variance component σ^2_{Error} represents the repeatability variance, and the sum of variance components $\left(\sigma^2_{\text{Oper}} + \sigma^2_{\text{Part} \times \text{Oper}} \right)$

Table 9.9 Expected values of the mean squares for a two-way random effects model

Mean square	Expected value
MS_{Part}	$\sigma^2_{Error} + n\sigma^2_{Part \times Oper} + on\sigma^2_{Part}$
MS_{Oper}	$\sigma^2_{Error} + n\sigma^2_{Part \times Oper} + pn\sigma^2_{Oper}$
$MS_{Part \times Oper}$	$\sigma^2_{Error} + n\sigma^2_{Part \times Oper}$
MS_{Error}	σ^2_{Error}

Table 9.10 Variance component estimates for a two-way random effects model

Variance component	Estimate
$\hat{\sigma}^2_{Part}$	$\dfrac{MS_{Part} - MS_{Part \times Oper}}{on}$
$\hat{\sigma}^2_{Oper}$	$\dfrac{MS_{Oper} - MS_{Part \times Oper}}{pn}$
$\hat{\sigma}^2_{Part \times Oper}$	$\dfrac{MS_{Part \times Oper} - MS_{Error}}{n}$
$\hat{\sigma}^2_{Error}$	MS_{Error}

represents the reproducibility variance. The part-to-part variance component σ^2_{Part} would not be of concern from the perspective of measurement uncertainty.

9.3.2 Mixed Effects Models

Mixed effects ANOVA models are used when both random and fixed effects are present in the measurement study. The difference between the two types of effects is sometimes subtle. For random effects, interest is in the component of variance associated with each effect. For fixed effects, interest is in the mean associated with each effect.

Consider a case where multiple calipers are being used to measure steel thickness and the goal is to quantify the difference between the calipers. Because the objective is to estimate differences between specific calipers, caliper would be considered a fixed factor. A two-factor mixed effects model would be used in the analysis, with one factor (Caliper) having fixed effects and one factor (Operator) having random effects. For simplicity, Part is excluded as a factor in this example, although it could easily be added as a second factor with random effects. The mixed effects model is thus defined as

$$y_{ijk} = \text{Ave} + \text{Operator}_i + \text{Caliper}_j + (\text{Operator} \times \text{Caliper})_{ij} + \text{Error}_{ijk} \quad (9.7)$$

with $i = 1, \ldots, o$ operators, $j = 1, \ldots, c$ calipers, and $k = 1, \ldots, n$ measurements. Here, Ave is again the overall average measurement. The Caliper_j effects are fixed, while the remainder of the terms are modelled as normal random variables with mean zero and constant variances. For the mixed model expressed as in Eq. (9.7), the fixed factor Caliper assumes a different average for each of its levels, while the random factor Operator assumes the same variance for each of its levels.

9.3 ANOVA Models in Metrology

Table 9.11 Expected values of the mean squares for a two-way mixed effects model

Mean square	Expected value
$MS_{Caliper}$	$\sigma^2_{Error} + n\sigma^2_{Caliper \times Oper} + on \frac{\sum Caliper^2_j}{c-1}$
MS_{Oper}	$\sigma^2_{Error} + cn\sigma^2_{Oper}$
$MS_{Caliper \times Oper}$	$\sigma^2_{Error} + n\sigma^2_{Caliper \times Oper}$
MS_{Error}	σ^2_{Error}

Table 9.12 Variance component estimates for a two-way mixed effects model

Variance component	Estimate
$\hat{\sigma}^2_{Oper}$	$\frac{MS_{Oper} - MS_{Error}}{cn}$
$\hat{\sigma}^2_{Caliper \times Oper}$	$\frac{MS_{Caliper \times Oper} - MS_{Error}}{n}$
$\hat{\sigma}^2_{Error}$	MS_{Error}

The expected mean squares and estimated variance components for the random factor and interaction are provided in Tables 9.11 and 9.12, respectively.

In the context of a measurement study, the primary focus is on the measurement-related variance terms σ^2_{Error}, σ^2_{Oper}, and $\sigma^2_{Caliper \times Oper}$. The random effect terms would become part of the Type A combined standard uncertainty, providing estimates of both repeatability and reproducibility.

The estimate for each level of the fixed factor (Caliper) can be calculated through ordinary least squares. The resulting equation is (Kutner et al. 2005):

$$\widehat{Caliper}_j = \bar{y}_{.j.} - \bar{y}_{...} \tag{9.8}$$

with variance equal to

$$\hat{\sigma}^2\{\widehat{Caliper}_j\} = \frac{MS_{Caliper \times Oper}}{on}. \tag{9.9}$$

The individual effect terms (9.8) and their corresponding variances (9.9) would be used to test for caliper differences that possibly introduce bias into the measurement.

9.3.3 Underlying ANOVA Assumptions

There are a few key assumptions to be aware of when performing an ANOVA:

1. Normality. Error terms are assumed to be normal, although slight deviations from normality are typically not critical.

2. Constant variance (homoscedasticity). This is a stronger assumption than normality. In a random effects or mixed effects model, it is assumed that the variance is the same at each level of a particular random effect.
3. Independence of the error terms. Plotting the residuals $e_{ijk} = (y_{ijk} - \bar{y}_{ij.})$ is typically a good way to assess whether this assumption has been met. Lack of independence can dramatically affect an analysis and its interpretation.

Guidance on checking ANOVA assumptions is provided in Kutner et al. (2005). If these assumptions are not met, there may be additional modelling options. See Arendacka (2014) for dealing with non-constant variance (heteroscedasticity) and correlations (non-independence) between the random effects.

9.3.4 Gauge R&R Study (Random Effects Model)

Gauge R&R studies are frequently used in manufacturing to assess the amount of variation in measured data that can be attributed to the measurement process. A typical Gauge R&R study consists of multiple parts being measured multiple times by multiple operators, to allow estimation of both repeatability and reproducibility. More details on the structure and analysis of Gauge R&R studies are provided in Vardeman and Jobe (2016). A case study is presented in this section of a standard Gauge R&R study. It is followed by discussion and a proposal for designing Gauge R&R studies to improve the estimate of reproducibility.

A Gauge R&R study was performed to assess the uncertainty of a high-precision optical CMM used to inspect dimensions of a component assembled within the NSE. The study consisted of three operators ($o = 3$) measuring four manufactured parts ($p = 4$) three times each ($n = 3$) for a total of 36 measurements. The data appear in Table 9.13.

The graphical analysis in Fig. 9.13 shows that Parts 1 and 3 have very different lengths than Parts 2 and 4, but that the Operators are similar and repeated measurements of the same part by the same operator are similar.

The ANOVA associated with the study appears in Table 9.14. In this analysis, Part and Operator were both treated as random factors, for which the equations in Table 9.10 apply.

Equating mean squares to their expected values (Table 9.10) and solving for the variance components, the estimates are

$$\hat{\sigma}^2_{\text{Error}} = 108.5$$

$$\hat{\sigma}^2_{\text{Part} \times \text{Oper}} = \frac{136.8 - 108.5}{3} = 9.4$$

9.3 ANOVA Models in Metrology

Table 9.13 CMM length measurements in a Gauge R&R study

Part	Operator	Replicate	Length × 10^4 (in)
1	1	1	9680.3
1	1	2	9655.2
1	1	3	9650.0
1	2	1	9670.8
1	2	2	9651.7
1	2	3	9635.6
1	3	1	9701.0
1	3	2	9680.9
1	3	3	9667.8
2	1	1	9764.0
2	1	2	9755.1
2	1	3	9754.5
2	2	1	9763.6
2	2	2	9752.1
2	2	3	9754.2
2	3	1	9764.0
2	3	2	9757.6
2	3	3	9763.7
3	1	1	9686.4
3	1	2	9662.3
3	1	3	9665.3
3	2	1	9670.4
3	2	2	9658.8
3	2	3	9662.9
3	3	1	9694.5
3	3	2	9690.7
3	3	3	9686.1
4	1	1	9768.1
4	1	2	9748.5
4	1	3	9750.4
4	2	1	9756.9
4	2	2	9753.7
4	2	3	9760.4
4	3	1	9768.1
4	3	2	9760.9
4	3	3	9760.8

$$\widehat{\sigma}^2_{\text{Oper}} = \frac{955 - 136.8}{12} = 68.2, \text{ and}$$

$$\widehat{\sigma}^2_{\text{Part}} = \frac{23418.8 - 136.8}{9} = 2586.9. \tag{9.10}$$

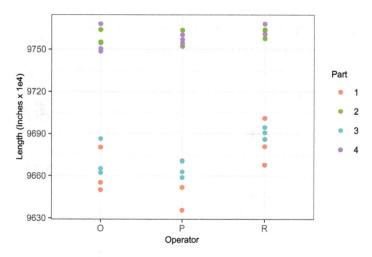

Fig. 9.13 Length measurement by operator and part

Table 9.14 ANOVA for random effects model from Gauge R&R

Analysis of variance					
Source	DF	Adjusted SS	Adjusted MS	F-value	P-value
Part	3	70256.3	23418.8	171.2	0.00
Operator	2	1910.1	955.0	6.9	0.03
Part*operator	6	820.6	136.8	1.3	0.31
Error	24	2603.8	108.5		
Total	35	75590.7			
Model summary					
S	R-squared				
10.416	96.5%				
Expected mean squares, using adjusted SS					
Source	Expected mean square for each term				
1 Part	(4) + 3.0000 (3) + 9.0000 (1)				
2 Operator	(4) + 3.0000 (3) + 12.0000 (2)				
3 Part*operator	(4) + 3.0000 (3)				
4 Error	(4)				

The Reproducibility variance term, combining the variance terms involving the random factor Operator, is

$$\sigma^2_{\text{Repr}} = \sigma^2_{\text{Oper}} + \sigma^2_{\text{Part} \times \text{Oper}}. \tag{9.11}$$

This expression is estimated by

9.3 ANOVA Models in Metrology

$$\widehat{\sigma}^2_{Repr} = \widehat{\sigma}^2_{Oper} + \widehat{\sigma}^2_{Part \times Oper}$$

$$= \frac{1}{12}\left(MS_{Oper} - MS_{Part \times Oper}\right) + \frac{1}{3}\left(MS_{Part \times Oper} - MS_{Error}\right)$$

$$= \frac{1}{12}\left(MS_{Oper}\right) + \frac{1}{4}\left(MS_{Part \times Oper}\right) - \frac{1}{3}\left(MS_{Error}\right)$$

$$= 79.6 + 34.2 - 36.2 = 77.6 \quad (9.12)$$

The estimate is expressed (Eq. 9.12) as a linear combination of independent mean squares so that the Welch–Satterthwaite approximation can readily be applied to yield an effective degrees of freedom for the reproducibility term.

The R&R variance, combining all the variance terms other than the *Part* term, is

$$\sigma^2_{R\&R} = \sigma^2_{Oper} + \sigma^2_{Part \times Oper} + \sigma^2_{Error}. \quad (9.13)$$

This expression is also estimated by a linear combination of independent mean squares,

$$\widehat{\sigma}^2_{R\&R} = \widehat{\sigma}^2_{Oper} + \widehat{\sigma}^2_{Part \times Oper} + \widehat{\sigma}^2_{Error}$$

$$= \frac{1}{12}\left(MS_{Oper} - MS_{Part \times Oper}\right) + \frac{1}{3}\left(MS_{Part \times Oper} - MS_{Error}\right) + MS_{Error}$$

$$= \frac{1}{12}\left(MS_{Oper}\right) + \frac{1}{4}\left(MS_{Part \times Oper}\right) + \frac{2}{3}\left(MS_{Error}\right)$$

$$= 79.6 + 34.2 + 72.3 = 186.1 \quad (9.14)$$

The *Part* variance term is not included as a focus of the Gauge R&R analysis because part-to-part variance is most likely due to variability in the manufacturing process, not variability in the measurement process. Reduction in the overall measurement error would thus depend on reducing the operator-to-operator variability as well as the repeated measurement variability.

Because each of the variance estimates can be expressed as a linear combination of independent mean squares, the Welch–Satterthwaite approximation can be used to calculate an effective degrees of freedom for each estimated variance.

The W-S approximation (see Satterthwaite 1946) applies when MS_1, MS_2, ..., MS_k are independent mean squares with degrees of freedom $v_1, v_2, \ldots, v_k$, and the linear combination

$$\widehat{V} = \sum_{i=1}^{k} a_i MS_i \quad (9.15)$$

is an estimate of a combined variance term. Given these conditions, the effective degrees of freedom of the approximating chi-square is given by

$$v_{\text{eff}} = \frac{\left(\sum_{i=1}^{k} a_i E(\text{MS}_i)\right)^2}{\sum_{i=1}^{k} \frac{(a_i E(\text{MS}_i))^2}{v_i}}. \quad (9.16)$$

This expression is evaluated by substituting the observed mean squares for their expected values, giving

$$v_{\text{eff}} = \frac{\left(\sum_{i=1}^{k} a_i \text{MS}_i\right)^2}{\sum_{i=1}^{k} \frac{(a_i \text{MS}_i)^2}{v_i}} \quad (9.17)$$

as the effective degrees of freedom for the combined variance term.

The effective degrees of freedom associated with the reproducibility variance estimate $\hat{\sigma}^2_{\text{Repr}}$, using this approximation and the mean squares in Eq. (9.12), is thus

$$v_{\text{Repr}} = \frac{(77.6)^2}{\left(\frac{(79.6)^2}{2} + \frac{(34.2)^2}{6} + \frac{(36.2)^2}{24}\right)} \cong 2. \quad (9.18)$$

The results of this analysis indicate that the Reproducibility portion of the Gauge R&R study, the σ^2_{Repr} term, is very poorly estimated because of the small number of operators (3) used in the study. The number of operators in such a study would need to be significantly more than the number typically recommended (2 or 3) to achieve a useful estimate of the Reproducibility term.

The effective degrees of freedom associated with the R&R variance $\hat{\sigma}^2_{\text{R\&R}}$, using the W-S approximation and the mean squares in Eq. (9.14), is

$$v_{\text{R\&R}} = \frac{(186.1)^2}{\left(\frac{(79.6)^2}{2} + \frac{(34.2)^2}{6} + \frac{(72.3)^2}{24}\right)} \cong 9.7 \quad (9.19)$$

The $\hat{\sigma}^2_{\text{R\&R}}$ term is estimated with more degrees of freedom because of the contribution of the error degrees of freedom to $v_{\text{R \& R}}$.

Discussion and Proposal for Gauge R&R Studies

It was pointed out by Satterthwaite (1946) that care must be taken when one or more of the terms in the expression for a variance component is negative, as is the case with the expression for $\hat{\sigma}^2_{\text{Repr}}$ above. If the value of $\hat{\sigma}^2_{\text{Repr}}$ can be negative with high probability, the approximating chi-square distribution determined via the W-S approximation will be a poor approximation. In this study, because MS_{Oper} is considerably larger than MS_{Error} in the expression for $\hat{\sigma}^2_{\text{Repr}}$, the approximation seems reasonable.

A second observation regards another aspect of the W-S approximation for this problem. Ballico (2000) showed that the W-S approximation is not a good approximation when an uncertainty term with many degrees of freedom is added to a dominant uncertainty term with a small number of degrees of freedom. This is often the case with Gauge R&R studies designed according to typical recommendations (see, for example, Down et al. 2010). The MS_{Oper} term in the calculation of v_{Repr} will often be larger, with fewer degrees of freedom, than the MS_{Error} term, which typically has more degrees of freedom.

Reproducibility is defined as the uncertainty in a measurement system over time, under conditions that are subject to change. The standard Gauge R&R study typically looks at variation of a measurement system over a relatively short period of time, with only the operator being varied. Other factors affecting the reproducibility of the measurement are often not included in these studies. The degrees of freedom analysis above indicates that the standard Gauge R&R study produces little information with respect to reproducibility. The degrees of freedom associated with repeatability, however, is much more likely to be adequate.

Our recommendation is therefore to use the Gauge R&R study, as typically designed, to estimate only the repeatability portion of uncertainty. An estimate of reproducibility should be based on data collected over time under different conditions. This could be achieved in various ways. First, the control chart methodology described in Sect. 11.1 could be used to monitor uncertainty in the measurement over time. Repeated measurements of a check standard over time would provide a good estimate of reproducibility. A handful of manufactured units (say 3–5) could also be set aside and repeatedly measured by multiple operators over time. These two potential sources of data would provide a far better estimate of reproducibility than that obtained from a short-term Gauge R&R study such as described above. After acceptable estimates of R&R are obtained, a complete uncertainty analysis can be performed, and test uncertainty ratios (see Sect. 5.2.1.1) can be used to assess gauge performance relative to product specifications.

With few operators in a short-term Gauge R&R study, Operators could be treated as a fixed effect. A simple check for differences between operators could then be made, and any significant differences could be addressed by reviewing measurement procedures and training.

It should be noted that the same design structure used for a Gauge R&R study can be used for comparing laboratories, with the factor "Operator" replaced with the factor "Laboratory," as in Suzuki et al. (2018).

9.3.5 Voltage Standard Uncertainty Analysis (Mixed Effects Model)

An uncertainty study was performed by measuring trigger output voltages (see Sect. 6.3.1) with an external standard Tektronix TDS 3024 digitizer. The uncertainty study was performed over a three-day period with 10 measurements taken each day

Table 9.15 Results from voltage standard uncertainty study

Day	Sample	Input trigger 60 V	Input trigger 150 V	Input trigger 350 V
1	1	59.6	153.0	354.0
1	2	59.2	152.0	350.1
1	3	59.7	152.0	354.0
1	4	59.2	153.0	352.2
1	5	59.3	152.0	354.0
1	6	59.9	152.0	352.0
1	7	59.9	153.0	350.0
1	8	59.9	151.0	352.2
1	9	59.6	152.0	352.2
1	10	59.2	152.0	358.0
2	1	59.6	152.0	356.0
2	2	59.2	153.0	354.0
2	3	59.2	152.0	358.0
2	4	58.8	153.0	354.0
2	5	59.2	152.0	358.0
2	6	60.0	152.0	356.0
2	7	59.2	152.0	356.0
2	8	59.2	152.0	354.0
2	9	59.6	153.0	356.0
2	10	59.6	154.0	356.0
3	1	59.6	150.0	354.0
3	2	60.0	151.0	353.0
3	3	59.7	153.0	352.0
3	4	59.6	151.0	350.0
3	5	59.6	152.0	352.0
3	6	59.3	152.0	352.0
3	7	59.9	152.0	352.0
3	8	59.6	151.0	350.0
3	9	59.7	151.0	350.0
3	10	59.6	154.0	354.0

at three different cardinal points (60 V, 150 V, and 350 V) to cover the range of possible voltages. The purpose of the study was to determine the uncertainty of the external voltage standard. The results from the study appear in Table 9.15.

Analyzed as a single experiment, the data can be modelled using a mixed effects model with Voltage as a fixed factor and Day as a random factor. Since each trigger voltage level was used on each day, the factors are crossed, with 10 repeated measurements at each combination of Voltage and Day.

Following the form of Eq. (9.6), the mixed effects model can be expressed as

9.3 ANOVA Models in Metrology

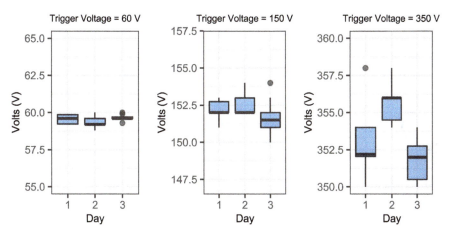

Fig. 9.14 Boxplots of measured voltage by trigger voltage level and day

$$y_{ijk} = \text{Ave} + \text{Voltage}_i + \text{Day}_j + (\text{Voltage} \times \text{Day})_{ij} + \text{Error}_{ijk} \qquad (9.20)$$

with $i = 1, \ldots, 3, j = 1, \ldots, 3$, and $k = 1, \ldots, 10$. The Ave term is the overall average. The Voltage term is treated as fixed, while the Day term, Voltage × Day term, and Error term are modelled as normal random variables with mean zero and constant variances. The fixed factor Voltage assumes a different average for each level of the trigger voltage, while the random factor Day assumes the same variance for each day.

Figure 9.14 shows the outcome of the experiment in boxplot form. The boxplots show the mean and range of the data for each day and trigger voltage level in the experiment. Clearly the different trigger voltage levels produce difference mean response. The different voltage trigger levels, however, also appear to produce greater uncertainty as the voltage level increases.

The ANOVA for the voltage standard study appears in Table 9.16. In this analysis, Voltage is treated as fixed and Day, Voltage × Day, and Error are treated as random. For this model, the equations in Tables 9.11 and 9.12 apply.

Note in the table of expected mean squares that Trigger Voltage has a quadratic term Q[1], indicating that the associated factor is fixed. The three random terms Day, Voltage × Day, and Error are all estimated using the formulas in Table 9.12. The variance components table indicates that most of the variation in the data, adjusted for voltage differences, is associated with the Voltage × Day and Error variance terms. This means that to reduce the uncertainty of measurement, emphasis would be placed on reducing the day-to-day and within-day variation.

Based on the graphical analysis above, it is of interest to investigate the performance of the standard across the range of input voltages. The presence of repeated measurements at each voltage level make it possible to investigate both bias and uncertainty at each level. Figure 9.15 shows the bias at each voltage level, with the bias of the kth measurement at the ith voltage level on the jth day defined as

Table 9.16 ANOVA for mixed effects model from voltage standard uncertainty study

Analysis of variance

Source	DF	Adjusted SS	Adjusted MS	F-value	P-value
Trigger voltage	2	1,355,705	677,853	51798	0.00
Day	2	34.0	17.0	1.29	0.37
Trigger voltage*day	4	52.0	13.0	9.55	0.00
Error	81	111.0	1.3		
Total	89	1,355,903			

Model summary

S	R-squared				
1.171	99.9%				

Expected mean squares, using adjusted SS

Source	Expected mean square for each term				
1 Part	(4) + 10.0 (3) + Q (1)				
2 Operator	(4) + 10.0 (3) + 30.0 (2)				
3 Part*operator	(4) + 10.0 (3)				
4 Error	(4)				

Variance components, using adjusted SS

Source	Variance	% of Total	St Dev		
Day	0.128	4.8	0.357		
Trigger voltage*day	1.172	43.9	1.082		
Error	1.370	51.3	1.171		
Total	2.670	100.0			

$$\text{bias}_{ijk} = y_{ijk} - \text{Voltage Level}_i. \tag{9.21}$$

The bias plot can be used to check the *linearity* of the measurement system (Kimothi 2002). In this context, linearity refers to the bias of the measurements over the range of input trigger voltages. To estimate linearity, repeated measurements are required at each voltage level over the range of the device. These data can also be used to evaluate the uncertainty at each voltage level. As is the case here, it is common in measurement processes for both the bias and uncertainty of a device to increase as the average of the measurement increases.

The fitted line (Fig. 9.15) can be used to predict the average bias over the range of the device. These data can also be used to construct confidence intervals (Fig. 9.16) for the standard deviations at each voltage level. These intervals show a clear difference in uncertainty at each level. The intervals represent the combined uncertainty due to day-to-day and within-day random differences.

9.3 ANOVA Models in Metrology

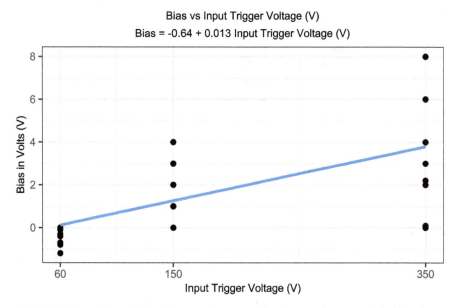

Fig. 9.15 Bias and linearity in voltage measurement

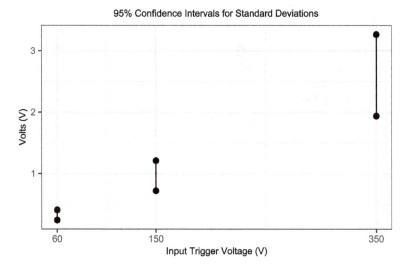

Fig. 9.16 Confidence intervals for the standard deviation at each voltage level

A mixed effects model was used to analyze data from a voltage standard uncertainty study. The fixed factor was Voltage, with input trigger voltages set at 60 V, 150 V, and 350 V. The random factor was Day, resulting in random terms for Day, Voltage × Day, and Error. The variance components analysis showed that the *Day* component was a small part of the total variation, while the Voltage × Day and

Error components together accounted for 95% of the total variation adjusted for voltage differences.

Bias and uncertainty were significantly different at each voltage level. As a result, a detailed uncertainty analysis would be done at each level of input trigger voltage. The uncertainty analysis would follow the approach given in Sect. 6.3.1 for the standard tested at 350 V. A separate expanded uncertainty (see Eq. 6.27), including the bias term, would be reported for each level.

9.3.6 Summary of ANOVA Method

ANOVA models, including random effects and mixed effects models, are useful for assessing the amount of variation that can be attributed to random components of a measurement process. ANOVA techniques provide a simple way to estimate variance components of the associated models. The relative magnitudes of the estimated variance components indicate where emphasis should be placed to reduce measurement uncertainty.

Sample sizes for ANOVA-based measurement studies should be based on a degrees of freedom argument. Each estimated variance component has a degrees of freedom associated with it, often computed using the Welch–Satterthwaite approximation. The degrees of freedom should always be reported along with the estimated variance component to assess the adequacy of the estimate.

A recommended goal in an ANOVA is to have 20–30 degrees of freedom for the error variance term, with 15 as minimally acceptable. Other variance components in an ANOVA will typically have far fewer degrees of freedom. It is recommended that the number of degrees of freedom for these components be built up over time through ongoing monitoring of the measurement system. More detailed recommendations regarding degrees of freedom and sample size are given in Sect. 11.5.

9.4 Related Reading

Metrology's guiding bodies have to date provided relatively little guidance on the use of design of experiments (DOEx) and ANOVA in uncertainty analyses. The JCGM 100 (GUM) has a brief discussion of basic ANOVA models and uses of ANOVA in metrology, but with no discussion of DOEx other than to say "a well-designed experiment can greatly facilitate reliable evaluations of uncertainty and is an important part of the art of measurement." The JCGM 101 and JCGM 102 have no mention of DOEx or ANOVA. The JCGM 103 "Developing and Using Measurement Models" (in draft form at this writing) discusses the use of experimental data to develop measurement models, but without discussion of DOEx.

ISO 3534-3, "Statistics - Vocabulary and Symbols - Part 3: Design of Experiments" defines the terms used in the field of design of experiments, although

9.4 Related Reading

it is not specifically for metrology applications. It includes a discussion of full factorial, fractional factorial, screening designs, randomized block designs, Latin Squares, Incomplete Block Designs, Split Plot Designs, Response Surface Designs, Mixture Experiments, Nested Designs, and Optimal Designs.

ISO 17025, ISO 17043, and ISO 13528 discuss proficiency testing, a simple experimental procedure that is used to compare the performance of individual laboratories on the same specific set of measurements. The statistic used to make the comparison has been referred to as the zeta score, z-prime score, or normalized error. This statistic corresponds to a two-sample t-test in the case of two laboratories being compared, or an ANOVA F-test if more than two laboratories are being compared.

ISO 5725-2 discusses design of experiments (at a high level) in the context of planning an accuracy study to compare laboratories. The choice of number of laboratories and replicates at each laboratory is based on a desired level of precision. ISO 23471, "Experimental Designs for Evaluation of Uncertainty – Use of Factorial Designs for Determining Uncertainty Functions" may in the future provide valuable guidance for DOEx in metrology. It is in the early stages of development at this writing. The NIST research library (https://www.nist.gov/nist-research-library) contains downloadable papers illustrating the use of DOEx and ANOVA in metrology studies, and metrology journals such as *Metrologia, Measurement Science and Technology,* and *Measurement* contain an occasional paper describing DOEx or ANOVA methods.

Discipline-specific journals also occasionally feature articles that use formal DOEx methodologies in measurement studies. Barari et al. (2014) used DOEx (2^6 full factorial) in a thermometry study to investigate the output of a mathematical model. The approach is used to compute interactions between input variables and to determine which input variables have a significant impact on measurement errors. Fong et al. (2009) also use DOEx methods to investigate uncertainties in finite element model (FEM) simulations. The technique they present relies on small factorial and fractional factorial designs to estimate uncertainties due to variability in input parameters such as material property constants, geometric inputs, and loading rates.

A number of papers have appeared in the recent literature illustrating the use of DOEx to evaluate the uncertainty of coordinate measurement machines (CMMs). Papers by Vrba et al. (2015), Papananias et al. (2017), Barini et al. (2010), Piratelli-Filho and Di Giacomo (2003), and Feng et al. (2007) all apply DOEx to the CMM measurement uncertainty problem. The authors use a variety of full factorial and fractional factorial designs, usually with multiple replications of each trial, to evaluate the effect of experimental factors and their two-way interactions on CMM uncertainty.

A detailed discussion of the design and analysis of Gauge R&R studies is given by Burdick et al. (2005). This book covers the basics of Gauge R&R studies, with emphasis on one-factor random models, two-factor random effects models, and two-factor mixed effects models. The analyses focus on constructing confidence intervals for variance components and ratios of variance components used to

9.5 Exercises

1. When might a full factorial design be used in a measurement study? When might a fractional factorial design be used?
2. What are the various ways experimental factors can be handled during an experiment?
3. What is "confounding" of factors in a two-level fractional factorial design? How would one determine the confounding pattern for such a design?
4. Design a $2^{(6-2)}$ fractional factorial design for six factors at two levels. What is the highest resolution that can be achieved? Suppose it is suspected that the AB and CE two-way interaction terms are both important. How would this affect the design strategy?
5. Table 9.17 presents measurement data from a full factorial experiment in which moisture Gauges were used to measure the dewpoint of sealed containers at sub-zero temperatures.
   The experimental factors are: 3 Moisture Gauges, 3 Dewpoint Levels, and 2 Operators, with 2 replicates of each measurement. The standard dewpoint value in the table can be considered the "actual" value. The total number of experimental observations is $3 \times 3 \times 2 \times 2 = 36$.

   Explore the data graphically, including a main effects plot. What does the graphical analysis suggest? Perform an ANOVA treating Moisture Gauge and Dewpoint Level as fixed factors and Operator as a random factor. Which factor has the greatest effect on the dewpoint measurement? Which factors contribute to the measurement uncertainty? What is the estimated repeatability standard deviation? What is the total Type A standard uncertainty associated with the measurement?

   Compute the differences Delta = (Measured Value − Standard Value). Graph the main effects associated with this calculated value. Is the measurement biased? Perform an ANOVA treating Moisture Gauge and Dewpoint Level as fixed effects and Operator as a random effect. Which factors contribute most to the measurement bias?
6. When might a variance components design (ANOVA) be used in a measurement study? What is a mixed ANOVA model? What is the difference between factors that are crossed and factors that are nested?
7. What is the Welch–Satterthwaite approximation used for in ANOVA studies?
8. Table 9.18 gives the data resulting from a Gauge R&R study in which the length of a component piece part was measured. The nominal length of the part was

9.5 Exercises 223

Table 9.17 Moisture gauge DOEx (3 × 3 × 2)

Obs	Moisture gauge	Dewpoint level	Operator	Rep	Standard value	Measured value
1	1	1	1	1	−26.1	−25.90
2	1	1	1	2	−26.7	−26.50
3	1	1	2	1	−27.4	−27.00
4	1	1	2	2	−27.8	−27.50
5	1	2	1	1	−35.6	−34.70
6	1	2	1	2	−36.2	−35.30
7	1	2	2	1	−36.7	−36.00
8	1	2	2	2	−37.3	−37.00
9	1	3	1	1	−44.9	−46.00
10	1	3	1	2	−45.5	−46.10
11	1	3	2	1	−46.0	−47.00
12	1	3	2	2	−46.6	−47.00
13	2	1	1	1	−23.7	−26.60
14	2	1	1	2	−24.4	−26.20
15	2	1	2	1	−25.1	−25.00
16	2	1	2	2	−25.7	−24.60
17	2	2	1	1	−33.4	−30.80
18	2	2	1	2	−34.0	−30.00
19	2	2	2	1	−34.7	−29.30
20	2	2	2	2	−35.3	−32.80
21	2	3	1	1	−43.4	−46.50
22	2	3	1	2	−43.9	−45.00
23	2	3	2	1	−44.5	−44.00
24	2	3	2	2	−45.1	−43.00
25	3	1	1	1	−24.8	−26.20
26	3	1	1	2	−25.5	−26.90
27	3	1	2	1	−23.6	−24.80
28	3	1	2	2	−24.2	−25.20
29	3	2	1	1	−34.4	−34.70
30	3	2	1	2	−35.0	−35.00
31	3	2	2	1	−33.2	−34.50
32	3	2	2	2	−33.8	−33.20
33	3	3	1	2	−44.4	−44.80
34	3	3	1	1	−42.8	−42.00
35	3	3	2	2	−43.3	−43.30
36	3	3	2	1	−43.9	−44.20

0.978. The experiment consisted of two operators measuring the length of two parts three times each on two separate days. The total number of observations is thus $2 \times 2 \times 3 \times 2 = 24$. Explore the data graphically. What does the graphical analysis suggest? Perform an ANOVA treating each factor as a random factor. Which factor has the largest variance component? Which factors contribute to the measurement uncertainty? What is the total Type A standard uncertainty

Table 9.18 Gauge R&R study data

Observation	Day	Operator	Part number	Replicate	Length (in)
1	1	1	1	1	0.970558
2	1	1	1	2	0.970183
3	1	1	1	3	0.969758
4	1	1	2	1	0.977658
5	1	1	2	2	0.977967
6	1	1	2	3	0.977933
7	1	2	1	1	0.970275
8	1	2	1	2	0.969833
9	1	2	1	3	0.969817
10	1	2	2	1	0.978125
11	1	2	2	2	0.978125
12	1	2	2	3	0.977908
13	2	1	1	1	0.970625
14	2	1	1	2	0.969758
15	2	1	1	3	0.970258
16	2	1	2	1	0.977833
17	2	1	2	2	0.978525
18	2	1	2	3	0.977600
19	2	2	1	1	0.970425
20	2	2	1	2	0.969667
21	2	2	1	3	0.970058
22	2	2	2	1	0.977833
23	2	2	2	2	0.978075
24	2	2	2	3	0.978292

associated with this measurement? What are the degrees of freedom for repeatability? Reproducibility?

9. What is an adequate number of degrees of freedom for estimating the error variance component (repeatability) in an ANOVA experiment? How can you increase the number of degrees of freedom to improve the estimate of reproducibility in a Gauge R&R study?
10. When would you use a Resolution III fractional factorial design?

References

Addelman, S., Kempthorne, O.: Orthogonal Main Effects Plans. Aeronautical Research Laboratory Office of Aerospace Research. Wright-Patterson Air Force Base, OH (1961)

Arendacka, B.: Linear mixed models: GUM and beyond. Meas. Sci. Rev. **14**(2), 52–61 (2014)

Barari, F., Morgan, R., Barnard, P.: A design of experiments approach to optimise temperature measurement accuracy in solid oxide fuel cell (SOFC). J. Phys. Conf. Ser. 547 (2014)

References

Ballico, M.: Limitations of the Welch-Satterthwaite approximation for measurement uncertainty calculations. Metrologia, 61–64 (2000)

Barini, E., Tosello, G., De Chiffre, L.: Uncertainty analysis of point-by-point sampling complex surfaces using touch probe CMMs. Precis. Eng. **34**, 16–21 (2010)

Box, G., Hunter, J., Hunter, W.: Statistics for Experimenters, 2nd edn. Wiley, New York (2005)

Burdick, R., Borror, C., Montgomery, D.: Design and Analysis of Gauge R&R Studies: Making Decisions with Confidence Intervals in Random And Mixed ANOVA Models. ASA-SIAM Series on Statistics and Applied Probability (2005)

Daniel, C.: Use of half-normal plots in interpreting factorial two-level experiments. Technometrics. **1**, 311–340 (1959)

Down, M., Czubak, F., Gruska, G., Stahley, S., Benham, D.: Measurement Systems Analysis Reference Manual. Automotive Industry Action Group (2010)

Feng, C., Saal, A., Salsbury, J., Ness, A., Lin, G.: Design and analysis of experiments in CMM measurement study. Precis. Eng. **31**, 94–101 (2007)

Fong, J., de Wit, R., Marcal, P., Filliben, J., Heckert, N.: A design-of-experiments plug-in for estimating uncertainties in finite element simulations. In: SIMULIA Customer Conference (2009)

Fong, J., Heckert, N., Filliben, J., Ma, L., Stupic, K., Keenan, K., Russek, S.: A design-of-experiments approach to FEM uncertainty analysis for optimizing magnetic resonance imaging RF coil design. In: Proceedings of the 2014 COMSOL Conference, Boston (2014)

ISO 13528: Statistical Methods for USE in Proficiency Testing by Interlaboratory Comparison. (2015)

ISO 17043: Conformity Assessment - General Requirements for Proficiency Testing. (2010)

ISO 3534-3: Statistics Vocabulary and Symbols - Part 3: Design of Experiments. (2013)

ISO 5725-1: Accuracy (Trueness and PRECISION) of Measurement Methods and Results - Part 2: Basic Method for the Determination of Repeatability and Reproducibility of a Standard Measurement Method. (1994)

ISO/IEC 17025: General Requirements for the Competence of Testing and Calibration Laboratories. (2017)

JCGM 100: Evaluation of Measurement Data - Guide to the Expression of Uncertainty in Measurement. (2008)

JCGM 101: Evaluation of Measurement Data - Supplement 1 to the Guide to the Expression of Uncertainty in Measurement - Propagation of Distributions Using a Monte Carlo Method. (2008)

JCGM 102: Evaluation of Measurement Data - Supplement 2 to the -Guide to the Expression of Uncertainty in Measurement - Extension to any Number of Output Quantities. (2011)

Kimothi, S.K.: The Uncertainty of Measurements: Physical and Chemical Metrology and Analysis. ASQ Quality Press, Milwaukee (2002)

Kutner, M.H., Nachtsheim, C.J., Neter, J., Li, W.: Applied Linear Statistical Models, 5th edn. McGraw-Hill, New York (2005)

Lenth, R.: Quick and easy analysis of unreplicated factorials. Technometrics. **31**, 469–473 (1989)

Montgomery, D.: Design and Analysis of Experiments, 9th edn. Wiley, New York (2017)

Papananias, M., Fletcher, S., Longstaff, A., Forbes, A.: Uncertainty evaluation associated with versatile automated gauging influenced by process variations through design of experiments approach. Precis. Eng. **49**, 440–455 (2017)

Piratelli-Filho, A., Di Giacomo, B.: CMM uncertainty analysis with factorial design. Precis. Eng. **27**, 283–288 (2003)

Satterthwaite, F.: An approximate distribution of estimates of variance components. Biom. Bull. **2**(6), 110– 114 (1946)

Searle, S.R., Gruber, M.H.J.: Linear Models. Wiley, New York (2016)

Suzuki, T., Takeshita, J., Ono, J., Lu, X.: Designing a measurement precision experiment considering distribution of estimated precision measures. IOP Conf. Series J. Phys. Conf. Series 1065 (2018)

Vardeman, S., Jobe, J.: Statistical Methods for Quality Assurance. Springer, New York (2016)

Vrba, I., et al.: Different approaches in uncertainty evaluation for measurement of complex surfaces using coordinate measuring machine. Meas. Sci. Rev. **15**(3), 111–118 (2015)

Zhang, S., et al.: Error analysis of data acquisition of reverse engineering process using design of experiment. In: Advances in Manufacturing Technology XVI. Professional Engineering Publishing Limited, London (2002)

Chapter 10
Determining Uncertainties in Fitted Curves

This chapter discusses methods of fitting curves to measured data while accounting for uncertainty in the data. We start with equations for the best linear fit and uncertainties in the slope and intercept. A short discussion of linear fits when there is uncertainty in the x-values is presented next. Confidence and prediction bands are presented for expressing uncertainty in the fitted line and uncertainty in predictions made using the fitted line. Goodness-of-fit metrics are used to assess model adequacy. A summary of techniques for fitting nonlinear equations to data is given, including formulas for nonlinear confidence and prediction bands. In this chapter "nonlinear equation" means an equation that is nonlinear in the independent variable. Finally, a case study applying curve fitting to a calibration drift problem is provided with a short discussion of how to choose optimal calibration intervals based on predicted drift.

10.1 The Purpose of Fitting Curves to Experimental Data

Previous chapters presented the GUM's propagation of uncertainty approach, in which uncertainties of multiple independent variables are propagated through an indirect measurement model $y = f(x_1, x_2, \ldots, x_N)$. In practice it is often useful to take measurements of a dependent variable y over a range of values of an independent variable x. Fitting a line or a curve to the data then provides a model that can be used to predict y, and its uncertainty, for any value of x. In this chapter, the measurement model

$$y_i = f(x_i; p) + \varepsilon_i$$

is used to address this prediction problem. In this expression, y_i is the measured dependent variable and x_i is the independent variable that could be either known (such as time) or measured with uncertainty (such as temperature).

The term ε_i represents the difference (the residual) between the fitted model and the observed dependent measurement. The goal of curve fitting as discussed in this chapter is to identify the form of the function f and estimate the parameters p that produce the least error (smallest $\varepsilon_i's$) in predicting y given x.

Measurements are often made over a range of physical settings. For example, density may be measured at various temperatures, radiation dose may be measured at various distances from the source, and resistance may be measured at various time intervals to observe drift or decay. All these cases involve characterizing the relationship between two variables. By fitting a curve to experimental data, the relationship between two variables can be described concisely with a single equation. The equation also allows prediction of points on the curve for which experimental data was not measured, that is, interpolating the expected value and uncertainty between measurement points, or in some cases, extrapolating beyond the region of measured values. Expressing the uncertainty in y as a function of x usually results in a more precise (and smaller) estimate of uncertainty. Rather than using a bound on uncertainty in the analysis, the uncertainty is based on the value of x.

Fitting a curve to experimental data generally involves minimizing the distance between a proposed curve and the data. While finding the best-fitting curve can be relatively straightforward, propagating uncertainties through the fitting process can be more complicated.

10.1.1 Resistance vs. Temperature Data

A resistance standard is nominally 100 ohms at a temperature of 20 °C. However, the resistor cannot always be used in a controlled environment and may be subject to a wide temperature range. As such, the resistance was measured at temperatures from 0 °C to 40 °C as shown in Table 10.1 and Fig. 10.1. When the resistor is in use, the temperature is monitored and used to determine the actual resistance value at the time of use. A curve fit may be appropriate in this case, allowing the user to predict the resistance when used at 23 °C, a value not explicitly measured in the table.

Table 10.1 Measured resistance versus temperature

Temperature (°C)	Resistance (Ω)
0	92.648
5	95.675
10	96.923
15	97.732
20	100.473
25	101.258
30	104.426
35	105.484
40	109.237

10.1 The Purpose of Fitting Curves to Experimental Data

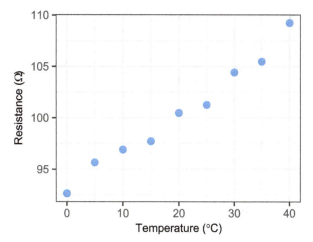

Fig. 10.1 Resistance versus temperature

The sections that follow below use the resistance versus temperature data to illustrate various cases of fitting curves and to describe the uncertainty in those fits.

10.1.2 Considerations When Fitting Models to Data

The first consideration when fitting a model to measured data is determining whether there is even a relationship between the two variables of interest. It is not uncommon to see attempts to fit a line to data that has no real trend. It can sometimes be difficult to distinguish between a relationship between two variables and random noise in the measurement data.

Once a clear relationship is established, the next consideration is choice of a model. In many cases, the simplest possible model is a line. The resistance vs. temperature example has data that appears to fit a line over the range of interest; however, a linear model cannot be extended up or down infinitely—it will eventually break down. For example, when the temperature gets too high, the resistor will melt. From an engineering perspective, a linear model works under reasonable assumptions, but care must be taken when extrapolating the model outside of its valid range. For example, a line fit to the data in Fig. 10.1 may make reasonable predictions in the range from 0 to 40 °C, and a few degrees outside this range, but the same line should not be used to predict resistance at 100 °C. Nonlinear models are often required. For example, an exponential model works well for problems involving decay. Sometimes quadratic or higher-order polynomials work well if the exact physical nature of the model is unknown.

Finally, it should be ensured that there are a sufficient number of data points to define the fitted curve. Two points determine a line exactly, but with no degrees of freedom for error, the uncertainty will be large. Three points determine a quadratic curve that passes exactly through all three points, yet is likely "overfit" resulting in the curve describing random fluctuations rather than the true physical relationship between the variables.

Several goodness-of-fit metrics are discussed in this chapter that can help quantitatively validate the relationship between variables and proper selection of a model.

In what follows, techniques for finding the best fit parameters (such as slope and intercept) and the uncertainties in these parameters are discussed. Then, formulas for finding the uncertainty in a prediction based on the fitted curve are given, along with some goodness-of-fit metrics. Finally, the cases of nonlinear curve fitting and curve fitting with uncertainties in both x and y values are discussed.

10.2 Methods for Fitting Curves to Experimental Data

The common method for fitting a curve to data is a technique called "least squares fit" or "regression." This method finds the curve that minimizes the vertical distance (technically the sum of the squares of the vertical distances), illustrated in Fig. 10.2, from each measured point to the curve. This can be done analytically for straight line models, but an iterative numerical solution is often required for more complex curves. Most modern computing packages (such as R, Matlab, Python/Scipy) have line and curve fitting functions. However, some of the functions do not consider known uncertainty in the measured values or assume an uncertainty based on scatter in the data.

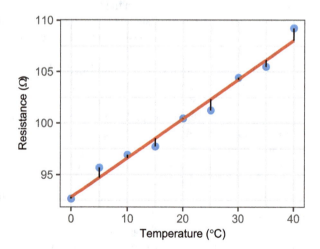

Fig. 10.2 Resistance versus temperature fit line

10.2.1 Linear Least Squares

The linear case of least-squares fitting can be solved analytically. The problem is to find the best fitting line of the form $y = A + Bx$ through the data points. The N measured data points are x_i and y_i.

The least-squares method compares differences between the fit line and each data point to minimize the sum of the squared residuals (the $\varepsilon_i's$), that is, the squared distances from a measured data point to the fitted line. The sum of squared residuals is

$$Q = \sum_{i=1}^{N} (y_i - A - Bx_i)^2. \tag{10.1}$$

This assumes that all the error in the measurement is in the y values. The x values here have no uncertainty. The residuals are minimized by taking the derivative of Q with respect to A and B, setting them equal to zero, and solving the system of equations for A and B. Direct expressions can be obtained:

$$A = \frac{\Sigma x_i^2 \Sigma y_i - \Sigma x_i \Sigma x_i y_i}{\Delta}$$

$$B = \frac{N \Sigma x_i y_i - \Sigma x_i \Sigma y_i}{\Delta}$$

$$\Delta = N \Sigma x_i^2 - (\Sigma x_i)^2. \tag{10.2}$$

See Kutner et al. (2005) for a full derivation of these equations. Any uncertainty in the y values is not needed to find the nominal slope and intercept but is used to find the uncertainties of these estimated parameters.

Resistor Example In the resistor vs. temperature data, the best fitting line can be computed using the linear least-squares formulas to obtain a slope of 0.381 $\Omega/°C$ and intercept of 92.8 Ω. The line, with the measured data points and residual distances (vertical lines), is shown in Fig. 10.2.

10.2.2 Uncertainty in Fitting Parameters

Uncertainty in the slope B and intercept A can be calculated with Taylor (1997):

$$\sigma_A = \sigma_y \sqrt{\frac{\Sigma x_i^2}{\Delta}}, \quad \sigma_B = \sigma_y \sqrt{\frac{N}{\Delta}}, \tag{10.3}$$

where σ_y is the uncertainty in the measured y values. If σ_y is not explicitly known based on the details of the measurement, it is common to estimate it using the residuals:

$$\sigma_y^2 = \frac{\sum_{i=1}^{N} \varepsilon_i^2}{N-2}$$

$$= \frac{\sum_{i=1}^{N} (y_i - A - Bx_i)^2}{N-2}. \tag{10.4}$$

This may or may not always be a valid assumption. First, it assumes every measurement point has the same uncertainty distribution, and second it assumes that a line is actually the correct model for the data. The basic line-fitting algorithms in many computing packages often make this assumption and may not provide a method to override the value of σ_y when more information is known from the measurement. If σ_y is known from independent sources, this known value should be used in place of the estimate. However, it is reasonable to expect the known σ_y to have similar magnitude as the σ_y estimated from residuals.

Resistor Example Using the residuals to estimate the uncertainty in resistance σ_y, the uncertainties of the slope and intercept in the resistor/temperature example (Eq. 10.1) are ($k = 1$)

$$B = 0.381 \, \Omega/°C \pm 0.021 \, \Omega/°C$$

$$A = 92.81 \, \Omega \pm 0.50 \, \Omega. \tag{10.5}$$

The "slope test" for testing the significance of the estimated slope B is described in Sect. 10.4.2. It is based on the output of the regression analysis in Eq. (10.5).

```
temperature <- c(0,5,10,15,20,25,30,35,40)
resistance <- c(92.648, 95.675, 96.923, 97.732, 100.473, 101.258,
104.426, 105.484, 109.237)
# Linear model fit
fit <- lm(resistance ~ temperature)
summary(fit)$coefficients
## Estimate Std. Error t value Pr(>|t|)
## (Intercept) 92.80744 0.50421269 184.06408 3.687337e-14
## temperature 0.38105 0.02118117 17.99004 4.051657e-07
```

10.2.3 Weighted Least Squares: Non-constant u(y)

In some practical cases, the uncertainty may change for different measurement points on the line. For example, an ohmmeter could change ranges as resistance increases, where each range has a different known uncertainty value. In this case, every measurement point has its own uncertainty σ_i. The above equations can be modified to weight the least squares appropriately using this value. Assigning weights as $w_i = 1/\sigma_i^2$, the weighted solution becomes (Taylor 1997)

$$A = \frac{\sum w_i x_i^2 \sum w_i y_i - \sum w_i x_i \sum w_i x_i y_i}{\Delta}$$

$$B = \frac{\sum w_i \sum w_i x_i y_i - \sum w_i x_i \sum w_i y_i}{\Delta}$$

$$\Delta = \Sigma w_i \Sigma w_i x_i^2 - (\Sigma w_i x_i)^2 \qquad (10.6)$$

and uncertainties:

$$\sigma_A = \sqrt{\frac{\sum w_i x_i^2}{\Delta}}$$

$$\sigma_B = \sqrt{\frac{\sum w_i}{\Delta}}. \qquad (10.7)$$

10.2.4 Weighted Least Squares: Uncertainty in Both x and y

When the measurements have uncertainty in both x and y, the problem becomes much more complicated. An analytical solution is no longer possible or straightforward, and an iterative numerical technique must be used to minimize the residuals. One good example is the algorithm described by York et al. (2004) and its pseudocode implementation by Wehr and Saleska (2017) which accounts for not only σ_x and σ_y but also the correlation between errors in x and y.

In some cases, it may be simpler to apply a Monte Carlo approach (Robert and Casella 2004) rather than implement a complex algorithm such as this.

Resistor Example Suppose the uncertainty in the recorded temperature is 0.75 °C and the uncertainty in resistance measurements follows 0.5 + 2.5% Ω. These uncertainties are shown as the error bars in Fig. 10.3. Accounting for these uncertainties using the York algorithm, the new slope and intercept values are

$$B = 0.373 \; \Omega/\text{cm} \pm 0.025 \; \Omega/\text{cm}$$

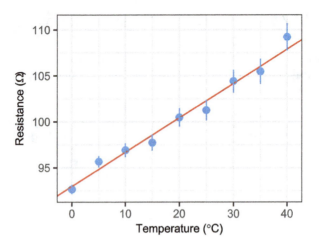

Fig. 10.3 Resistance vs. temperature with uncertainties in both axes

$$A = 92.96\,\Omega \pm 0.43\,\Omega \tag{10.8}$$

```
# Uncertainty in x and y
sdx <- 0.75
sdy <- temperature/40+0.5

# Best-fit straight line
fit <- bfsl(temperature, resistance, sd_x = sdx, sd_y = sdy)
fit
## Best-fit straight line
##
## Estimate Std. Error
## Intercept 92.96163 0.42860
## Slope 0.37299 0.02541
##
## Goodness of fit:
## 0.6418
```

Notice that with σ_x and σ_y accounted for, σ_A is a bit smaller, but σ_B is a bit larger because σ_y is no longer assumed from residuals only.

10.3 Uncertainty of a Regression Line

Two types of uncertainties can be calculated based on fitted lines: the uncertainty in the slope and intercept parameters, and the uncertainty in the line itself.

10.3.1 Uncertainty of Fitting Parameters

The above least-squares methods provide the slope, intercept, and the uncertainty in each, but not the uncertainty in the overall line, i.e., the uncertainty in the predicted value of y for a particular value of x. In the resistor problem, for example, the resistance (y) and uncertainty at a given temperature (x) cannot be obtained directly from the slope and intercept and their uncertainties.

10.3.2 Confidence Bands

Just as the mean of a measurement sample is an estimate of a variable's true mean, the fitted line $y = A + Bx$ is an estimate of the true line relating the two variables. Analogous to the standard error of the mean, the confidence band is a measure of error in the fit line. Because it encompasses all the possible lines that might fit the data, and the line is best defined near the center of the data points, the confidence band is a function of x and is typically smallest near the center of the line and largest on the ends. The confidence band as a function of x is

$$u_{\text{conf}}(x) = \sigma_y \sqrt{\frac{1}{N} + (x - \bar{x})^2 \left(\frac{\sigma_B}{\sigma_y}\right)^2}, \tag{10.9}$$

and the true line, with k-factor applied, is expected to fall in the range $f(x) \pm k \cdot u_{\text{conf}}(x)$ with high confidence.

10.3.3 Prediction Bands

A prediction band is used to determine the uncertainty in a prediction associated with an individual measurement (x) on the fitted curve. This band includes both uncertainty in the fitted line (i.e., the confidence band) and uncertainty in the individual measurement. The prediction band can be calculated from the RSS of the confidence band with the σ_y uncertainty estimated from the residuals:

$$u_{\text{pred}}(x) = \sqrt{\sigma_y^2 + u_{\text{conf}}^2(x)} \tag{10.10}$$

Doing a little algebra:

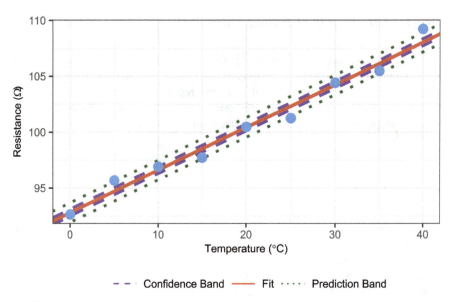

Fig. 10.4 Confidence and prediction bands

$$u_{\text{pred}}(x) = \sigma_y \sqrt{1 + \frac{1}{N} + (x - \bar{x})^2 \left(\frac{\sigma_B}{\sigma_y}\right)^2}. \qquad (10.11)$$

A particular value of x is substituted into u_{pred}, and the confidence factor k can be applied to estimate the total uncertainty when making a measurement at that value of x on the curve (Fig. 10.4):

$$f(x) \pm k \cdot u_{\text{pred}}(x). \qquad (10.12)$$

The coverage factor k is determined based on level of confidence and degrees of freedom, similar to traditional uncertainty calculations presented in Chaps. 6 and 7. However, the degrees of freedom for a fitted line is $N - 2$ rather than $N - 1$ due to estimating two parameters, the slope and intercept.

Resistor Example During a particular measurement, the resistor standard was used at a temperature of 27.5 °C, a value not measured in the original data set, so interpolation was needed. Using the fitted line, the resistance at this operating temperature can be calculated:

$$R = A + BT = 92.96 + 0.373(27.5) = 103.29\ \Omega. \qquad (10.13)$$

Uncertainty in this value can be found from the confidence band (Eq. 10.9):

10.3 Uncertainty of a Regression Line

$$u_{\text{conf}}(27.5) = 0.32\ \Omega. \tag{10.14}$$

However, if one needs to predict the value of the resistor during use, the uncertainty in the measurements must be included by using the prediction band (Eq. 10.11).

$$u_{\text{pred}}(27.5) = 0.88\ \Omega \tag{10.15}$$

Therefore, the resistance with uncertainty at 27.5 °C is

$$R = 103.29\ \Omega \pm 0.88\ \Omega\ (k = 1) \tag{10.16}$$

```
# Fit linear model
fit <- lm(resistance ~ temperature)

# Estimate line using fit coefficients
lineT <- fit$coefficients[[1]] + fit$coefficients[[2]]*temperature

# Calculate sigma_b and sigma_y
N <- length(temperature)
sigB <- sqrt(diag(vcov(fit)))[2]
sigy <- sqrt(sum((resistance - lineT)^2)/(N-2))

# Define functions for confidence and prediction bands
uconf <- function(sigy, xx, xbar, sigB) {
 return(sigy*sqrt(1/N + (xx - xbar)^2*(sigB/sigy)^2))
}
upred <- function(sigy, xx, xbar, sigB) {
 return(sigy*sqrt(1 + 1/N + (xx - xbar)^2*(sigB/sigy)^2))
}

# Calculate confidence and prediction uncertainty values at 27.5 degrees
xbar <- mean(temperature)
uc <- uconf(sigy, 27.5, xbar, sigB)
uc[[1]]
## [1] 0.3162432
up <- upred(sigy, 27.5, xbar, sigB)
up[[1]]
## [1] 0.8791887
```

10.4 How Good Is the Model?

A simple line may fit the data reasonably well and still not be the best model for the data. Other models may fit the data better or make more physical sense. Some qualitative and quantitative tests can be used to determine if the most appropriate model was chosen. Those tests are discussed in this section.

10.4.1 Residual Analysis

The residual plot (plotting the $\varepsilon_i's$ vs the $x_i's$) is one of the key diagnostic tools for assessing a fitted model. For any reasonable fit of the data, the individual measurements should fall randomly above and below the curve. The residuals (distances from each point to the fitted curve) should also be roughly normally distributed with mean zero. This assumption can be tested via a simple histogram or normal probability plot. A good linear fit and residual plot is shown in Fig. 10.5a.

Now consider Fig. 10.5b. The data show a slight curvature when compared to the linear fit. The curvature is not always as obvious as in this example, so a better qualitative test is checking the residual plot. Note the order of the residuals, with positive values occurring near the ends and negative values in the center. This indicates a straight line model may not be the best fit for this data. Perhaps a polynomial or exponential curve should be considered.

The data in Fig. 10.5c show an example with a single large outlier. The large outlier greatly influences the slope of the fitted line, and the majority of the data are not well fit by the line. Again, using a residual plot shows clearly that the data are not well fit, and the residuals are not randomly distributed about zero.

10.4.2 Slope Test

In fitting simple lines, another good check of the fit is the slope test. This test can be used to confirm a linear relationship between the x and y variables. If there is indeed a linear relationship, the slope of the line will have a statistically significant positive or negative slope. To apply the slope test, check the 95% coverage interval of the slope and its uncertainty (Eq. 10.5). If this interval contains zero, the test fails, and one cannot conclude there is a linear relationship between the x and y variables (Fuller 1987). The slope test is equivalent to testing that the (x, y) correlation coefficient is statistically different from zero (Glantz and Slinker 2001). The validity of these tests relies on the assumptions regarding the residuals as stated in Sect. 10.4.1.

10.4 How Good Is the Model?

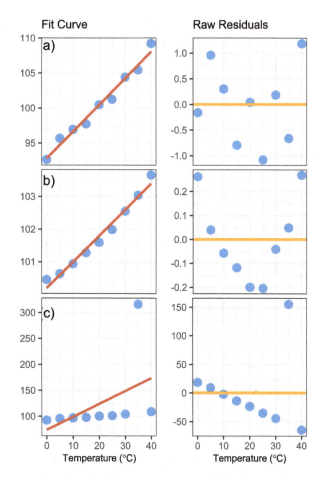

Fig. 10.5 Residual analysis examples

Similar tests can be used to check the validity of certain nonlinear models. For example, a test of the parameter A in the model $y = \exp(Ax)$ can be used to confirm an exponential relationship between the x and y variables. On the log scale, the parameter A becomes a slope parameter, and the slope test applies.

10.4.3 Quantitative Residual Analysis

In some cases, a more quantitative approach is desired to assess the goodness-of-fit. Many statistical tests for goodness-of-fit are described by Glantz and Slinker (2001), but one of the most common is the coefficient of determination, or r^2 value. We define r^2 by

$$r^2 = \frac{\text{(Regression Sum of Squares)}}{\text{(Total Sum of Squares corrected for the mean)}}.$$

It is calculated from

$$r^2 = \frac{SS_{reg}}{SS_{tot}} = 1 - \frac{SS_{res}}{SS_{tot}}, \qquad (10.17)$$

where

$$SS_{reg} = \sum_{i=1}^{N} (\hat{y}_i - \bar{y})^2$$

$$SS_{res} = \sum_{i=1}^{N} (y_i - A - Bx_i)^2,$$

and

$$SS_{tot} = \sum_{i=1}^{N} (y_i - \bar{y})^2. \qquad (10.18)$$

In the expression for SS_{reg}, $\hat{y}_i$ is the predicted value of y_i. The total sum of squares is equal to the regression sum of squares plus the residual sum of squares:

$$SS_{tot} = SS_{reg} + SS_{res}.$$

Given these definitions, r^2 can be interpreted as the proportion of the total variation that is explained by the regression model.

SS_{res}, the residual sum of squares, is the same quantity that is minimized (Eq. 10.1) to find the best fit. SS_{tot}, the total sum of squares, computes the difference between each data point and the overall mean. When $r^2 = 0$ ($SS_{reg} = 0$), the model has not explained any of the variation in the data, and the fitted model is of no value. If $r^2 = 1$ ($SS_{reg} = SS_{tot}$), the fitted model has explained all of the variation in the data, and the model is a perfect fit ($SS_{res} = 0$).

While an r^2 value close to one is a sign of a good fit, it is difficult to interpret the actual r^2 value or assign a threshold for what qualifies as "good enough" for a particular problem. Another metric is the residual mean square, often denoted S:

$$S = \sqrt{\frac{SS_{res}}{(N-2)}} \qquad (10.19)$$

which is an estimate of the standard deviation of the residuals about the fitted line.

10.5 Uncertainty in Nonlinear Regression

Under the assumption that the residuals are normally distributed with constant variance, approximately 95% of the measured values (the $y_i's$) should fall within $\pm 2S$ of the fitted line (the corresponding $\hat{y}_i's$). The value of S can be used to compare competing candidate models, with models having lower values of S generally being preferred.

In the resistance vs. temperature example from Table 10.1, the r^2 value is 0.97, indicating a very good linear fit. The value of S is 0.82 Ω, meaning that 95% of the measurements are expected to fall within 1.64 Ω of the fitted line.

10.5 Uncertainty in Nonlinear Regression

The previous sections examined fitting lines (slope and intercept) to measurement data. This section generalizes the method above to fitting nonlinear curves. In this chapter "nonlinear" means nonlinear in the independent variable x.

10.5.1 Nonlinear Least Squares

The same least-squares methods can be adapted to nonlinear curve fitting. Recall that the sum of squared residuals (Eq. 10.1) in the linear case has the form:

$$Q = \sum_{i=1}^{N} (y_i - A - Bx_i)^2. \quad (10.20)$$

In the nonlinear case, the distance between the measurements and fitted curve is still used, but the minimization problem takes the more general form:

$$Q = \sum_{i=1}^{N} [y_i - f(x_i; p)]^2, \quad (10.21)$$

where f is the nonlinear function relating x to y, and p represents a vector of parameters that are estimated to minimize Q. Typically, an iterative numerical solution is required to minimize Q using an algorithm such as gradient descent, Gauss–Newton Method, or the Levenberg–Marquardt Method (Gavin 2017).

Resistor Example It was decided to calibrate the resistance standard over a wider temperature range. After taking new measurements up to 100 °C, a new line was fit, shown in Fig. 10.6; however, the residual analysis indicated problems with the linear model.

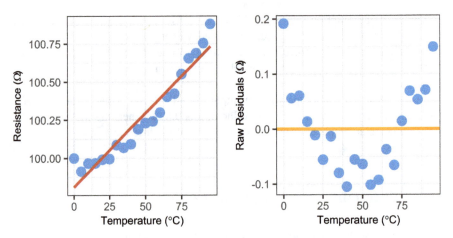

Fig. 10.6 Resistor example: poor linear fit

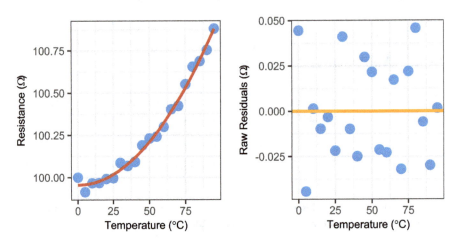

Fig. 10.7 Resistor example: better quadratic fit

Instead of using a linear model that would result in larger SS_{res}, a quadratic function $y = A + Bx + Cx^2$ was used and values of A, B, and C were found, resulting in a fitted curve of

$$y = 100.0 - 0.00254x + 0.000125x^2 \qquad (10.22)$$

with standard uncertainties in A, B, and C of 0.023, 0.0011, and 0.000011, respectively (Figs. 10.7).

10.5 Uncertainty in Nonlinear Regression

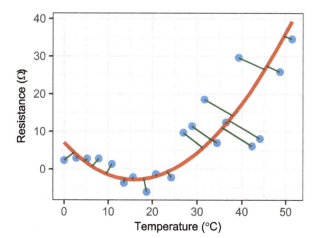

Fig. 10.8 Orthogonal distance regression

```
# Generate data
T2 <- seq(0, 95, by=5)
R2 <- 100 + (T2)**2 / 10000 - .04
R2 <- rnorm(length(T2), R2, 0.04)
# Fit quadratic model
fit <- lm(R2 ~ poly(T2,2, raw = T))
summary(fit)$coefficients
## Estimate Std. Error t value Pr(>|t|)
## (Intercept) 1.000112e+02 2.298894e-02 4350.404352 7.658464e-53
## poly(T2, 2, raw = T)
1 -2.538013e-03 1.121652e-03 -2.262745 3.703469e-02
## poly(T2, 2, raw = T)
2 1.250023e-04 1.139747e-05 10.967549 3.933882e-09
```

10.5.2 Orthogonal Distance Regression

An alternative to least-squares regression, useful when fitting a nonlinear curve with uncertainty in both *x* and *y*, is Orthogonal Distance Regression (ODR). It takes the same approach as least-squares regression but instead of minimizing the sum of vertical distances between each point and the curve, it minimizes the sum of shortest orthogonal distances (Fig. 10.8).

```
# Generate data
xo <- seq(0, 50, length.out = 20)
yo <- .0005 * xo^3 -.01 * xo^2 - .05 * xo + 3
uxo <- seq(.1,1, length.out = length(xo))
uyo <- seq(1,10, length.out = length(yo))
yo <- yo + rnorm(length(yo), 0, uyo)
```

```
xo <- xo + rnorm(length(xo), 0, uxo)
dat <- data.frame(xo, yo)
# Fit orthogonal nonlinear least-squares regression model
fit <- onls(yo ~ b1 + b2*xo + b3*xo^2 + b4*xo^3, data = dat, start = list
(b1=.05, b2=-.05, b3=.2, b4=3))
summary(fit)$coefficients
## Estimate Std. Error t value Pr(>|t|)
## b1 2.7102953151 4.3367491296 0.6249601 0.5408084
## b2 -0.1242620821 0.7614549688 -0.1631903 0.8724120
## b3 -0.0105391979 0.0355942465 -0.2960927 0.7709678
## b4 0.0005111474 0.0004619429 1.1065162 0.2848629
```

Unlike ordinary least-squares regression, ODR produces the same curve regardless of which axis is considered the independent variable. In other words, using ODR to fit a line to data (x_i, y_i), regressing x on y, will generate an equivalent result as fitting a line to (y_i, x_i), regressing y on x. Ordinary least squares in the first case would minimize residuals in the y direction and in the second case minimize residuals in the x direction, leading to different results. This is illustrated in Fig. 10.9.

For details of the algorithm, see Boggs and Rogers (1990). Implementations of ODR exist in computing packages such as *R*, Matlab, and Python/Scipy.

10.5.3 Confidence and Prediction Bands in Nonlinear Regression

The confidence and prediction bands given above were derived for the special case of linear regression. In the nonlinear case, the confidence interval may be found by approximating the function using a Taylor series expansion and using the delta method as described in Cox and Ma (1995):

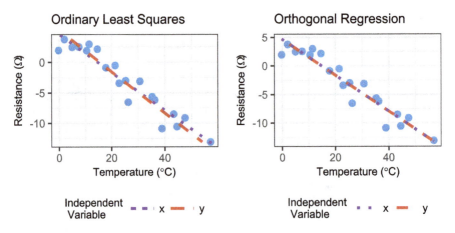

Fig. 10.9 Independent variable in ODR compared to ordinary least-squares regression

10.6 Using Monte Carlo for Evaluating Uncertainties in Curve Fitting

$$u_{\text{conf}}(x) = \sqrt{\nabla f^T \cdot C \cdot \nabla f}, \qquad (10.23)$$

where ∇f is the gradient matrix of the function at a particular x and C the covariance matrix of the input variables.

Prediction can then be found as before by combining the confidence band with the uncertainty in an individual measurement:

$$u_{\text{pred}}(x) = \sqrt{u_{\text{conf}}^2 + \sigma_y^2}. \qquad (10.24)$$

Note that σ_y may also be a function of x and may not be explicitly known for all possible x values.

10.6 Using Monte Carlo for Evaluating Uncertainties in Curve Fitting

Sometimes it is easier to apply a simple least-squares approach to find the best fit parameters, and then use a Monte Carlo method to propagate the uncertainties through the fitted curve. This becomes especially true when a programming language or computing package has a least-squares method that does not allow the input of uncertainties into the independent variables, or only allows a fixed uncertainty to apply to all measurement values. Monte Carlo can also be used to relax the assumption that each measurement value is normally distributed.

10.6.1 Monte Carlo Approach

In the Monte Carlo approach to estimating uncertainty in a fitted curve, each data point (x_i) is considered an input to the model. For each Monte Carlo sample, each data point is independently sampled from its distribution. The distributions for each data point do not have to be the same and can be drawn from any probability distribution. Perhaps there is greater uncertainty when measuring low values of resistance than measuring high values of resistance, for example. Then, with this set of sampled data points, one possible fitted curve is found using the least-squares fit (with no uncertainty input). The sampling process is repeated N times to generate N possible fitted curves and the parameters are estimated for each curve. This results in a Monte Carlo distribution and summary statistics for each parameter (e.g., the mean and standard deviation of possible slopes and intercepts) (Fig. 10.10).

Using Monte Carlo methods for curve fitting reduces the complexity of the calculation. The curve-fitting algorithm does not need to include uncertainty as an input and does not need to produce uncertainties as outputs. If developing code

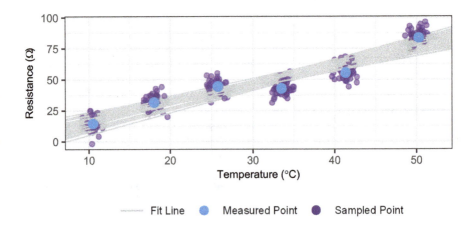

Fig. 10.10 Monte Carlo for curve fitting

from scratch, a basic least-squares fit with no uncertainty consideration can be much easier to implement. The Monte Carlo method does not require the uncertainties in data points to be normally distributed or even identically distributed. Little additional effort is required to account for uncertainty in both the x and y values of each point, a factor that makes analytical methods very complex.

An obvious disadvantage is computation time. A least-squares method on a nonlinear model often requires an iterative (slow) solution, and with Monte Carlo, the slow solution must be computed many times to converge on a representative set of fit parameters. However, depending on the number of data points, "slow" may mean computing the solution in tens of seconds rather than milliseconds, still perfectly acceptable for offline data analysis.

10.6.2 Markov-Chain Monte Carlo Approach

In certain curve fitting problems where the model is highly nonlinear or the variance is unknown, a Bayesian approach called "Markov-Chain Monte Carlo" (MCMC) may be more suitable. With this approach the data points ($x_i's$) remain fixed but the actual fitting parameters (such as slope and intercept for a fitted line) are varied during each Monte Carlo sample (Brooks et al. 2011). A slope and intercept value are randomly generated from an expected distribution and the probability that those parameters produce the best-fitting curve is calculated. New samples of slope and intercept are generated based on probabilities, generating a "random-walk" through the expected parameter space. After N trials, a distribution of slope and intercept parameters can be constructed and summary statistics for each parameter can be evaluated.

10.7 Case Study: Contact Resistance

A common way of measuring the resistance between a metal contact and semiconductor starts by making a series of identical contacts spaced at different lengths along the semiconductor. By measuring the total resistance between each pair of contacts and extrapolating back to zero length, the resistance due to only a single pair of contacts can be estimated (Fig. 10.11).

Table 10.2 lists resistance measurements made between various Titanium-Gold contacts at different lengths along a silicon nanowire. The goal is to find the intercept of the best-fit line.

As displayed in Fig. 10.12, extrapolating back to zero length should provide the combined resistance of two contacts. It is straightforward to calculate A and B for the least-squares line by hand using Eq. (10.2).

Fig. 10.11 Schematic for measuring contact resistance

Fig. 10.12 Nanowire resistance vs. length for contact resistance determination

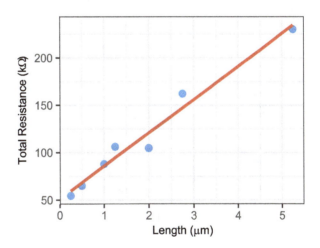

Table 10.2 Contact resistance data

Length (μm)	Resistance (kΩ)
0.25	54.6
0.50	65.1
1.00	88.1
1.25	106.2
2.00	104.9
2.75	162.3
5.25	230.3

$$\Delta = 125.0$$
$$A = 50.90 \text{ k}\Omega$$
$$B = 35.00 \text{k}\Omega/\mu\text{m}. \tag{10.25}$$

Because uncertainty in the resistance measurement is not given, σ_y can be estimated using the residuals (Eq. 10.4):

$$\sigma_y = 11.66 \text{ k}\Omega. \tag{10.26}$$

And this estimate is used to find uncertainty in the slope and intercept (Eq. 10.3):

$$\sigma_A = 6.76 \text{ k}\Omega$$
$$\sigma_B = 2.76 \text{k}\Omega/\mu\text{m}. \tag{10.27}$$

Because the total resistance with zero length includes two contacts, the contact resistance is half of A and its uncertainty is half of σ_A:

$$R_c = 25.45 \text{ k}\Omega \pm 3.38 \text{ k}\Omega. \tag{10.28}$$

The slope, B, is related to the sheet resistance of the semiconductor.

```
L <- c(.25, .5, 1, 1.25, 2, 2.75, 5.25)
R <- c(54.6, 65.1, 88.1, 106.2, 104.9, 162.3, 230.3)

# Fit linear model
fit <- lm(R~L)
summary(fit)$coefficients
## Estimate Std. Error t value Pr(>|t|)
## (Intercept) 50.91 6.755270 7.536338 6.514436e-04
## L 35.01 2.757827 12.694776 5.391758e-05
# Calculate sigma_y, sigma_b, and sigma_a
N <- length(L)
lineT <- fit$coefficients[[1]] + fit$coefficients[[2]]*L
sigy<- sqrt(sum((R - lineT)^2)/(N-2))
sigy
## [1] 11.65395
sigB <- sqrt(diag(vcov(fit)))[2]
sigB[[1]]
## [1] 2.757827
sigA <- sqrt(diag(vcov(fit)))[1]
sigA[[1]]
## [1] 6.75527
```

The measurements were made using an Agilent B1500 Semiconductor Parameter Analyzer. Resistance measurements were made by applying a voltage and measuring the current. According to the B1500 data sheet, under the ranges used for this measurement, the voltage accuracy is 0.018% + 150 μV and the current accuracy is 0.05% + 100 pA. Using the GUM method described in Chap. 6 for the measurement model $R = V/I$ with these uncertainty values, the total resistance uncertainty is 470 Ω.

This value can be used in place of the σ_y calculated above from the residuals. Now, solving for σ_A gives 270 Ω. This value is much smaller than the 6.7 kΩ calculated before. In this case, the residuals predict a much higher uncertainty than simply accounting for the published uncertainty of the measurement instrument. It is likely there are other uncertainties to consider here in addition to the 470 Ω of the measurement instrument.

Suppose a nanowire device is to be constructed with 4 μm separation between contacts. What is the expected total resistance? The solution involves predicting a value for a new measurement from the fit line; thus, the prediction band must be used. Calculating u_{pred} with $\sigma_B = 2.76$ kΩ, $S_{xy} = \sigma_y = 11.66$ kΩ, and $N = 7$:

$$u_{\text{pred}} = 13.79 \text{ k}\Omega, \tag{10.29}$$

and the total resistance of a 4 μm length of semiconductor with two contacts, to 95.4% confidence ($k = 2$), is estimated to be

$$R_c = 190 \text{ k}\Omega \pm 28 \text{ k}\Omega. \tag{10.30}$$

10.8 Drift and Predicting Future Values

There are additional issues to consider when calibrating devices that drift or decay predictably over time. The same line or curve fitting techniques discussed in this section apply, with the x-axis being time. In typical applications, the fitted curve is extrapolated into the future to predict a value and uncertainty during the time of use.

Consider a resistor on which 15 measurements have been made over the course of 15 years and in which the resistance value appears to be drifting linearly. As always, caution must be used when choosing a model. The resistor cannot have downward linear drift forever or the resistance would eventually become negative! However, in a reasonable lifespan of the device, the linear model likely is sufficient.

Using the methods previously described, a fitted line and its prediction band can be computed for the drifting resistance (Fig. 10.13).

10.8.1 Uncertainty During Use

The resistance value and uncertainty for the latest measurement (Year 14) are presently known, but the resistor will be used over the course of one more year

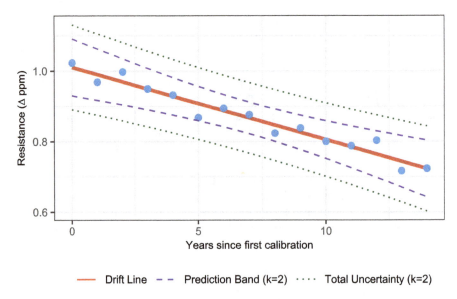

Fig. 10.13 Drift with time

before its next calibration. There are three common methods for generating an uncertainty expression that applies to the entire calibration interval.

Method 1: Function of Time The most accurate method of recording the predicted value and uncertainty is to specify its value and uncertainty as functions of time, valid over the upcoming interval, by applying a coverage factor k to the prediction band:

$$R(t) = A + Bt$$

$$U_R(t) = k \cdot u_{\text{pred}}(t) = k \sqrt{\sigma_y^2 \left[1 + \frac{1}{N} + \left(\frac{\sigma_B}{\sigma_y}\right)^2 (t - \bar{t})^2\right]}, \quad (10.31)$$

where t is time elapsed (in years) since the first calibration. It may make sense to convert the time into days since the last calibration, so the user does not need to work with fractional years or deal with the offset between the present time and the first ever calibration. Alternatively, time t could be represented as an absolute ordinal date, such as the number of seconds since the epoch of January 1, 1970 (a common time reference in many computer systems), in which case the reference point of the first calibration time is not required. Of course, when working in an absolute time representation, the intercept A may not be as useful since it gives the predicted value at the time reference, not at the time of first calibration.

Method 2: Conservative Worst-Case Estimate In some cases, calculating the value and uncertainty at each time of use is impractical. Another option is to report a single value and uncertainty that will cover the entire interval of use. This results in

10.8 Drift and Predicting Future Values

a more conservative uncertainty estimate, although the number is valid at any point during the interval. This uncertainty is the prediction band at the time of the next calibration, with the magnitude of the expected drift over the interval added on assuming linear drift continues.

$$R = R(t_{\text{now}})$$

$$U_R = k\sqrt{\sigma_y^2 + \sigma_y^2\left[\frac{1}{N} + \left(\frac{\sigma_B}{\sigma_y}\right)^2 (t_{\text{next}} - t)^2\right]} + \left|\text{rate} \times \left(t_{\text{next}} - t_{\text{now}}\right)\right|, \quad (10.32)$$

with t_{now} the current calibration time and t_{next} the time when the next calibration is due. While this method is simple to calculate, the drift term does not have a coverage factor applied. With k calculated for 95% coverage, the actual out of tolerance (OOT) rate is much less than 5% for typical measurements.

Method 3: Mean Values As a compromise between reporting worst-case uncertainty and using a full function of time, mean values can be computed to cover the calibration interval. Section F.2.4.5 of the GUM describes use of calibration curves and how to report uncertainty when the curve is not applied. The following analysis could be adapted to reduce the predicted uncertainty of any curve fitting problem to a single value valid over a range (for example, a resistor uncertainty over a large temperature range), but this section considers the specific case where the x value is time and the drift is linear.

In the GUM terminology, the correction factor to apply is simply a function of the expected drift, where r is the drift rate:

$$b(t) = r \times t, \quad (10.33)$$

with an average correction over the interval from t_1 to t_2 of

$$\bar{b} = \frac{1}{t_2 - t_1} \int_{t_1}^{t_2} b(t)\, dt = \frac{r(t_1 + t_2)}{2}. \quad (10.34)$$

The uncertainty used over the entire calibration interval is the combination of three components. First, uncertainty in the average correction $\bar{b}$:

$$\sigma_b^2 = \frac{1}{t_2 - t_1} \int_{t_1}^{t_2} (b(t) - \bar{b})^2\, dt = \frac{r^2}{12}(t_1^2 - 2t_1 t_2 + t_2^2), \quad (10.35)$$

the mean uncertainty in the fitted line (confidence band; see Sect. 10.3.2) over the interval:

$$\sigma_c^2 = \frac{1}{t_2 - t_1} \int_{t_1}^{t_2} u_{\text{conf}}^2(t) \, dt \tag{10.36}$$

and mean uncertainty in $y(t)$ over the interval:

$$\sigma_{\hat{y}}^2 = \frac{1}{t_2 - t_1} \int_{t_1}^{t_2} \sigma_{\hat{y}}^2(t) \, dt, \tag{10.37}$$

leading to a total reported uncertainty, valid for the full calibration interval, of

$$u_R = \sqrt{\sigma_{\hat{b}}^2 + \sigma_c^2 + \sigma_{\hat{y}}^2}. \tag{10.38}$$

Using this method, the resistance value is reported at the average time during the interval:

$$R = R(t_{\text{avg}}). \tag{10.39}$$

Often the rate has units of drift per year, and the interval is 1 year ($t_1 = 0$, $t_2 = 1$), which simplifies this expression further. (For example, $\sigma_{\hat{y}}^2 = \sigma_y^2$ and $\sigma_{\hat{b}}^2 = r^2/12$).

This method produces a result not as accurate as reporting the values as functions of time (Method 1), but not as conservative as using worst-case estimates (Method 2). It will produce an average OOT rate of 5% when u_R is expanded to 95% coverage under normal rate and uncertainty conditions (Delker et al. 2020).

When using any of the above methods to account for drift, the method and assumptions must be explicitly stated.

10.8.2 Validating Drift Uncertainty

Every time a new calibration is made, the fitted line and uncertainty bands must be recomputed to include the new data point. The model itself should be evaluated after each new measurement to ensure a linear fit is still adequate. After many calibration cycles, the prediction band can become quite large due to the $(t - \bar{t})$ term increasing.

Exercise caution when applying linear drift to devices that may not be predictably drifting. Data that may look like drift could also be random fluctuations due to other uncertainties. Use the slope test to validate whether the drift rate (slope) is significantly different than zero. Attempting to fit a drift line is especially problematic when a device has only been calibrated a few times. Generally, there is much larger uncertainty in the slope of a line going through 3 data points than a line going through 50, due to less spread in the t values and smaller degrees of freedom for estimating the expanded uncertainty. As a general rule, a minimum of 5 measurements are necessary for the slope test to reliably indicate whether the device is drifting.

10.8 Drift and Predicting Future Values

Consider the measurements in Fig. 10.14. At first glance, they appear to be drifting downward with time. However, the slope with 95% confidence is $b = -0.46 \pm 0.89$, or $-1.35 < b < 0.43$, a range that includes zero. This means that we cannot conclude that the device has a slope significantly different than zero. This can be illustrated graphically by finding the slope of all possible lines that fit within the 95% confidence band. If it is possible to draw lines with both positive and negative slopes, the test fails.

Case Study: Thomas 1-Ohm Resistor with Drift

This case study presents a complete uncertainty analysis on a Thomas-type 1-Ohm resistor, accounting for Type A and Type B uncertainties as well as an uncertainty due to drift over time.

Resistance standards are calibrated periodically using a resistance bridge. Historical data on a particular set of resistors shows a slow linear drift over time, but the amount and even direction of drift vary from device to device. While an ISO 17025 accredited calibration does not account for change over time, some customers require a "certification uncertainty" that includes any predicted drift or other affects to ensure the resistor stays within its uncertainty throughout the calibration interval.

The three major components of a standard resistor uncertainty budget are the Type B uncertainty, the Type A measurement uncertainty, and the uncertainty due to predicted drift over the next calibration interval.

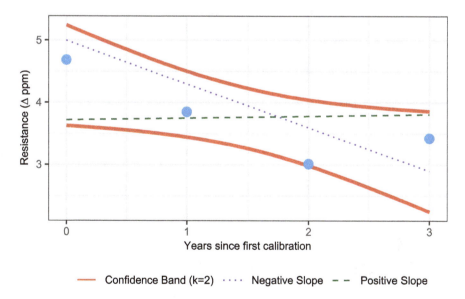

Fig. 10.14 Failure in slope test

10.8.2.1 Type B Uncertainty

The Type B uncertainty comes from the resistance bridge used for the measurement, a 1-ohm reference standard calibrated by NIST, and fluctuation in environmental conditions. Other uncertainties, such as uncertainty due to leakage current, were deemed insignificant compared to the other components and were omitted from this analysis. The Type B uncertainty components are listed in Table 10.3.

All components in this table have been normalized and converted to $k = 1$ standard uncertainties. For example, the bridge uncertainties as reported by the instrument specifications were interpreted as uniform distributions and were standardized by dividing by the square root of 3.

10.8.2.2 Type A Measurement Uncertainty

After waiting 30 days for a resistor under test to stabilize in the oil bath under the laboratory environmental conditions, multiple repeated measurements were made over the course of 5 days spread over a 2-week period. Each day's measurement consisted of 40 individual measurements averaged together.

The combined standard deviation is found by pooling the standard deviations of the 5 times 40 individual measurements, grouped by day:

Table 10.3 Type B uncertainty components

Uncertainty	Value $(k = 1)$ μΩ/Ω	Description
Reference standard resistor	0.01	Time-of-test uncertainty as calibrated by NIST
Reference standard drift allowance	0.006	Determined using same drift allowance method as described below, but on the reference standard
Resistance bridge uncertainty	0.013	1 Ω specifications for MI-6010C bridge
Resistance bridge linearity	0.0029	1 Ω specifications for MI-6010C bridge
Resistance bridge resolution	0.00029	1 Ω specifications for MI-6010C bridge
Temperature fluctuation	0.0021	Includes oil bath stability and temperature coefficient of standards
Pressure fluctuation	0.0017	From pressure coefficients of standards and typical atmospheric changes during test
Combined type B uncertainty	0.0179	RSS of above components ($k = 1$)

10.8 Drift and Predicting Future Values

$$u_A = \sqrt{\frac{\sum_{i=1}^{m}(n_i-1)s_i^2}{\sum_{i=1}^{m}(n_i-1)}} \qquad (10.40)$$

where n_i is the number of measurements made on day i, and s_i the standard deviation of the m measurements on that day. Degrees of freedom for the Type A uncertainty is $\sum_{i=1}^{m}(n_i-1)$.

For the data shown in Table 10.4, the Type A uncertainty ($k=1$) is 0.00862 $\mu\Omega/\Omega$ with 195 degrees of freedom.

10.8.2.3 Drift Uncertainty

When determining an uncertainty component to account for drift over the next calibration interval, the methodology described in Sect. 10.7 can be used. Historical data on the resistor since 1990 is plotted and a linear regression is performed (Fig. 10.15). For ease of calculation in this case study, the time-of-test uncertainty was assumed to be a constant $\sigma_y = 0.020$, (RSS of the Type A and Type B components found above) for all years. Complete analysis would include a varying uncertainty for each year to account for equipment changes, etc. and would require a weighted linear regression.

The regression results in a drift rate of 0.031 $\mu\Omega/\Omega$/year, with uncertainty $\sigma_B = 0.0011$ $\mu\Omega/\Omega$/year. With $N = 19$ historical data points and $\bar{t} = 732168$ (the average of all historical calibration dates), substituting the values into the prediction band formula (see Sect. 10.3.3), and simplifying gives the $k = 1$ uncertainty at any point:

$$u_{\text{pred}}(t) = \sigma_y \sqrt{1 + \frac{1}{N} + (t-\bar{t})^2 \left(\frac{\sigma_B}{\sigma_y}\right)^2}$$

Table 10.4 Type A uncertainty measurements

Day	Average deviation from nominal ($\mu\Omega/\Omega$)	Standard deviation	Number of measurements
1	0.197	0.0075	40
2	0.217	0.0062	40
3	0.198	0.0095	40
4	0.199	0.0101	40
5	0.196	0.0092	40

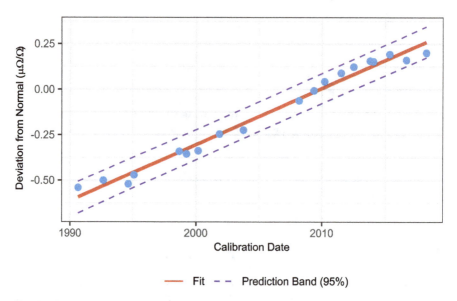

Fig. 10.15 Linear Regression of Thomas 1-Ohm resistor drift

$$= 0.03647\sqrt{6.379 \times 10^{-9}(t - 732168)^2 + 1.053} \ \mu\Omega/\Omega, \quad (10.41)$$

where t is the Gregorian ordinal date (number of days since January 1 of the year 1, another commonly used date format in software). Since the customer will not want to deal with this expression on every use of the resistor, the methodology in Sect. 10.8.1 (Method 3) is used to reduce the prediction band uncertainty to a single value over the next calibration cycle. Following this method, with $r=0.0000847$ (drift per day), $t_1 = 736782$ (calibration date 29-March-2018), and $t_2 = 737147$ (expiration date 29-March-2019), the correction factor (fit drift line), $b(t)$, and the average correction, $\bar{b}$, over the interval are

$$b(t) = rt,$$

$$\bar{b} = \frac{1}{t_2 - t_1} \int_{t_1}^{t_2} b(t)dt = \frac{r(t_1 + t_2)}{2} = 0.01542. \quad (10.42)$$

The variance in the average correction is

$$\sigma_{\bar{b}}^2 = \frac{1}{t_2 - t_1} \int_{t_1}^{t_2} (b(t) - \bar{b})^2 dt = \frac{r^2}{12}(t_1^2 - 2t_1 t_2 + t_2^2) = 7.960 \times 10^{-5} \quad (10.43)$$

and the variance in the fit line is

10.8 Drift and Predicting Future Values

$$\sigma_c^2 = \frac{1}{t_2 - t_1} \int_{t_1}^{t_2} u_{conf}^2(t) dt = \sigma_B^2 \left(\frac{t_1^2}{3} + \frac{t_1 t_2}{3} - t_1 \bar{t} + \frac{t_2^2}{3} - t_2 \bar{t} + \bar{t}^2 \right) = 0.000195. \quad (10.44)$$

The drift uncertainty, u_R, is then given as the RSS combination of the two components:

$$u_R = \sqrt{\sigma_b^2 + \sigma_c^2} = 0.0141 \, \mu\Omega/\Omega. \quad (10.45)$$

The time-of-test uncertainty $\sigma_{\bar{y}}$, as listed in Sect. 10.8.1, is omitted from the drift uncertainty u_R and included in the next section. The degrees of freedom for a fitted line are $N - 2$ where N is the number of measurement values ($y_i's$), and 2 is subtracted to account for the 2 parameters that were fit (slope and intercept). In this case, degrees of freedom are 17.

10.8.2.4 Expanded Uncertainty

The total combined uncertainty is the RSS combination of Type A and Type B time-of-test uncertainties and the drift uncertainty. The effective degrees of freedom are calculated via the W-S formula and come out large enough in this case to use $k = 2.0$ for a 95.45% confidence interval (Table 10.5).

Therefore, the reported resistance deviation from nominal is the mean of the 5 days of measurements, and the total expanded uncertainty is

$$0.201 \, \mu\Omega/\Omega \pm 0.048 \, \mu\Omega/\Omega (95.45\%, k = 2.0). \quad (10.46)$$

However, at the time of the next calibration, the resistor will be evaluated for OOT conditions based on the prediction band formula calculated on the expiration date, not the single interval uncertainty value of 0.048.

Table 10.5 Combined uncertainty components

Component	$k = 1$ Uncertainty ($\mu\Omega/\Omega$)	Degrees of freedom
Type B	0.0179	∞
Type A	0.00862	195
Drift allowance	0.0141	17
Combined (RSS)	0.024	2506

10.9 Calibration Interval Analysis

Interval analysis, determining the optimal length of a calibration interval, is another application of curve fitting in metrology. Many calibrations expire after 1 year. While convenient from a calendar perspective, this may not be the optimal interval from a risk perspective. If the calibration interval is set too long, a device is more likely to drift out of tolerance before the next calibration. If too short, there will be increased costs due to extra calibrations and equipment downtime.

Typically, interval adjustment is performed on devices that are tolerance tested to a specification limit such as a manufacturer's specification. The goal of interval analysis is to find the longest interval that results in the device remaining within its limits to some given probability during the interval. A good interval analysis requires a sufficient amount of historical data. Frequently, data from an individual device will not be sufficient, but when the data from all devices of the same model are combined, better results can be obtained.

Depending on what type of historical records are available for class of devices, there are three common methods for determining an interval. Sometimes, only the pass/fail status is recorded in a calibration database without keeping any information on exactly where a device measured within or outside the limits. In this case, if all the calibration intervals in the historical record were approximately the same length, the Interval Test method can be used. It is commonly referred to as "Method A3" due to its designation in NCSLI Recommended Practice 1 (RP-1) (NCSLI 2010). When the historical data recorded only pass/fail status but the intervals were of varying length, the Binomial Method (often referred to as "Method S2" from RP-1) can be used for better results. The Variables Method can be used if historical data contains the actual value measured in addition to pass/fail status, providing the most accurate results (Castrup 2005). The Variables Method and Method S2 both involve fitting a curve to the historical data. Methods A3 and S2 are fully described in RP-1, so the remainder of this section will focus on the Variables Method.

The first step in performing an interval analysis using the Variables Method is to convert the historical data from time of calibration (t) and measured value (y) into deltas Δt and Δy, the differences between subsequent calibrations. Note that if any adjustments are made during the calibration, Δy values should be calculated as the difference between as-found value of one calibration and as-left value of the previous calibration.

Once the historical data is in (Δt, Δy) pairs, fit a curve to the points. In theory, the curve could be any model, such as polynomial or exponential, but it is typically a straight line with intercept forced to zero. This assumes the device is adjusted back to the center of the specification limits after each calibration. Determining an optimal interval then becomes a matter of calculating the prediction band at the desired confidence level and solving for the time at which the prediction band exceeds the specification limits.

10.9 Calibration Interval Analysis

Consider a voltage source which has been calibrated 15 times at varying intervals. Fourteen pairs of Δt, Δy values were calculated from the calibrations as shown in Table 10.6.

Using the methods in Sect. 10.5.1 for the fitted model, $y = bx$, the slope b was found to be 0.000278 V/day with uncertainty $u_b = 8.811 \times 10^{-6}$ V/day. From the residuals, σ_y is computed to be

$$\sigma_y = \sqrt{\sum_{i=1}^{N} \frac{(\Delta y_i - b\Delta t)^2}{N-1}} = 0.0137. \tag{10.47}$$

Note the denominator is $N - 1$ due to having only 1 free parameter (slope) rather than 2 free parameters (slope and intercept) in the typical linear regression which divides by $N - 2$. The prediction band can be written using Eqs. (10.20) and (10.21), as a function of time. Because there is only one estimated parameter, the prediction band simplifies to

$$u_{\text{pred}}(t) = \sqrt{t^2 u_b^2 + \sigma_y^2}. \tag{10.48}$$

Multiplying u_{pred} by $k = 2$, for approximately 95% confidence, and adding to the expected y value, the prediction band can be plotted as a function of time, shown in Fig. 10.16. With an upper specification limit of 0.2 V above nominal, the intersection of the prediction band with 0.2 determines the maximum calibration interval, in this case 614 days. A significant cost savings may be realized by using this interval rather than a standard 365 days that is often blindly applied.

Table 10.6 Historical calibration data on a voltage source

Days since last calibration	Deviation from prior calibration (V)
117	0.0480
203	0.0760
213	0.0756
260	0.0517
272	0.0730
289	0.0681
295	0.0591
309	0.0818
467	0.1292
470	0.1207
554	0.1698
591	0.1595
638	0.1767
650	0.1897

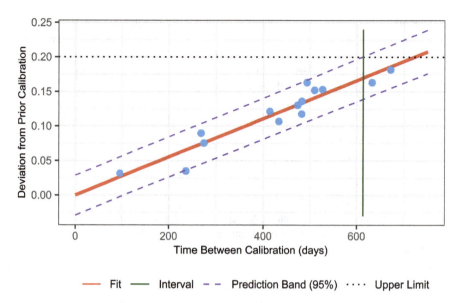

Fig. 10.16 Interval analysis for a voltage source

10.10 Summary

Fitting a curve to data is a common exercise in metrology, yet a proper evaluation of uncertainty in the fitted curve is often neglected. This chapter has provided general guidance on how uncertainties can be determined in both the fitted parameters and in the curve as a whole. The benefits of a careful assessment of fitted curve uncertainties can be great. For example, uncertainties involving equipment with temperature dependencies can be significantly reduced when the temperature dependency is fit to a curve. The temperature can then be measured during each use of the equipment, and the uncertainty at that temperature can be used rather than a worst-case uncertainty. Using curve fitting techniques, in combination with the risk analysis of Chap. 5, also provides a means of predicting the probability of future OOT conditions.

It should always be verified that the model being fit to the data (linear, polynomial, exponential, etc.) makes physical sense and represents the data without overfitting. Sample sizes should also be chosen to ensure an adequate number of degrees of freedom are available to estimate overall uncertainty. Additional points along the fitted curve not only increase total degrees of freedom, but they reduce the uncertainties in fitted parameters. For more recommendations on degrees of freedom and sample size, see Sect. 11.5.

10.11 Related Reading

While the GUM says very little about curve fitting, the BIPM has announced a document in preparation titled "Applications of the least-squares method" which, once released, promises to be a definitive resource on curve fitting as applied to metrology. ISO/TS 28037:2010 provides guidance on use of straight-line fits for calibration curves, and ISO/TS 28038:2018 extends that guidance to polynomial calibration curves. NCSLI's Recommended Practice 1 (RP-1) applies regression analysis to the determination of suitable calibration intervals for managing risk.

Most nonlinear curve fitting problems apply a numerical minimization technique to find the best fit parameters. The Gauss–Newton method is described in Hartley (1961), and the Levenberg–Marquardt algorithm described by Moré (1978). Both are implemented in modern computer code by Press et al. (2002).

Concerns over selecting the correct model (polynomial vs. exponential fit, appropriate order of polynomial, etc.) to prevent under- or overfitting, and validating the adequacy of the model itself are discussed by Harrell (2015). Izenman (2013) provides an in-depth discussion of how to extend the concepts introduced in this chapter to the multivariate case—when there are more dimensions than just x and y in the regression problem. Details of the Orthogonal Distance Regression algorithm are given in Boggs and Rogers (1990). Carroll and Ruppert (1996) cautions the user on use of ODR when the measurement equation introduces its own error. An analysis of least-squares fitting applied to coordinate measuring machines (CMMs) is given by Srinivasan (2012), which considers fitting measured CMM points to lines, planes, and other curved surfaces to evaluate the dimensions of the object under test. A treatment of the often-ignored effects of autocorrelation (correlation in time between uncertainties in all x values, for example) is provided in King and Giles (2018).

Regression problems can also be approached from a Bayesian perspective, accounting for a priori information. Klauenberg (2015) gives a tutorial on a Bayesian approach to regression, and Elster and Wübbeler (2016) provides an example and comparison to traditional least-squares regression. Markov-Chain Monte Carlo also applies a Bayesian analysis. For the statistics behind the algorithm as applied to regression problems, see Elster and Toman (2011). For an engineering application of MCMC to fitting X-ray diffraction spectra, see Iamsasri et al. (2017).

10.12 Exercises

1. A pressure gauge has been calibrated five times using a known 180 psi reference with results shown in Table 10.7.
   (a) Assuming the gauge is drifting in a straight line, find the best fit slope and intercept using Eq. (10.2).
   (b) Use Eq. (10.3) to determine the uncertainty in this slope and intercept assuming each measurement had a total ($k = 1$) uncertainty of 0.2 psi.

Table 10.7 Pressure gauge calibration data

Days since first calibration	Indicated pressure (psi)
0	179.9
701	181.4
1364	182.2
1721	183.1
2378	184.1

Table 10.8 Power meter resistance versus temperature

Temperature °C	Power meter resistance Ω
16.2	200.0416
17.3	200.0433
18.3	200.0454
19.3	200.0478
20.4	200.0491
21.4	200.0513

(c) The next calibration is due on day 2743 (1 year after the last calibration). Calculate the predicted pressure reading on this date.

2. An electrical power meter has an internal resistance that changes with temperature. To increase the accuracy of calibrations made by the meter, the resistance was measured over a range of temperatures as shown in Table 10.8. This data can then be used to predict the actual resistance when the meter is used at any lab temperature.

   (a) Assuming the data follows a straight line model, use a line fit function in your favorite statistical software to find the slope and intercept for the best fit line. Write an expression to predict the resistance at any temperature, and plot the data along with the fit line.
   (b) Compute the uncertainty in slope and intercept of the fit line and find σ_y (estimated from residuals). Then write an expression to calculate the predicted uncertainty in resistance at any temperature.
   (c) Plot the residuals of the power meter resistance curve fit. Does the linear fit appear to model the data well? Attempt to fit a quadratic model ($y = a + bx + cx^2$) to the data and replot the residuals. Compute the r^2 value for each fit. Which model is a better fit to the data?

3. Radiation detectors are calibrated by placing the detector at fixed distances from a known radiation source and reading the output signal. When in use, the reading and uncertainty of the detector must be adjusted for the distance if the actual distance falls between calibration points. The radiation intensity is known to follow a power law equation of the form $y = ax^b$, where x is the distance between detector and source, y is the detector reading, and a and b are fit parameters determined during the calibration. Given the data in Table 10.9, find the coefficients a and b and their uncertainties by fitting the power law equation to the data using a nonlinear curve fitting function. Plot the measured data and the fit curve.

10.12 Exercises

Table 10.9 Radiation detector calibration data

Distance (mm)	Reading (R/hr)
281.0	882.5
300.0	774.3
320.0	678.0
340.0	600.1
360.0	535.3

Note 1: Many nonlinear curve fitting solvers require a good starting guess for the coefficients before converging on the best solution. A reasonable guess for this problem is $a = 7 \times 10^7$ and $b = -2.0$.

Note 2: Frequently the uncertainties in fit parameters are returned in the form of a covariance matrix. The square root of the diagonal of the covariance matrix provides uncertainties of the individual coefficients.

4. Now account for calibration uncertainties in the radiation detector of exercise 3. Assume the reading values were calibrated with ($k = 1$) uncertainty of 5.0 R/hr., and the distances were placed with ($k = 1$) uncertainty of 0.5 mm. This is a nonlinear curve fit with uncertainty in both x and y, which cannot be solved by many typical curve fitting algorithms. It could be done with ODR, but instead, take a Monte Carlo approach. For each of the 5 distances, generate a random sample from the normal distribution with mean of the distance and standard deviation 0.5. For each of the 5 readings, generate a random sample from the normal distribution with mean of the reading and standard deviation 5.0. Then find a and b fit to this set of points, assuming they have no uncertainty. Repeat the process at least 5000 times to collect a distribution of a and b values. Plot a histogram of the a and b samples, and report the mean and standard uncertainty of a and b.

5. One method of measuring the oxide thickness t_{ox} of a metal-oxide-semiconductor structure used in transistor gates is to measure the gate capacitance C per unit area as a function of gate voltage $V_g - V_{fb}$. (Schroder 2006). A plot of $1/C$ versus $1/(V_g - V_{fb})$ is approximately a straight line with an intercept of $1/C_{ox}$. Then, the oxide thickness can be calculated using $t_{ox} = \epsilon_{ox}/C_{ox}$ where ϵ_{ox} is the permittivity of the oxide material.

   (a) For the measured $1/C$ and $1/(V_g - V_{fb})$ values in Table 10.10, find $1/C_{ox}$ and $u(1/C_{ox})$ by finding the intercept of the fit line.
   (b) Use a GUM uncertainty propagation to find the oxide thickness and uncertainty in nanometers assuming $\epsilon_{ox} = 3.7\epsilon_0 \cong 3.276 \times 10^{-13}$ F/cm with uncertainty $u(\epsilon_{ox}) = 0.2\epsilon_0 \cong 1.771 \times 10^{-14}$ F/m.

6. Consider the x, y data in Table 10.11. Fit a straight line to the 5 points and calculate coefficient of determination r^2 and the standard error of the estimate S. Repeat the fit with a second-order (quadratic) polynomial, and continue up to a fourth order polynomial. Based only on r^2 and S, what can you conclude about the best polynomial order to use to fit this data? Without context, it can be difficult to answer such a question. Suppose the x values are temperature

Table 10.10 Oxide thickness measurement data

$1/C$ (cm²/μF)	$1/(V_g - V_{fb})$
2.212	0.667
2.215	0.725
2.223	0.794
2.221	0.877
2.233	0.980
2.248	1.111
2.261	1.282
2.288	1.515
2.312	1.852
2.330	2.381
2.408	3.333

Table 10.11 Example x, y data for polynomial order determination

x	y
10	79.0
12	91.5
14	127.9
16	210.8
18	260.0

Table 10.12 Leak rate standard calibration data

Days since first calibration	Leak rate (cc/s @ STP)	Leak rate uncertainty (cc/s @ STP)
0	6.64×10^{-8}	7.4×10^{-10}
380	6.53×10^{-8}	7.4×10^{-10}
794	6.35×10^{-8}	6.8×10^{-10}
1247	6.26×10^{-8}	3.6×10^{-10}
1673	6.21×10^{-8}	4.0×10^{-10}
2031	5.97×10^{-8}	4.0×10^{-10}
2747	5.91×10^{-8}	3.9×10^{-10}

and y values are resistance, then plot each fit curve in the range from 0 to 30. Does assuming these data points represent resistance versus temperature and looking at the fit projection outside the range of measured data change your best-order decision?

7. Standards used for calibrating gas leak rate detectors are measured in cm³/sec at standard temperature and pressure (STP). Leak rate standards are known to decay exponentially over time, in the form $y = a \exp(-bt)$, where a and b are fitting parameters. A particular leak rate standard has been calibrated over several years, with data shown in Table 10.12. Because each calibration incorporated Type A and Type B uncertainties, the total uncertainty at each calibration varied. Accounting for the uncertainties in each measurement, find the fit parameters a and b and their uncertainties $u(a)$ and $u(b)$. (A reasonable initial guess is $a = 6 \times 10^{-8}$ and $b = 4 \times 10^{-5}$).

References

Boggs, P.T., Rogers, J.E.: Orthogonal Distance Regression. National Institute of Standards and Technology, Applied Computational Mathematics Division, Gaithersburg, MD: U.S. Department of Commerce (1990)

Brooks, S., Gelman, A., Jones, G.L., Meng, X.: Handbook of Markov Chain Monte Carlo. Chapman & Hall/CRC, Boca Raton (2011)

Carroll, R.J., Ruppert, D.: The use and misuse of orthogonal regression in linear errors-in-variables models. Am. Stat. **50**(1), 1–6 (1996)

Castrup, H.: Calibration Intervals from variables data. NCSLI Workshop and Symposium, Washington DC (2005)

Cox, C., Ma, G.: Asymptotic confidence bands for generalized nonlinear regression models. Biometrics. **51**(1), 142–150 (1995)

Delker, C.J., Auden, E., Solomon, O.: Calculating interval uncertainties for calibration standards that drift with time. NCSLI Measure [accepted] (2020)

Elster, C., Wübbeler, G.: Bayesian regression versus application of least squares – an example. Metrologia. **53**, S10–S16 (2016)

Elster, C., Toman, B.: Bayesian uncertainty analysis for a regression model versus application of GUM supplement 1 to the least-squares estimate. Metrologia. **48**, 233–240 (2011)

Fuller, W.A.: Measurement Error Models. Wiley, New York (1987)

Gavin, H.P.: The Levenberg-Marquardt Method for Nonlinear Least Squares Curve-Fitting Problems. Duke University, Department of Civil and Environmental Engineering (2017)

Glantz, S., Slinker, B.: Primer of Applied Regression and Analysis of Variance, 2nd edn. McGraw-Hill, New York (2001)

Harrell, F.E.: Regression Modeling Strategies, 2nd edn. Springer, New York (2015)

Hartley, H.O.: The modified Gauss-Newton method for the fitting of non-linear regression functions by least squares. Technometrics. **3**, 269–280 (1961)

Iamsasri, T., et al.: A Bayesian approach to modeling diffraction profiles and application to ferroelectric materials. J. Appl. Crystallogr. **50**, 211–220 (2017)

Izenman, A.J.: Modern Multivariate Statistical Techniques – Regression, Classification, and Manifold Learning. Springer, New York (2013)

King, M.L., Giles, D.E.: Specification Analysis in the Linear Model. Routledge, New York (2018)

Klauenberg, K., Wübbeler, G., Mickan, B., Harris, P., Elster, C.: A tutorial on Bayesian normal linear regression. Metrologia. **52**, 878–892 (2015)

Kutner, M.H., Nachtsheim, C.J., Neter, J., Li, W.: Applied Linear Statistical Models, 5th edn. McGraw-Hill, New York (2005)

Moré, J.: The Levenberg-Marquardt Algorithm. Numerical Analysis: Lecture Notes in Mathematics, vol 630. Springer, Berlin (1978)

National Conference of Calibration and Standards Laboratories International: Recommended Practice 1 - Establishment and Adjustment of Calibration Intervals. (2010)

Press, W.H., Teukolsky, S.A., Vetterling, W.T., Flannery, B.P.: Numerical Recipes in C - The Art of Scientific Computing, 2nd edn. Cambridge University Press, New York (2002)

Robert, C.P., Casella, G.: Monte Carlo Statistical Methods, 2nd edn. Springer, New York (2004)

Schroder, D.K.: Semiconductor Material and Device Characterization, 3rd edn. Wiley, Hoboken (2006)

Srinivasan, V., Sharkarji, C.M., Morse, E.P.: On the enduring appeal of least-squares fitting in computational coordinate metrology. J. Comput. Inform. Sci. Eng. 12(1), 1–15 (2012)

Taylor, J.R.: An Introduction to Error Analysis: the Study of Uncertainties in Physical Measurements, 2nd edn. University Science Books, Sausalito, CA (1997)

Wehr, R., Saleska, S.R.: The long-solved problem of the best-fit straight line: application to isotopic mixing lines. Biogeosciences. **14**, 17–29 (2017)

York, D., et al.: Unified equations for the slope, intercept, and standard errors of the best straight line. Am. J. Phys. **72**(3), 367–375 (2004)

Chapter 11
Special Topics in Metrology

Metrology guides such as the JCGM 100 and JCGM 101 present in detail the most commonly used methods for evaluating measurement uncertainty. Additional guides for less commonly used methods are in development. In this chapter we present an overview of methods that we find valuable and deserving of more coverage. These methods include statistical process control (SPC), binary measurement systems (BMS), destructive measurements, sample size determination, and Bayesian analysis in metrology.

11.1 Introduction

This chapter discusses several special topics in metrology that have not received a great deal of emphasis in the metrology literature, but are considered important enough to cover here. The first special topic is statistical process control (SPC), in which the *process* of making a measurement is monitored using the tools of SPC. The second special topic is the evaluation of Binary Measurement Systems (BMS), in which the outcome of a measurement, such as with a go/no-go gauge, is binary. The third special topic is quantifying uncertainty when the measurement is destructive. The fourth special topic is sample size determination and allocation in metrology studies, and the final topic is Bayesian analysis in metrology.

11.2 Statistical Process Control (SPC)

Statistical process control (SPC) has traditionally been used in manufacturing to monitor and improve production processes. The goal of SPC is to quickly detect when a process upset has occurred and to eliminate the "assignable" cause. The same goal applies to the process of making measurements. Any significant departure of a measurement process from nominal performance should be quickly detected and mitigated. In this section we apply the tools of SPC to this problem.

The many potential applications of SPC for monitoring measurement processes include monitoring testers involved in manufacturing, from product design to production to product acceptance testing. Whenever measurement data can be collected and ordered over time, the performance of the tester can be monitored via an SPC control chart. The tools of SPC are ideally suited to this task. A brief overview of SPC is given here, followed by a case study involving the SPC of a battery tester. See Montgomery (2013) for a detailed treatment of statistical process control.

SPC in metrology recognizes that measurement process variability is unavoidable. Variation can be separated into two categories: variation due to natural causes and variation due to unnatural (assignable) causes. Natural variation is the inherent variability of the process that cannot be eliminated, but it may be reducible. Examples in metrology include the inherent variation in test equipment, cables, connectors, and in the configuration of a test setup. Unnatural variation is identifiable, assignable, and is the by-product of causes that can be found and eliminated. Examples in metrology include operator error, tester bias, tester drift, and unusual environmental conditions that affect the measurement. To identify variability and remove its causes, both subject matter expertise and statistical expertise are required.

SPC tools useful for identifying and eliminating unnatural variation include boxplots, histograms, time-ordered plots, control charts, and DOEx (Chap. 9). The SPC tool most often used for this purpose is the control chart. Standard control charts are based on 3-sigma control limits, for which normal distribution theory implies a roughly 3 in 1000 chance of a measurement falling outside these limits when the process is at nominal. Control charts are used to detect departures from nominal, either in past data (retrospective analysis) or in present data (ongoing analysis) with previously established control limits.

The $\overline{X} - R$ chart (Average-Range) and the $X - MR$ chart (Individuals-Moving Range) are two of the most popular control charts that can be used to monitor measurement processes. The $\overline{X} - R$ chart usually is constructed from a minimum of 25–30 subgroups of 3–5 observations each. Two separate charts are plotted, the $\overline{X}$ chart and the R chart. The $\overline{X}$ chart is used to detect changes in the process mean and the R chart to detect changes in the process standard deviation.

When it is not possible to collect data in subgroups, the $X - MR$ chart can be used to monitor the measurement process. The moving range is based on the differences between successive observations. If the ordered observations are represented as $x_1, x_2, \ldots, x_n$, the ith moving range is defined as (see Montgomery 2013)

$$MR_i = |x_{i+1} - x_i|, i = 1, 2, \ldots, (n-1).$$

The $X - MR$ chart should be initially constructed from a minimum of 50 individual measurements when possible. Two charts are usually plotted, one for the individual values (X) and one for the moving ranges (MR). The individuals chart detects changes in the process mean and the moving range chart detects changes in

11.2 Statistical Process Control (SPC)

the process standard deviation. Additional details for both of these charts are presented in Crowder (1987a) and Montgomery (2013).

Common control chart patterns (departures from nominal) can be detected with these two charts. The most common patterns include gradual changes in mean level, single unusual values, instability, sudden shifts in mean level, sustained trends up or down, and cyclical behavior. Each of these patterns may point to specific causes associated with the measurement process. It is in identifying causes of unusual behavior that subject matter expertise and statistical expertise come together.

In the case study in Sect. 11.2.1, an X – MR chart was used to monitor the performance of a battery tester used for product acceptance.

11.2.1 Case Study: Battery Tester Uncertainty and Monitoring Via SPC

In manufacturing applications, a check standard (Sect. 3.6.2) is often used to monitor tester performance over time. The check standard should be constructed as close as possible to the manufactured part being measured, and it should be a stable device that can replicate measurements over time. Measurements of the check standard are made on a frequent basis and can be plotted on an SPC control chart to verify ongoing tester performance. Any assignable causes of variation in the measurement process can then be detected via the control chart and eliminated to the extent possible.

The tester under investigation in this case study was a battery product acceptance tester that measured voltage output from four different channels of a battery. Because the testing of manufactured batteries is destructive, a check standard was used to simulate nominal voltages from the four channels. These voltages were measured by the tester at the rate of 10 readings per second for 15 s. The data retrieval was performed 50 times to establish a baseline for each of the four channels, resulting in 7500 ($10 \times 15 \times 50$) total measurements for each channel.

An $\overline{X}$ – R chart was first considered for monitoring tester performance. Prior to implementing this control chart for ongoing control, an uncertainty analysis of the tester was performed to establish baseline control limits. The initial control limits were based on the combined standard uncertainty of the battery tester determined from measurements of the check standard. Data from just the 10-volt channel were used in this case study.

The combined standard uncertainty of the battery tester was based on the Type A uncertainty of the tester combined with the Type B uncertainties of both the tester and the check standard. The standard deviation of the initial 7500 measurements of the 10-volt channel was $u_A(V) = 0.0141$ V. Prior independent analysis of the check standard yielded $u_B(V) = 0.000171$ V, while the Type B uncertainty of the tester was considered negligible. The combined standard uncertainty of the battery tester was thus

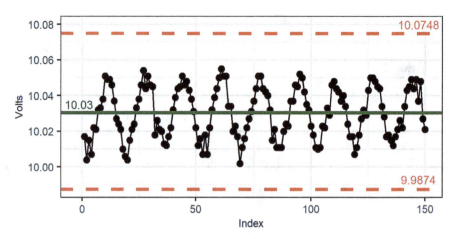

Fig. 11.1 First 150 voltage measurements

$$u_c(V) = \sqrt{u_A^2(V) + u_B^2(V)}$$
$$= \sqrt{(0.0141V)^2 + (0.000171V)^2} = 0.0141V. \quad (11.1)$$

Note that the uncertainty of the check standard contributes little additional uncertainty to the battery tester. The expanded uncertainty was then calculated using the coverage factor determined from the *t*-distribution for a 99.73% level of confidence. The total degrees of freedom were significantly more than 100.

$$U_V^{99.73} = t_{99.73}(100)u_c(V)$$
$$= (3.1) \times (0.0141)V = 0.0437 \ V. \quad (11.2)$$

Before implementing an $\overline{X} - R$ chart, the first 150 observations (collected over 15 s) were plotted to look for possible short-term patterns in the data. The resulting plot appears in Fig. 11.1, with limits based on the expanded uncertainty of the battery tester.

Subject matter experts determined that the pattern seen in the data was due to an AC-induced sine wave caused by AC-powered isolating transformers used to condition the voltage signals. This variation was due to the frequency of sampling and was not considered due to assignable causes that could be eliminated from the tester measurements. As a result of this pattern, however, it was decided to construct an X − MR control chart based on the *averages* of the 150 measurements from each of the 50 independent data sets. This focused the analysis on long-term characteristics of the data and removed the autocorrelation seen in the plot above.

The results of the baseline study are summarized in Table 11.1. The "Baseline Data" column lists the means of the 50 data sets that were generated as a baseline for ongoing control. The "Ongoing Data" column lists the means of the subsequent

11.2 Statistical Process Control (SPC)

Table 11.1 Voltage data summarized for 50 baseline and 50 additional data sets

Data Set	Baseline data	Ongoing data
1	10.0303	10.0312
2	10.0302	10.0311
3	10.0308	10.0308
4	10.0309	10.0312
5	10.0310	10.0315
6	10.0309	10.0314
7	10.0310	10.0314
8	10.0309	10.0311
9	10.0306	10.0315
10	10.0307	10.0313
11	10.0306	10.0310
12	10.0311	10.0313
13	10.0313	10.0308
14	10.0312	10.0309
15	10.0311	10.0313
16	10.0312	10.0310
17	10.0305	10.0311
18	10.0307	10.0309
19	10.0310	10.0309
20	10.0308	10.0312
21	10.0315	10.0309
22	10.0310	10.0308
23	10.0313	10.0305
24	10.0311	10.0303
25	10.0315	10.0299
26	10.0318	10.0297
27	10.0313	10.0303
28	10.0312	10.0304
29	10.0314	10.0300
30	10.0314	10.0305
31	10.0312	10.0302
32	10.0311	10.0295
33	10.0316	10.0303
34	10.0311	10.0295
35	10.0307	10.0301
36	10.0312	10.0300
37	10.0307	10.0300
38	10.0308	10.0300
39	10.0305	10.0305
40	10.0313	10.0303
41	10.0307	10.0301
42	10.0314	10.0300

(continued)

Table 11.1 (continued)

Data Set	Baseline data	Ongoing data
43	10.0311	10.0299
44	10.0308	10.0309
45	10.0313	10.0302
46	10.0317	10.0299
47	10.0306	10.0302
48	10.0315	10.0304
49	10.0317	10.0305
50	10.0319	10.0305
Mean	10.0311	10.0306

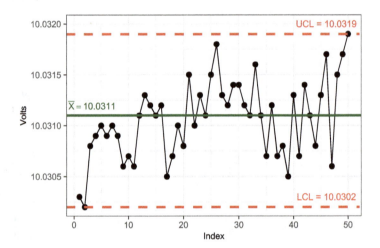

Fig. 11.2 Plot of baseline voltage data

50 data sets that were generated after initial control chart limits were established. The control charts for the baseline analysis and the ongoing monitoring of the battery tester also appear below. Figure 11.2 shows that the baseline data were relatively stable and the resulting control limits were thus appropriate for ongoing control. The autocorrelation that appeared in the short-term data was mostly removed. The control limits were set at

$$\overline{X} \pm 3\hat{\sigma},$$

where $\hat{\sigma}$ is the estimate of the standard deviation σ based on the average moving range:

11.2 Statistical Process Control (SPC)

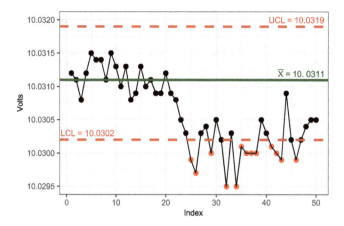

Fig. 11.3 Individuals control chart of ongoing voltage data

$$\widehat{\sigma} = \frac{\sum_{i=1}^{49} |x_{i+1} - x_i|/49}{1.128}.$$

The estimate $\widehat{\sigma}$ is based on the average of 49 moving ranges (from 50 observations) divided by the constant 1.128. This constant is the value that makes $\widehat{\sigma}$ an unbiased estimate of σ when the data are normally distributed (see Montgomery 2013). These baseline control limits were then used for an ongoing X − MR control chart of the subsequent 50 observations (Fig. 11.3). The control chart in Fig. 11.3 shows that initially the ongoing performance of the battery tester was consistent with baseline performance. However, around observation 20, the mean voltage quickly dropped to a new level, triggering "out-of-control" alarms denoted by the red observations. This shift was statistically significant, but was not considered practically significant, given the voltage specification limits (10 ± 0.1 V). Because the shift was not considered important, the battery tester continued to be monitored using the check standard and the technique described here without adjustment to the check standard or tester. If this shift in mean level was considered important, an investigation would be needed to identify the cause of the shift and necessary adjustments to the measurement process. The moving range chart, constructed for the baseline data, did not indicate any significant increase in tester variability. Construction of the moving range chart for the ongoing data is left as an exercise at the end of the chapter.

11.2.2 Discussion

The case study in Sect. 11.2.1 illustrates the use of control charts to monitor tester performance over time. Data from a well-designed, stable check standard were used to establish baseline control limits for a battery tester. Ongoing data collected over time from the same tester measuring the same check standard were then monitored using a control chart with limits determined from the baseline data. The control chart used for ongoing control was an individuals chart with control limits based on the moving range (X − MR chart). Other control charts could be used for this problem, including a cumulative sum chart (Lucas 1982) or an exponentially weighted moving average chart (Crowder 1987b, 1989). These alternative SPC control charts are more sensitive to subtle process changes than the $\overline{X}$ − R or X − MR chart.

The control chart in this example was used to monitor the measurement process, not the manufacturing process. It should be noted that the measurement process can be stable and in-control even if the manufacturing process is out of control. These two processes can behave independently of each other, although the measurement uncertainty affects whether the manufacturing process is viewed as meeting specification limits (Chap. 5). If the measurement process remains in a state of statistical control over an extended period, the results could be used to assess tester drift and to establish calibration intervals for the tester (Chap. 10).

Many applications of SPC in metrology, such as the one above, involve the use of a check standard. When possible, it is preferable to measure manufactured parts over time, if the parts are stable and can be measured multiple times without degrading their output. The data collected over time can also be used to improve baseline estimates of repeatability and reproducibility (Sect. 9.3.4). In this case, the destructive nature of the test prevented repeated measures of actual batteries over time.

11.3 Binary Measurement Systems (BMS)

Binary Measurement Systems (BMS) are systems where the measurements can be classified into one of two categories. In engineering applications, these classifications are commonly 0/1, Pass/Fail, or go/no-go and are usually determined via a simple gauge or visual inspection. While BMS studies do not provide as much information as studies where the response is a continuous variable, they can be more efficient and are often used in practice Zhao et al. (2008). This section gives an overview of BMS studies and how to analyze the results, as well as guidance for choosing a sample size when designing a new study.

11.3.1 BMS Overview

For clarity, several aspects of a BMS study are defined below. A part is defined as the object being measured, an operator is the person taking the measurement, and a repetition is a repeated measurement of a single part. In a BMS study, several operators typically measure several parts, and each operator might measure each part multiple times. The total number of measurements, N, can be calculated as

11.3 Binary Measurement Systems (BMS)

$$N = o \times u \times m, \qquad (11.3)$$

where o is the number of operators, u is the number of parts, and m is the number of repetitions of the measurement. For each inspection, the operator classifies the part using a binary outcome, such as "go" or "no-go."

There are two goals that are common in BMS studies: (1) assessing the agreement of operators by estimating the within-operator and between-operator agreement and (2) assessing the accuracy (correctness) of operators by estimating misclassification rates. Methods for addressing both goals will be discussed in the following sections. Guidance on choosing a sample size for a BMS study is provided in Sect. 11.3.4.

11.3.2 BMS Case Study Introduced

A case study is used to illustrate the concepts associated with evaluating a BMS. In the study, 3 operators measured 6 components at 20 different locations. For the purpose of illustration, each location on each component was considered an independent measurement, resulting in (effectively) 120 different parts. Each operator measured each part three times and provided a classification of Pass or Fail. Thus, the total number of inspections, N, was 360 for each operator. Table 11.2 provides a subset of the data for the component with serial number 2170.

11.3.3 Evaluation of a BMS

Assessing agreement and correctness is an important part of a BMS analysis as it provides a measure of the consistency of operators and the overall performance of the BMS. We will focus on the following three characteristics of a BMS:

- Within-operator agreement: the agreement within a single operator when repeated measurements are taken on the same part. Also known as intra-rater agreement.
- Between-operator agreement: the agreement between multiple operators when measurements are taken on the same part. Also known as inter-rater agreement.
- Correctness: the error rates (correctness rates) of an operator's classification of a part.

These metrics provide an indication of how the BMS is performing. If within-operator agreement or correctness is low for a given operator, that may be an indication that the operator requires additional training. Similarly, if the between-operator agreement or overall correctness is low, multiple operators may require additional training, or there may be issues with the gauge or the measurement process that need to be addressed.

Table 11.2 Subset of binary measurement system case study data

Serial Number	Location	Rep	Operator 1	Operator 2	Operator 3
2170	1	1	Pass	Pass	Pass
2170	1	2	Pass	Pass	Pass
2170	1	3	Pass	Pass	Pass
2170	2	1	Pass	Pass	Pass
2170	2	2	Pass	Pass	Pass
2170	2	3	Pass	Pass	Pass
2170	3	1	Fail	Fail	Fail
2170	3	2	Fail	Fail	Fail
2170	3	3	Fail	Fail	Fail
2170	4	1	Fail	Fail	Fail
2170	4	2	Fail	Fail	Fail
2170	4	3	Fail	Fail	Fail
2170	5	1	Fail	Fail	Fail
2170	5	2	Fail	Fail	Fail
2170	5	3	Fail	Fail	Fail
2170	6	1	Pass	Pass	Pass
2170	6	2	Pass	Pass	Pass
2170	6	3	Pass	Pass	Pass
2170	7	1	Pass	Pass	Pass
2170	7	2	Pass	Pass	Pass
2170	7	3	Pass	Pass	Pass
2170	8	1	Pass	Pass	Pass
2170	8	2	Pass	Pass	Pass
2170	8	3	Pass	Pass	Pass
2170	9	1	Pass	Pass	Pass
2170	9	2	Pass	Pass	Pass
2170	9	3	Pass	Pass	Pass
2170	10	1	Pass	Pass	Fail
2170	10	2	Fail	Fail	Fail
2170	10	3	Fail	Fail	Fail
2170	11	1	Pass	Pass	Pass
2170	11	2	Pass	Pass	Pass
2170	11	3	Pass	Pass	Pass
2170	12	1	Pass	Pass	Pass
2170	12	2	Pass	Pass	Pass
2170	12	3	Pass	Pass	Pass
2170	13	1	Pass	Pass	Pass
2170	13	2	Pass	Pass	Pass
2170	13	3	Pass	Pass	Pass
2170	14	1	Pass	Pass	Pass
2170	14	2	Pass	Pass	Pass
2170	14	3	Pass	Pass	Pass

(continued)

11.3 Binary Measurement Systems (BMS)

Table 11.2 (continued)

Serial Number	Location	Rep	Operator 1	Operator 2	Operator 3
2170	15	1	Fail	Fail	Pass
2170	15	2	Pass	Pass	Pass
2170	15	3	Pass	Pass	Pass
2170	16	1	Pass	Pass	Pass
2170	16	2	Pass	Pass	Pass
2170	16	3	Pass	Pass	Pass
2170	17	1	Pass	Pass	Pass
2170	17	2	Pass	Pass	Pass
2170	17	3	Pass	Pass	Pass
2170	18	1	Pass	Pass	Pass
2170	18	2	Pass	Pass	Pass
2170	18	3	Pass	Pass	Pass
2170	19	1	Pass	Pass	Pass
2170	19	2	Pass	Pass	Pass
2170	19	3	Pass	Pass	Pass
2170	20	1	Fail	Fail	Fail
2170	20	2	Fail	Fail	Fail
2170	20	3	Fail	Fail	Fail

Table 11.3 Within-operator agreement for the BMS case study

Operator number	Number of parts inspected	Number in agreement	Proportion of agreement
Operator 1	120	113	94.2
Operator 2	120	116	96.7
Operator 3	120	114	95.0

11.3.3.1 Within-Operator Agreement

A straightforward method for assessing within-operator agreement is to tabulate the number of parts that agree (i.e., have the same classification for all repetitions) for each operator. This is referred to as the proportion of agreement within operators, and can be formally defined as

$$p_i = \frac{u_{ia}}{u}, \qquad (11.4)$$

where u_{ia} is the number of parts that agree for the ith operator and u is the number of inspected parts. Table 11.3 provides the proportion of agreement for each operator in the BMS case study.

In this example, measurements agree if all three repetitions for a part were given the same classification (i.e., all Pass or all Fail) by an operator. Here, the operators have a high level of agreement across repetitions, implying that there is relatively high precision for each operator. The acceptable level of within-operator agreement is problem specific, however, and may vary depending on the objectives of the analysis.

If more than three repetitions were taken, the requirement for agreement could be relaxed so that only a certain percentage (e.g., three out of four) of the classifications would need to be the same for the part to be in agreement, resulting in a less conservative estimate of the within-operator agreement.

11.3.3.2 Between-Operator Agreement

A basic method for assessing between-operator agreement is to determine how often the classification of different operators agrees across inspections. This proportion is defined as

$$p = \frac{n_a}{N}, \qquad (11.5)$$

where n_a is the number of inspections that received the same classification from all operators and N is the total number of inspections. Table 11.4 shows the overall number of times all operators agreed on the classification of the part. Here, it appears that the operators are generally in agreement, with 345 of the inspections having the same classification across operators. While this method provides a high-level measure of between-operator agreement, it does not tell us if certain operators are more in agreement than others.

Cross-Tabulation The cross-tabulation method as described in Down et al. (2010) provides a more in-depth method for quantifying between-operator agreement. The first step is to assess the between-operator agreement for each pairwise combination of the operators. Table 11.5 shows the format of the cross-tabulation matrix that can be calculated for two operators. The first value in each cell, denoted n_{xy}, represents the number of times Operator A classified an inspection as x and Operator B classified an inspection as y. For example, n_{00} provides the number of times Operators A and B both classified an inspection as 0, a "Fail."

Table 11.4 Proportion of agreement for the BMS case study

Number of inspections	Number in agreement	Proportion of agreement
360	345	95.8

Table 11.5 Format of cross-tabulation matrix

			Operator B		
			Fail	Pass	Total
Operator A		Fail	n_{00} E_{00}	n_{01} E_{01}	$n_{0\cdot}$
		Pass	n_{10} E_{10}	n_{11} E_{11}	$n_{1\cdot}$
		Total	$n_{\cdot 0}$	$n_{\cdot 1}$	N

11.3 Binary Measurement Systems (BMS)

Table 11.6 Cross-tabulation for operators 1 and 2

	Op. 2: fail	Op. 2: pass	Total
Op. 1: Fail	29 (2.98)	0 (26.02)	29
Op. 1: Pass	8 (34.02)	323 (296.98)	331
Total	37	323	360

Table 11.7 Cross-tabulation for operators 1 and 3

	Op. 3: fail	Op. 3: pass	Total
Op. 1: Fail	25 (2.42)	4 (26.58)	29
Op 0.1: Pass	5 (27.58)	326 (303.42)	331
Total	30	330	360

Table 11.8 Cross-tabulation for operators 2 and 3

	Op. 3: fail	Op. 3: pass	Total
Op. 2: Fail	27 (3.08)	10 (33.92)	37
Op. 2: Pass	3 (26.92)	320 (296.08)	323
Total	30	330	360

The second value in each cell, denoted E_{xy}, represents the expected count for Operator A classifying an inspection as x and Operator B classifying it as y. The expected count is calculated as

$$E_{xy} = N \times P_{x.} \times P_{.y}, \quad (11.6)$$

where N is the total number of inspections, and $P_{x.}$ and $P_{.y}$ are given by

$$P_{x.} = \frac{n_{x.}}{N}$$

$$P_{.y} = \frac{n_{.y}}{N}. \quad (11.7)$$

The table counts can be used to calculate the proportion of agreement between each pairwise combination of the operators. This proportion, p_0, is defined as

$$p_0 = \frac{(n_{00} + n_{11})}{N} = \frac{n_a}{N}. \quad (11.8)$$

The expected counts are used to calculate a kappa value, discussed below.

Tables 11.6, 11.7 and 11.8 provide the cross-tabulation results for all three operators. In each pairwise comparison, the number of times both operators agreed (i.e., both said 'Fail' or both said 'Pass') is given. For example, Table 11.6 shows

Table 11.9 Kappa values for each pairwise combination of operator

	Op. 1	Op. 2	Op. 3
Op. 1	–	0.867	0.834
Op. 2	0.867	–	0.786
Op. 3	0.834	0.786	–

that Operator 1 and Operator 2 both classified an inspection as Fail 29 times, and both classified an inspection as Pass 323 times, resulting in a proportion of agreement of $(29 + 323)/360 \times 100 = 97.8\%$. The expected counts are included in parenthesis underneath the observed counts. Note that the expected marginal totals equal the observed marginal totals.

Cohen's Kappa Value Within the cross-tabulation method, Cohen's kappa value (Cohen 1960) can be calculated to assess the level of agreement between operators. It can be advantageous over the basic proportion of agreement metric because it accounts for the possibility that operators may have randomly chosen the same classification. The kappa value generally falls between 0 and 1, where 0 indicates that there is no agreement between operators beyond what is expected by chance, and 1 indicates that there is total agreement. The kappa value can also be negative, implying that the agreement between operators is worse than what would be expected by chance. The kappa value between two operators is defined as

$$\kappa \equiv \frac{p_0 - p_e}{1 - p_e}, \tag{11.9}$$

where p_0, the proportion of agreement between two operators, is equal to $(n_{00} + n_{11})/N$ and p_e, the probability of random agreement using the observed data, is equal to $(E_{OO} + E_{11})/N$.

Table 11.9 shows the pairwise kappa values for all three operators. Operators 1 and 2 have the highest kappa value, implying they have the best between-operator agreement. Generally, kappa values larger than 0.75 suggest good between-operator agreement and that their agreement is much higher than what would be expected by chance. Again, the adequate kappa value may be problem specific (Down et al. 2010).

There are limitations to consider when using a kappa value. The p_e value in the kappa formula has several assumptions, including that the operators' classifications are statistically independent and that the operators randomly guessed for each classification. These assumptions are generally not met, so the kappa value will usually provide a conservative estimate of between-operator agreement (Ubersax 2010). In practice, it is recommended that both the percent agreement and the kappa values be evaluated to determine if there is an adequate level of between-operator agreement for the intended application.

11.3.3.3 Assessing BMS Correctness

Often in a BMS study it is of interest to know the accuracy (correctness) of classifications. Correctness refers to whether an operator's classification of a part is equal to the "true state" of that part. At times, the "true state" of a part can be determined using a gold standard system (GSS). This section describes methods for assessing correctness when a GSS is available.

To formally define the methods for assessing correctness, let

$$X_{ijh} = \begin{cases} 1 \text{ if pass} \\ 0 \text{ if fail} \end{cases}$$

for $i = 1, \ldots, o$ operators, $j = 1, \ldots, u$ parts and $h = 1, \ldots, m$ repetitions. The "true state" can similarly be defined for the ith part as

$$Y_i = \begin{cases} 1 \text{ if conforming} \\ 0 \text{ if nonconforming} \end{cases}.$$

If a GSS is available, the "true state" is simply the classification provided by the GSS. It is assumed that the GSS value always classifies correctly. To assess BMS correctness, it is often of interest to estimate how often the system is incorrectly classifying parts, as well as what type of misclassifications are occurring. This can be accomplished by estimating the following two error rates:

$$\alpha = \Pr(X_{ijh} = 1 | Y_i = 0)$$

and

$$\beta = \Pr(X_{ijh} = 0 | Y_i = 1).$$

Here, α is the probability the BMS passes a nonconforming part, while β is the probability the BMS rejects a conforming part. The misclassification rates α (also known as miss rate or false positive rate) and β (also known as false alarm or false negative rate) are often used to assess performance of a BMS. In many applications, having a low false positive rate is more important than a low false negative rate as it could be worse to pass a nonconforming part than to fail a conforming part. However, it is important to assess which rate will have a larger impact for a given application.

To estimate these rates, two roughly equal-sized samples (each of size N) of conforming and nonconforming parts (determined by the GSS) are taken and evaluated by having multiple operators classify the parts multiple times using the BMS under investigation. Each classification is then assigned one of these four labels:

Table 11.10 False positive and false negative rates for all operators in BMS case study

	α (False positive rate)	β (False negative rate)
Op. 1	0.022	0.003
Op. 2	0.006	0.008
Op. 3	0.019	0.003
Total	0.016	0.005

1. True positive—$X_{ijh} = 1$ and $Y_i = 1$.
2. True negative—$X_{ijh} = 0$ and $Y_i = 0$
3. False positive—$X_{ijh} = 1$ and $Y_i = 0$
4. False negative—$X_{ijh} = 0$ and $Y_i = 1$

The sample misclassification rates can then be calculated by summing the number of false positives (of N negatives) and false negatives (of N positives) over the total number of inspections in Eq. (11.10):

$$\alpha = \frac{\sum \text{False Positives}}{N}$$
$$\beta = \frac{\sum \text{False Negatives}}{N}. \qquad (11.10)$$

Table 11.10 provides the false positive and false negative rates for the three operators, as well as an overall value across the operators. Operator 2 has a higher false negative rate, while Operators 1 and 3 have higher false positive rates. Overall, the operators have a higher false positive than false negative rate.

11.3.4 Sample Sizes for a BMS Study

This section addresses choice of sample sizes when designing a BMS study. With any estimate of agreement or correctness there is uncertainty due to the limited sample size used to generate the estimate. Consequently, a large enough sample size is needed to adequately assess a BMS given the constraints (time, money, etc.) that may limit the number of samples.

To assess how many classifications are needed, an analyst can quantify the uncertainty in the estimate of agreement and correctness for varying sample sizes. This section will describe how to calculate a confidence interval for the proportion of agreement p_0 between two operators. A similar analysis can also be done for a kappa value confidence interval (Blackman and Koval 2000) or a false positive rate confidence interval using the Binomial distribution.

A $100 \times (1 - \alpha)\%$ confidence interval for the proportion of agreement describes the uncertainty due to a finite sample size and can be calculated as

11.3 Binary Measurement Systems (BMS)

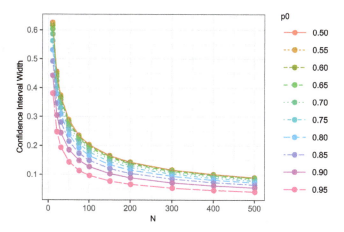

Fig. 11.4 Confidence interval width as a function of the total sample size

Table 11.11 Confidence interval width as a function of the total sample size

		\multicolumn{9}{c	}{p0}
		0.50	0.55	0.60	0.65	0.70	0.75	0.80	0.85	0.90	0.95
	10	0.626	0.623	0.616	0.604	0.586	0.562	0.531	0.492	0.442	0.381
	20	0.456	0.454	0.448	0.438	0.424	0.404	0.379	0.347	0.305	0.247
	30	0.374	0.372	0.367	0.359	0.347	0.330	0.309	0.281	0.244	0.193
	50	0.289	0.288	0.284	0.277	0.267	0.254	0.237	0.214	0.185	0.143
	75	0.236	0.234	0.231	0.225	0.217	0.206	0.192	0.173	0.148	0.113
N	100	0.203	0.202	0.199	0.195	0.187	0.178	0.165	0.149	0.127	0.096
	150	0.165	0.164	0.162	0.158	0.152	0.144	0.134	0.120	0.103	0.077
	200	0.143	0.142	0.140	0.136	0.131	0.124	0.115	0.104	0.088	0.066
	300	0.116	0.115	0.114	0.111	0.107	0.101	0.094	0.084	0.071	0.053
	400	0.100	0.100	0.098	0.096	0.092	0.087	0.081	0.072	0.061	0.045
	500	0.089	0.089	0.088	0.085	0.082	0.078	0.072	0.065	0.055	0.040

$$\left(\frac{\nu_{1L} \times F(\nu_{1L}, \nu_{2L}, \alpha/2)}{\nu_{2L} + \nu_{1L} \times F(\nu_{1L}, \nu_{2L}, \alpha/2)}, \frac{\nu_{1U} \times F(\nu_{1U}, \nu_{2U}, 1 - \alpha/2)}{\nu_{2U} + \nu_{1U} \times F(\nu_{1U}, \nu_{2U}, 1 - \alpha/2)} \right). \quad (11.11)$$

Here, $\nu_{1L} = 2n_a$, $\nu_{2L} = 2(N - n_a + 1)$, $\nu_{1U} = 2(n_a + 1)$, $\nu_{2U} = 2(N - n_a)$, and $F(\nu_{1*}, \nu_{2*}, \alpha/2)$ is the $\alpha/2$th percentile of the F distribution with ν_{1*} and ν_{2*} degrees of freedom (Zhao et al. 2008). Recall that n_a is the number of inspections that agree.

To understand the effect of sample size, a confidence interval can be calculated for a range of sample sizes and potential proportions of agreement. Figure 11.4 and Table 11.11 show the widths of 95% confidence intervals for varying sample sizes and proportions of agreement. Note that p_0 and $(1 - p_0)$ result in the same confidence interval width across sample sizes.

If the proportion of agreement is close to 0 or 1, fewer samples are needed to precisely estimate the proportion of agreement. Alternatively, as the proportion of agreement moves closer to 50%, many more samples are needed to reduce the estimate's uncertainty. Based on the results in Fig. 11.4 and Table 11.11, the total number of inspections, N, should be at least 400 to ensure that the confidence interval width is less than 0.10 for all proportions of agreement. This sample size will be large enough to bound the confidence interval width of the correctness measure (described in Sect. 11.3.3.3) by 0.10, as well. To achieve this level of uncertainty, the number of operators, o, parts, u, and repetitions, m, should thus be chosen so that the product $o \times u \times m$ is at least 400. This number could be achieved, for example, using 50–60 parts with 2–3 operators and 4–5 repetitions of each part.

11.4 Measurement System Analysis with Destructive Testing

Many manufactured parts have properties that can be measured only once due to the destructive nature of the test. Examples include parts for which break strength, impact strength, or burst pressure may be measured. Examples within the NSE include neutron sources, detonators, and thermal batteries, in which both inputs that activate the device and outputs from the device can only be measured once. In cases like these, assessing the repeatability and reproducibility of the measurement becomes more difficult because the part is not available for repeated measurements. We have used two different approaches to this measurement problem, each with important assumptions.

The first approach uses an alternative source to produce the same type of output produced by an actual part. This approach was used in both the battery tester case study (Sect. 11.2.1) and the lead probe case study (Sect. 6.4.1). In the lead probe uncertainty study, an alternative neutron source was used to assess measurement uncertainty. In Fig. 11.5, the alternative source (with the yellow sticker) produces a neutron stream that is measured by a lead probe standard, located to the right of the source. Repeated shots of the source produced the data that were used in the lead

Fig. 11.5 Alternative neutron source under test

probe uncertainty analysis. The key assumptions with this approach are that the alternative source is repeatable and that its output adequately mimics that of an actual test unit.

The second approach to the destructive test problem relies on samples of manufactured parts being relatively homogeneous. Although not an ideal approach, it may be the best option available. Each part is measured, and resulting values are analyzed with the understanding that variation in the data includes part-to-part manufacturing variation. The resulting uncertainty analysis will be conservative in that part-to-part variation will inflate the estimate of measurement uncertainty. The same Gauge R&R methodology described earlier (Sect. 9.3.4) can be used, with repeated measurements of the same parts replaced by single measurements of different (homogeneous) parts. The key assumption with this approach is that the test units are homogenous, or that any significant non-homogeneity can be modelled and removed from the analysis. An example of this approach is given in Benham (2002). De Mast and Trip (2005) discuss a variety of approaches and important assumptions associated with Gauge R&R studies in which measurements are destructive. Both of these papers include a handful of additional references taken from a relatively small literature on this topic.

11.5 Sample Size and Allocation of Samples in Metrology Experiments

The expression for expanded uncertainty (Eq. 6.7)

$$U = t_p(v_{\text{eff}})u_c(y) \qquad (11.12)$$

consists of two terms that are both greatly affected by the choice of sample size and sample allocation in a measurement uncertainty study. The first term is the multiplier $t_p(v_{\text{eff}})$, the $p \times 100$th percentile of the t-distribution with effective degrees of freedom v_{eff}, often computed using the W-S approximation. The second term is the estimated combined standard uncertainty that is typically a combination of several Type A and Type B variance terms.

The GUM provides little guidance regarding how to choose an appropriate sample size and how to allocate those samples in a measurement experiment. This section presents simple approaches to determining the effect of sample size and sample allocation on the value of $t_p(v_{\text{eff}})$ and the precision in $u_c(y)$. The results provide objective criteria for assessing the relative value of a given sample size and allocation. These criteria can be used to determine a desirable sample size for most of the measurement experiments described in this book.

The impact of the degrees of freedom v_{eff} on the multiplier $t_p(v_{\text{eff}})$, and hence on the expanded uncertainty, is easy to demonstrate graphically. For a 95% confidence interval, the value of $t_{95}(v_{\text{eff}})$ is shown as a function of v_{eff} in Fig. 11.6.

Fig. 11.6 The value of $t_{95}(v_{\text{eff}})$ as a function of v_{eff}

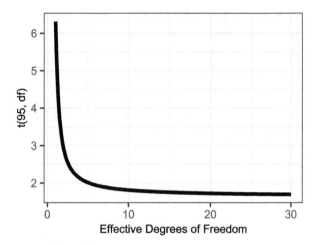

This simple plot shows that little is to be gained, in terms of reducing $t_p(v_{\text{eff}})$, by requiring v_{eff} to be greater than 30. Much is to be gained, however, from increasing v_{eff} from 3 to 10 degrees of freedom as this increase cuts the value of $t_p(v_{\text{eff}})$ by 30%. In practice, 30 degrees of freedom may be difficult to obtain, depending on the complexity of the measurement, but a minimum of 15 degrees of freedom is necessary to reach the flattest part of the curve for $t_{95}(v_{\text{eff}})$.

The second part of the sample size analysis quantifies the precision in $u_c(y)$ as a function of the effective degrees of freedom v_{eff}. The derivation of the W-S approximation for v_{eff} assumes that the quantity

$$\frac{v_{\text{eff}} \times \sum_{i=1}^{k} (c_i u(x_i))^2}{\sum_{i=1}^{k} (c_i \sigma_i)^2} \tag{11.13}$$

has approximately a Chi-square distribution with v_{eff} degrees of freedom. Here the variance term σ_i^2 is equal to $E(u^2(x_i))$, the expected value of the ith estimated variance term. It follows from the Chi-square distribution that

$$\text{Prob}\left(\chi^2_{(1-p)/2}(v_{\text{eff}}) \leq \frac{v_{\text{eff}} \times \sum_{i=1}^{k} (c_i u(x_i))^2}{\sum_{i=1}^{k} (c_i \sigma_i)^2} \leq \chi^2_{(p+1)/2}(v_{\text{eff}}) \right) = p, \tag{11.14}$$

where $\chi^2_p(v_{\text{eff}})$ represents the $p \times 100$th percentile of the Chi-square distribution with v_{eff} degrees of freedom. Using this ratio as a pivotal quantity, a $p \times 100\%$ confidence interval for $\sigma_c^2(y) = \sum_{i=1}^{k} (c_i \sigma_i)^2$ is

11.5 Sample Size and Allocation of Samples in Metrology Experiments

Table 11.12 Constants $A(v_{\text{eff}})$ and $B(v_{\text{eff}})$

v_{eff}	$A(v_{\text{eff}})$	$B(v_{\text{eff}})$
2	0.52	6.26
3	0.57	3.69
5	0.62	2.45
7	0.66	2.04
10	0.70	1.75
15	0.74	1.55
20	0.77	1.44
25	0.78	1.38
30	0.80	1.34
35	0.81	1.30
40	0.82	1.28
50	0.84	1.24
100	0.88	1.16
500	0.94	1.07

$$\left(\frac{v_{\text{eff}} \times \sum_{i=1}^{k}(c_i u(x_i))^2}{\chi^2_{(p+1)/2}(v_{\text{eff}})} \right) \leq \sum_{i=1}^{k}(c_i \sigma_i)^2 \leq \frac{v_{\text{eff}} \times \sum_{i=1}^{k}(c_i u(x_i))^2}{\chi^2_{(1-p)/2}(v_{\text{eff}})} \tag{11.15}$$

and a $p \times 100\%$ confidence interval for $\sigma_c(y) = \left(\sum_{i=1}^{k}(c_i \sigma_i)^2\right)^{1/2}$ is

$$\left(A(v_{\text{eff}}) \times u_c(y) \leq \left(\sum_{i=1}^{k}(c_i \sigma_i)^2\right)^{1/2} \leq B(v_{\text{eff}}) \times u_c(y) \right), \tag{11.16}$$

where $A(v_{\text{eff}}) = \left(\frac{v_{\text{eff}}}{\chi^2_{(p+1)/2}(v_{\text{eff}})}\right)^{1/2}$, $B(v_{\text{eff}}) = \left(\frac{v_{\text{eff}}}{\chi^2_{(1-p)/2}(v_{\text{eff}})}\right)^{1/2}$, and $u_c(y) = \left(\sum_{i=1}^{k}(c_i u(x_i))^2\right)^{1/2}$ is the estimated combined standard uncertainty. Table 11.12 shows the values of $A(v_{\text{eff}})$ and $B(v_{\text{eff}})$ as a function of v_{eff} for the special case $p = 0.95$. These values are the upper and lower multipliers used to construct a 95% confidence interval for the combined standard uncertainty $\sigma_c(y) = \left(\sum_{i=1}^{k}(c_i \sigma_i)^2\right)^{1/2}$. The resulting confidence interval is a measure of the precision associated with using $u_c(y)$ to estimate $\sigma_c(y)$ for a given effective degrees of freedom v_{eff}.

In Fig. 11.7, the $A(v_{\text{eff}})$ sequence converges to 1.0 from below, while the $B(v_{\text{eff}})$ sequence converges to 1.0 from above. With an "infinite" sample size, the confidence interval for $\sigma_c(y) = \left(\sum_{i=1}^{k}(c_i \sigma_i)^2\right)^{1/2}$ would thus be the point estimate

Fig. 11.7 Constants used to construct 95% confidence intervals for $\sigma_c(y)$

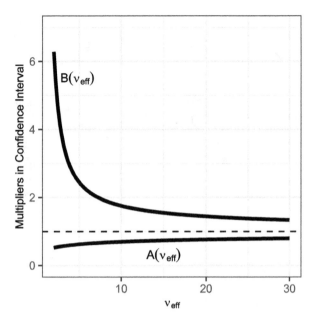

$u_c(y)$. Table 11.12 can be used to choose the sample size necessary to estimate the combined standard uncertainty within a given percent of its actual value. Figure 11.7 graphically indicates the point of diminishing returns, the point beyond which increasing the sample size does not produce much additional precision in the estimate. That point is approximately for v_{eff} in the range 20–30. Table 11.12 and Fig. 11.7 can also be used to quantify the loss of precision with very small sample sizes.

The expression for v_{eff} can be examined in greater detail to inform the allocation of test quantities, that is, to decide which variables in the measurement equation should be tested the most. A general expression for the Type A effective degrees of freedom obtained from the W-S approximation is

$$v_{\text{eff}} = \frac{u_c^4(y)}{\sum_{i=1}^{k} \frac{u_i^4(y)}{v_i}} \qquad (11.17)$$

In this expression, $u_i(y) = |c_i| u(x_i)$, the estimated uncertainty is associated with the input variable x_i for $i = 1, 2, \ldots, k$. It is easy to show, using the Cauchy–Schwarz inequality (Marsden and Tromba 2012), that v_{eff} is bounded by

$$\min_i (v_i) \leq v_{\text{eff}} \leq \sum_i^k v_i. \qquad (11.18)$$

11.5 Sample Size and Allocation of Samples in Metrology Experiments

If the experimental test plan results in $\min_i (v_i) \geq 15$, the minimum recommended effective degrees of freedom based on the above criteria will be achieved. If it is not possible to perform enough tests so that $\min_i (v_i) \geq 15$, the allocation of the total number of tests to the variables in the measurement equation can be used to optimize v_{eff}.

As an example of the gains that can be made through strategic allocation of available tests (when unequal allocation is possible), suppose that $k = 4$ variables are part of the measurement equation and that one of the uncertainties $u_i(y) = |c_i| u(x_i)$ is much larger than the other three. To illustrate, suppose

$$u_1(y) = u_2(y) = u_3(y) = \sigma \text{ and } u_4(y) = 2\sigma. \tag{11.19}$$

Then

$$v_{\text{eff}} = \frac{\left(u_1^2(y) + u_2^2(y) + u_3^2(y) + u_4^2(y)\right)^2}{\sum_{i=1}^{k} \frac{u_i^4(y)}{v_i}}$$

$$= \frac{\left(\sigma^2 + \sigma^2 + \sigma^2 + (2\sigma)^2\right)^2}{\frac{\sigma^4}{v_1} + \frac{\sigma^4}{v_2} + \frac{\sigma^4}{v_3} + \frac{(2\sigma)^4}{v_4}}$$

$$v_{\text{eff}} = \frac{\left(u_1^2(y) + u_2^2(y) + u_3^2(y) + u_4^2(y)\right)^2}{\sum_{i=1}^{k} \frac{u_i^4(y)}{v_i}} \tag{11.20}$$

The allocation of degrees of freedom, via allocation of number of tests to input variables, can have a significant effect on the resulting Type A effective degrees of freedom v_{eff}. For example, suppose 28 tests across the 4 variables are available, and equal allocation is chosen. Then each variable obtains $v_1 = v_2 = v_3 = v_4 = (7-1) = 6$ degrees of freedom. The effective degrees of freedom are then

$$v_{\text{eff}} = \frac{49}{\left(\frac{1}{6} + \frac{1}{6} + \frac{1}{6} + \frac{16}{6}\right)} = 15.5.$$

If instead the allocation $v_1 = v_2 = v_3 = 5$ and $v_4 = 9$ was possible, the effective degree of freedom are

$$v_{\text{eff}} = \frac{49}{\left(\frac{1}{5} + \frac{1}{5} + \frac{1}{5} + \frac{16}{9}\right)}$$

$$= 20.6,$$

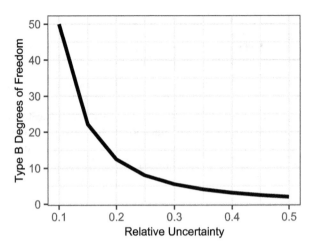

Fig. 11.8 Relative uncertainties and Type B degrees of freedom

an increase of five degrees of freedom gained by strategic allocation of resources. In general, if the individual uncertainties are roughly the same, equal allocation will be optimal and v_{eff} will be approximately equal to $\sum_{i=1}^{k} v_i$. However, if the uncertainties are significantly different, more tests should be allocated to the variable(s) with the largest uncertainty. Various allocation strategies can be explored prior to experimentation if the relative uncertainties (the $u_i(y)'s$) are known. If these are not known, it may be possible to estimate them by performing a small initial experiment. Engineering judgment may also provide valuable insight into the relative uncertainties. The total Type A degrees of freedom are thus determined by the total sample size and allocation of samples to variables in the measurement experiment.

Different suggestions have been made regarding how to assign the Type B degrees of freedom. One approach recommended in the GUM (G.4.2) is to approximate the Type B degrees of freedom for each variable using an equation introduced earlier (Eq. 6.16):

$$v_B \cong \frac{1}{2} \left[\frac{\Delta u_B(y)}{u_B(y)} \right]^{-2}. \tag{11.21}$$

This expression uses the relative uncertainty in $u_B(y)$, the ratio inside the brackets, to approximate v_B. Figure 11.8 shows how the relative uncertainty in $u_B(y)$ affects the Type B degrees of freedom via this approximation.

Note that if the relative uncertainty is as small as 10% (0.10), the Type B degrees of freedom are 50, while if the relative uncertainty is as large as 50% (0.50), the degrees of freedom are just 2. The GUM states (G.4.2) that the relative uncertainty "is a subjective quantity whose value is obtained by scientific judgement based on the pool of available information." Based on our experience and engineering judgement, we believe the relative uncertainty will rarely be greater than 20% (0.20), resulting in v_B no less than 12.5. Using this conservative lower bound for the Type B degrees of freedom will not by itself decrease the overall v_{eff} to an unacceptable level if the experiment has been properly designed.

The second approach discussed in the GUM (G.4.3) is to set the Type B degrees of freedom equal to $+\infty$. The GUM justifies this assumption by claiming it is equivalent to choosing a conservatively wide interval for the Type B uncertainty. If $+\infty$ is used, however, the effective degrees of freedom v_{eff} will be artificially inflated whenever the Type B uncertainty term dominates the combined standard uncertainty. As an example, suppose $u_A(y) = 1$ with degrees of freedom v_A and $u_B(y) = 2$ with degrees of freedom $v_B = +\infty$. Applying the W-S approximation,

$$v_{\text{eff}} = \frac{(1^2 + 2^2)^2}{\left(\frac{1^4}{v_A} + \frac{2^4}{+\infty}\right)}$$

$$= \frac{(5)^2}{\left(\frac{1}{v_A}\right)}$$

$$= 25 v_A. \tag{11.22}$$

Setting the Type B degrees of freedom to $+\infty$ in this simple example significantly increases the effective degrees of freedom, exaggerating the actual amount of information obtained from the measurement experiment.

Hall and Willink (2001) have written in greater detail regarding the performance of the W-S approximation in cases like this. They show that the W-S approximation does not perform well whenever $u_B(y) \geq 2u_A(y)$ and the Type B term is assigned infinite degrees of freedom. When a Type B term dominates the combined standard uncertainty, the relative uncertainty approach (Eq. 11.21) should be used to assign Type B degrees of freedom. Assigning degrees of freedom $+\infty$ should only be done with caution, taking into account the impact on the total effective degrees of freedom.

11.6 Summary of Sample Size Recommendations

Based on the demonstrated effect of sample size on the value of $t_p(v_{\text{eff}})$ and on the precision of the estimate $u_c(y)$, a reasonable goal for a measurement uncertainty study is to achieve Type A effective degrees of freedom ($v_{A,\text{eff}}$) in the 20 to 30 range, with $v_{A,\text{eff}} = 15$ as minimally acceptable. Allocating the number of tests to each variable in the measurement equation based on the value of the individual uncertainties (the $u_i(y)$'s) can be used to optimize $v_{A,\text{eff}}$. Properly assigning degrees of freedom to each Type B term via Eq. (11.21) will result in acceptable yet realistic values for both $v_{B,\text{eff}}$ and the overall v_{eff}. This is assured by the relationship

$$\min\left(v_{A,\mathit{eff}}, v_{B,\mathit{eff}}\right) \leq v_{\mathit{eff}} \leq \left(v_{A,\mathit{eff}} + v_{B,\mathit{eff}}\right),$$

where $v_{A,\mathit{eff}}$ and $v_{B,\mathit{eff}}$ are the effective degrees of freedom obtained from Type A and Type B uncertainty evaluations, respectively. These guidelines can be applied to sample size selection for basic uncertainty analyses (Chap. 6), test plans for factorial designs and ANOVAs (Chap. 9), and test plans for curve fitting (Chap. 10). Figures 11.6, 11.7, and 11.8 can also be used to investigate the increase in uncertainty associated with much smaller sample sizes than recommended here.

11.7 Bayesian Analysis in Metrology

The purpose of this section is to introduce the Bayesian approach to uncertainty analysis and to provide references for further study.

Since the publication of the GUM in 1995, Bayesian analysis has appeared regularly in the literature as an alternative to the classical frequentist uncertainty analysis emphasized in the GUM. The goal of a Bayesian uncertainty analysis is to incorporate prior belief regarding the unknown quantities of interest into the analysis. Experimental data is used to update the prior belief distribution to form a post-experiment, or posterior belief distribution about the unknown quantities. While the frequentist approach views the measurand as an unknown constant that is estimated from experimental data, the Bayesian approach views the measurand as a random variable whose distribution is derived from both prior knowledge and experimental data.

Prior knowledge is ideally available in the form of what is called an *informative* prior distribution. An informative prior can be developed from expert opinion, previous experimentation, or even from a previously derived posterior distribution. It is a distribution that reflects available knowledge prior to the collection of new data. When little or no prior information is available, Bayesian analysis may still be performed using what is called a *noninformative* prior. Noninformative priors are often taken to be simple constant values on the entire real line, although there is no universally agreed upon way to choose such a prior (Berger 1985, 2006). These priors are *improper* priors in that they cannot be normalized to integrate to one, a characteristic of proper probability density functions. This type of prior provides no information regarding the measurand or inputs to a measurement equation. A noninformative prior is used in the case of no prior information to preserve the mathematical framework of the Bayesian approach. Although the prior is improper, the resulting posterior distribution is proper.

A simple example of Bayesian inference for a normal distribution with known variance illustrates how the prior, sampling distribution (likelihood), and posterior distributions are related. Assume that a single measurement (x) is normally distributed with sampling distribution $f(x|\mu, \sigma^2)$, and that the prior distribution for the measurand μ is also normal with PDF $p(\mu|\mu_0, \sigma_0^2)$. Because the sampling distribution

11.7 Bayesian Analysis in Metrology

is normal and the prior distribution is normal, the posterior distribution will also be normally distributed (De Groot 1970). By Bayes theorem, the posterior distribution for μ is

$$p(\mu|x, \sigma^2) \propto f(x|\mu, \sigma^2) \cdot p(\mu|\mu_0, \sigma_0^2). \quad (11.23)$$

That is, the posterior distribution for μ is proportional to the product of the likelihood function and the prior distribution. Straightforward calculations show that the posterior mean is

$$\mu_1 = \left(\frac{\sigma^2}{\sigma^2 + \sigma_0^2}\right)\mu_0 + \left(\frac{\sigma_0^2}{\sigma^2 + \sigma_0^2}\right)x \quad (11.24)$$

and the posterior variance is

$$\sigma_1^2 = \left(\frac{1}{\sigma_0^2} + \frac{1}{\sigma^2}\right)^{-1}. \quad (11.25)$$

The posterior mean μ_1 can be thought of as the conditional expected value of the measurand, given prior information and sampled data. From expression (11.24), the posterior mean μ_1 is a weighted average of the estimate of μ coming from the prior distribution (μ_0) and the estimate of μ coming from the sample data (x). Examining the weights in (11.24), if the variance σ^2 from the sampling distribution is much larger than the prior variance σ_0^2, the prior mean μ_0 receives most of the weight and the data has little influence on the posterior mean. Conversely, if the prior variance σ_0^2 is much larger than the variance σ^2, the measurement data x receives most of the weight and the prior mean has little influence on the posterior mean. In addition, the posterior variance σ_1^2 in expression (11.25) will be less than either σ^2 or σ_0^2. The effect of the Bayesian approach in this example is to *discount* the value of sampled data if it is imprecise (large σ^2) and to *increase* the value of prior information if it is precise (small σ_0^2). Note that if n measurements $x_1, x_2, \ldots, x_n$, are used in the estimation of the posterior mean, the single measurement x would be replaced by the sample average $\bar{x}$, and the variance σ^2 would be replaced by $\frac{\sigma^2}{n}$ in these expressions.

Given a posterior distribution for μ, a Bayesian "credible" interval, based on lower and upper percentiles of this distribution, is used as an uncertainty interval for μ. This interval replaces the classical statistics confidence interval.

This special case of a Bayesian analysis is illustrated below using the Tek TDS 3024 (standard) data from the voltage measurement uncertainty study (Sect. 6.3.1). The Type A uncertainty information and Type B uncertainty information appear in Tables 11.13 and 11.14 below.

The Type B uncertainty information that appears in Table 11.14 can be used to define a plausible prior distribution for the measurand. If we assume that the standard is unbiased ($\mu_0 = 350$), with standard uncertainty $u_B = \sigma_0 = 4.05$, the prior

Table 11.13 Data from voltage measurement uncertainty study conducted at 350 V

Sample	Standard (V)
1	354.0
2	350.1
3	354.0
4	352.2
5	354.0
6	352.0
7	350.0
8	352.2
9	352.2
10	358.0
11	356.0
12	354.0
13	358.0
14	354.0
15	358.0
16	356.0
17	356.0
18	354.0
19	356.0
20	356.0
21	354.0
22	353.0
23	352.0
24	350.0
25	352.0
26	352.0
27	352.0
28	350.0
29	350.0
30	354.0
Avg	353.5
Std Dev	2.43

Table 11.14 Excerpt from calibration record providing Type B standard uncertainty

Digitizer	Assumed PDF	u_B ($k = 1$)
Tek TDS 3024 (standard)	Normal	4.05 V

distribution $p(\mu|\mu_0, \sigma_0^2)$ for a *single* voltage measurement is $N(\mu_0 = 350, \sigma_0^2 = (4.05)^2 = 16.4)$. The sampling distribution for a single measurement x is $f(x| \mu, \sigma^2)$, which can be approximated by a $N(\mu \cong 353.5, \sigma^2 \cong (2.43)^2 = 5.9)$ distribution. Using the sample average $\bar{x}$ as a possible value of a *single* measurement

11.7 Bayesian Analysis in Metrology

x and the sample variance s^2 as the "known" value of σ^2, the posterior mean (see Eq. 11.25) of the measurand μ is

$$\mu_1 = \left(\frac{5.9}{5.9 + 16.4}\right)350 + \left(\frac{16.4}{5.9 + 16.4}\right)353.5$$
$$= (0.26)(350) + (0.74)(353.5)$$
$$= 352.6 \text{ V}$$

and the posterior variance (see Eq. 11.25) is

$$\sigma_1^2 = \left(\frac{1}{16.4} + \frac{1}{5.9}\right)^{-1}$$
$$= 4.34 \text{ V}^2.$$

A resulting 95% uncertainty interval can be determined from the 2.5th and 97.5th percentiles of the posterior distribution. Note that the Bayesian credible interval does not rely on degrees of freedom, the Welch–Satterthwaite approximation, or the t-distribution. Given the posterior mean and variance above, a 95% credible interval for the measurand μ is

$$\mu_1 \pm z_{0.975} \cdot \sigma_1$$
$$= 352.6 \pm (1.96)(4.34)^{1/2}$$
$$= 352.6 \pm 4.1 \text{ V}.$$

In this expression $z_{0.975}$ is the 97.5th percentile of the standard normal distribution.

The resulting Bayesian credible interval has a different interpretation than a classical statistics confidence interval. The interpretation of this interval is that with probability 0.95, the random outcome of the measurand (a single voltage measurement) will fall in this interval. Figure 11.9 below illustrates graphically how the posterior distribution combines information from the prior distribution and sampling distribution (likelihood). The mean of the posterior distribution falls between the mean of the prior distribution and the mean of the sampled data, and the standard deviation of the posterior distribution is less than the standard deviation of either the prior distribution or the sampling distribution. In metrology applications the smaller posterior standard deviation results in a narrower uncertainty interval. It should be mentioned that the narrower interval is the result of the additional assumptions that have been made regarding prior information, and in practice these assumptions may be difficult to validate.

In most realistic metrology examples, the posterior distribution will not have a closed form expression and can only be determined numerically. Interpretation of the resulting distribution will be more difficult, but the basic idea is that the posterior

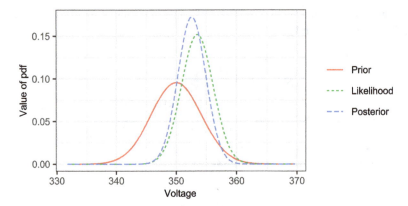

Fig. 11.9 Comparison of prior, likelihood, and posterior distributions

distribution will be a "weighted average" of the prior distribution, based on prior belief, and the sampling distribution, based on the measurement data. The weights given to the two distributions will depend on both the precision of the prior belief regarding the measurand, and the precision of the measurement data.

Lira and Grientschig (2010) present an assessment of Bayesian analysis in metrology and include several tutorial examples. The prior distribution is explained as a "state-of-knowledge" distribution carrying prior information about any supplementary and any input variables, *before* any data have been collected. The information provided by the measurement data is expressed in the form of a sampling distribution, and the resultant posterior distribution is expressed as proportional to the product of these two distributions. For measurement equations with even a few input quantities, the posterior distribution must be determined by a combination of mathematical analysis and numerical integration, but the availability of numerical analysis software makes this approach feasible. Several realistic metrology examples are presented, and uncertainty analyses are performed using this approach. This paper also provides numerous references from the literature that demonstrate Bayesian analysis in metrology.

11.8 Related Reading

Statistical process control (SPC) is not formally covered in the JCGM's guiding documents although the JCGM 100 (GUM, p. 7) states that "the use of check standards and control charts that can indicate if a measurement is under statistical control should be part of the effort to obtain reliable evaluations of uncertainty."

Guidelines for applying SPC to measurement problems appear in a variety of discipline-specific standards and regulations. These documents include ASTM D6299 (analytical measurements), ASTM E882 (chemical analysis), and ASTM

11.8 Related Reading

MNL7 (engineering data). NIST Handbook 143 provides performance standards and formalized procedures for the voluntary recognition of State legal metrology laboratories.

The Multi-Agency Radiological Laboratory Analytical Protocols (MARLAP) manual (2004) contains an entire chapter on laboratory quality control, with a significant treatment of the use of control charts to evaluate laboratory performance. This document cites the NIST/SEMATECH e-Handbook of Statistical Methods, http://www.itl.nist.gov/div898/handbook/ as an online source of instruction in the statistical control of a measurement process. Recent textbooks for general instruction in SPC include De Vor et al. (2007), Montgomery (2013), and Oakland and Oakland (2019).

Case studies involving the application of SPC methods to the monitoring and control of measurement processes appear in the metrology literature from time to time. Pereira da Cunha and Marcolino (2001) illustrate the use of ASTM D6299 guidelines to monitor the measurement of sulfur content in a standard oil supplied by NIST. Their SPC methods include sample selection, data collection, and use of an exponentially weighted moving average (EWMA) control chart. Pruckler (2016) assesses the stability of pipette calibration using $\overline{X}$ and S control charts, and Song and Vorburger (2007) use control charts with dynamic control limits to verify measurement uncertainty. The dynamic control limits are calculated from updated estimates of the mean and standard deviation using historical data from a check standard. Kuhl et al. (2003) use Individuals (X) and Moving Range (MR) charts to assess the state of control of a pressure transducer calibration process.

For BMS assessments, several papers have provided guidance for estimating misclassification rates when there is no gold standard system (GSS) available. Boyles (2001) illustrates a latent-class approach to this problem that estimates the misclassification rates while treating the "true state" of the parts as a latent variable. This method also allows the evaluation of reproducibility across multiple testers or test periods. Van Wieringen and Van den Heuvl (2005) give a comparison of Boyles' method with alternative approaches such as Cohen's kappa. Van Wieringen and De Mast (2008) later expand on the work of Boyles by developing an extended latent-class model that allows the modeling of both reproducibility and repeatability across multiple testers and replicate measures.

A number of papers focus on BMS assessment methods when the pass rate of a system is known, which is typically the case with 100% inspection of parts from a production process. Danila et al. (2008) provide two plans for estimating misclassification rates when the pass rate of a system is known and a GSS is available. Danila et al. (2010) extend this work to cases where no GSS is available. Danila et al. (2012, 2013) also address the case of misclassification rates varying from part to part both with and without a GSS available. Severn et al. (2016) extends the results of Danila et al. (2013) by providing a cost-effective approach to estimating varying misclassification rates by using a GSS on only a subsample of parts. The idea behind this approach is to repeatedly measure parts with the BMS and to only use the GSS on parts that have roughly an equal number of passes and failures. This approach has comparable statistical performance to using the GSS on all parts, but it is much more cost-effective. ASME Y14.43 (2011) provides guidance on the design, dimensioning and tolerancing of go/no-go gauges for BMSs. Ongoing assessment and control of a BMS that performs binary measurements over time can be accomplished using an SPC chart known as a Bernoulli CUSUM (Crowder 2017). Degradation in the correctness of a BMS could be monitored using stable check standards known to be passing or failing units. The Bernoulli CUSUM would provide a moving window of BMS performance.

Several of the JCGM guides mention or allude to Bayesian analyses. The JCGM 100 (GUM) does not mention Bayesian analysis, but it does state that probability can be thought of as a measure of the "degree of belief" that an event will occur (p. 7). This language is consistent with the Bayesian philosophy of probability and is in contrast with the frequentist philosophy that considers probability to be the "relative frequency" with which an event occurs. The JCGM 101 (GUM-S1) presents Bayes' theorem and shows for a simple normal case, using a *noninformative* prior (p. 25), that Bayesian analysis results in essentially the same uncertainty interval as does classical statistical analysis. However, the use of Bayesian analysis for a general measurement equation, with *informative* priors, is not presented in the GUM-S1. The JCGM 102 (GUM-S2) presents Bayes' theorem and shows for a simple multivariate normal case, using a multivariate *noninformative* prior (p. 12), that Bayesian analysis results in essentially the same uncertainty interval as does the classical analysis. The use of a full Bayesian analysis for a general multivariate measurement equation, with *informative priors*, is not presented in the GUM-S2.

Several authors have discussed (see Kackar and Jones 2003; Elster et al. 2015) the difference in the approaches to evaluating uncertainty in the GUM and the GUM-S1. In particular it has been noted that the GUM takes a classical (frequentist) approach to uncertainty analysis, while the GUM S1 takes a "Bayesian-like" approach to uncertainty analysis. In the former case, the measurand is treated as an unknown constant to be estimated. The uncertainty analysis results in an estimated mean with a confidence interval constructed about the mean, based on the expanded uncertainty. In the latter case, the measurand is treated as a random variable for which a distribution is derived via Monte Carlo simulation methods. This analysis results in a probability distribution for the measurand, and an uncertainty interval is derived from the percentiles of that distribution. While this treatment of the measurand is Bayesian-like, the resulting interval does not correspond to an interval derived purely by Bayesian analysis. The authors go on to show that a Bayesian analysis using particular *noninformative* priors can result in uncertainty intervals essentially the same as that derived using purely classical statistics in the GUM or using Monte Carlo methods in the GUM-S1. Elster et al. (2015) conclude, however, that the parallels between Bayesian approaches and approaches found in the GUM and GUM-S1 are very limited. In general, Bayesian approaches to uncertainty analysis will not correspond to the analyses in either the GUM or GUM-S1. When full Bayesian analyses with *informative* priors are used, the analyses and resulting distributions will be much more complicated than with either the GUM or GUM-S1 approaches.

Case studies applying Bayesian analysis to metrology problems appear in the National Institute of Standards (NIST) research library (https://www.nist.gov/nist-research-library) and in the metrology journals *Metrologia, Measurement Science and Technology,* and *Measurement*. The JCGM Working Group 1 News Note of May 4, 2019, states that JCGM 108, to be developed at a later time, will cover "Bayesian Methods." No time frame was given for its development and release. While Bayesian analysis has not been presented in this book in detail, the authors recognize it as a promising methodology to use in measurement studies when important prior knowledge is available. We believe that as the methodologies are formalized for applications in metrology, and presented in JCGM guidelines, Bayesian analysis will move into the mainstream as a tool for practitioners of metrology.

11.9 Exercises

1. Give examples of both natural ("common cause") and unnatural ("assignable cause") variations present in a measurement process.
2. Construct a moving range (MR) chart for the "Ongoing Data" in Table 11.1. What do these MR charts say about the measurement process stability?
3. What is the difference between "statistical" significance and "practical" significance when interpreting an out-of-control condition on a control chart?
4. What is a binary measurement system? When might one be used?
5. What is the difference between within-operator agreement and between-operator agreement?
6. Use the cross-tabulation matrix provided in Table 11.15 to calculate the expected counts in each cell. Then calculate the between-operator proportion of agreement and Cohen's kappa value. Why might the Cohen's kappa value be advantageous over the between-operator proportion of agreement?
7. What is the difference between the false positive rate and the false negative rate? Which rate is worse?
8. Table 11.16 gives the results from a study that was performed to assess agreement and correctness in a BMS. In this study, three operators measured 20 parts each three times and provided a classification of 1 (Pass) or 0 (Fail). A GSS was available and used for all 20 parts.
   Calculate the proportion of agreement within operators. What does this proportion suggest? Next, calculate the proportion of agreement between operators. How does this compare to the within-operator proportion of agreement?

Table 11.15 Cross-tabulation matrix for two operators

	Op. 2: Fail	Op. 2: Pass	Total
Op. 1: Fail	40	4	44
Op. 1: Pass	12	120	132
Total	52	124	176

Table 11.16 BMS study using 3 operators with 3 replicates each

Part	Op 1 Rep 1	Op 1 Rep 2	Op 1 Rep 3	Op 2 Rep 1	Op 2 Rep 2	Op 2 Rep 3	Op 3 Rep 1	Op 3 Rep 2	Op 3 Rep 3	GSS
1	1	1	1	1	1	0	1	1	1	1
2	1	0	1	1	1	1	0	0	0	1
3	1	1	1	0	1	1	1	1	1	1
4	0	0	0	0	0	0	0	0	0	0
5	1	1	1	1	1	0	1	1	1	1
6	1	1	0	1	1	1	1	1	1	1
7	1	1	1	1	1	1	0	0	0	1
8	0	0	0	0	1	0	0	0	0	0
9	0	1	0	0	0	1	0	0	0	0
10	0	0	0	0	0	0	0	0	1	0
11	1	1	1	1	1	1	1	1	1	1
12	1	1	1	1	1	0	1	1	1	1
13	0	0	0	0	0	0	1	1	1	0
14	1	1	1	1	1	1	0	0	1	1
15	1	1	0	1	1	1	1	1	1	1
16	1	1	1	1	1	0	1	1	1	1
17	0	0	0	0	0	0	0	0	0	0
18	0	0	0	0	0	1	0	0	0	0
19	1	1	1	1	1	1	1	1	1	1
20	0	0	0	0	0	0	0	0	0	0

Use the cross-tabulation method to calculate the pairwise proportion of agreement between each pair of operators. How can you interpret the results of the cross-tabulation method? Additionally, calculate the pairwise Cohen's kappa value for each pair of operators. Is the kappa value consistent with the between-operator proportion of agreement?

Finally, calculate the false positive and false negative rates for each operator. How do these rates compare?

9. Calculate a confidence interval for the between-operator proportion of agreement with a total number of inspections $N = 200$ and the number of inspections that agrees $n_a = 60$. What could be changed to narrow the width of this confidence interval?

10. Describe how sample allocation can be used to maximize the information obtainable in an uncertainty experiment. Suppose there are three independent variables in a measurement equation with $u(x_1) = 1$, $u(x_2) = 2$, and $u(x_3) = 3$ (assume the sensitivity coefficients $= 1.0$). With a total of 36 tests to allocate among these three variables, what would an optimal allocation be?

11. Give an example of how "prior" information could be used in a measurement uncertainty analysis. What are the key assumptions associated with a Bayesian analysis?

12. Assume that measurements of a counting process, $X_1, X_2, \ldots, X_n$, are independent and identically distributed (i.i.d.) *Poisson* (λ) (Eq. 4.10). What is the likelihood function $L(\lambda | x_1, x_2, \ldots, x_n)$ for λ, given these data? If a *Gamma*(α, β) distribution is chosen as the prior for λ, what is the resulting posterior distribution for $\pi(\lambda | x_1, x_2, \ldots, x_n)$?

(Note: The Gamma prior distribution for λ has the form: $p(\lambda) = \frac{\beta^\alpha}{\Gamma(\alpha)} \lambda^{\alpha-1} e^{-\beta\lambda}$.)

How would the posterior distribution for λ be used to construct a 95% uncertainty interval?

References

ASME Y14.43: Dimensioning and Tolerancing Principles for Gages and Fixtures. (2011)
ASTM D6299: Standard Practice for Applying Statistical Quality Assurance and Control Charting Techniques to Evaluate Analytical Measurement System Performance. (2013)
ASTM E882: Standard Guide for Accountability and Quality Control in the Chemical Analysis Laboratory. (2016)
ASTM MNL7: Manual on Presentation of Data and Control Chart Analysis. (2018)
Benham, D.: Non-Replicable Gauge R&R Study. Automotive Industry Action Group (2002)
Berger, J.: Statistical Decision Theory and Bayesian Analysis. Springer Series in Statistics. Springer, New York (1985)
Berger, J.: The case for objective Bayesian analysis. Bayesian Anal. **1**, 385–402 (2006)
Blackman, N.J.-M., Koval, J.J.: Interval estimation for Cohen's kappa as a measure of agreement. Stat. Med. **19**, 723–741 (2000)
Boyles, R.A.: Gauge capability for pass-fail inspection. Technometrics. **43**(2), 223–229 (2001)
Cohen, J.: A coefficient of agreement for nominal scales. Educ. Psychol. Meas. **20**, 37–46 (1960)
Crowder, S.V.: Computation of ARL for combined individual measurement and moving range charts. J. Qual. Technol. **19**(2), 98–102 (1987a)
Crowder, S.V.: A simple method for studying run length distributions of exponentially weighted moving average charts. Technometrics. **29**(4), 401–407 (1987b)
Crowder, S.V.: Design of exponentially weighted moving average schemes. J. Qual. Technol. **21**(2), 155–162 (1989)
Crowder, S.: An introduction to the Bernoulli CUSUM. In: Proceedings of the IEEE Annual Reliability and Maintainability Symposium, Orlando (2017)
Danila, O., Steiner, S.H., McKay, R.J.: Assessing a binary measurement system. J. Qual. Technol. **40**, 310–318 (2008)
Danila, O., Steiner, S.H., McKay, R.J.: Assessment of a binary measurement system in current use. J. Qual. Technol. **42**, 152–164 (2010)
Danila, O., Steiner, S.H., McKay, R.J.: Assessing a binary measurement system with varying misclassification rates using a latent class random effects model. J. Qual. Technol. **44**, 179–191 (2012)
Danila, O., Steiner, S.H., McKay, R.J.: Assessing a binary measurement system with varying misclassification rates when a gold standard is available. Technometrics. **55**, 335–345 (2013)
De Groot, M.: Optimal Statistical Decisions. McGraw-Hill, New York (1970)
De Mast, J., Trip, A.: Gauge R&R studies for destructive measurement. J. Qual. Technol. **37**(1), 40–49 (2005)
De Vor, R., Chang, T., Sutherland, J.: Statistical Quality Design and Control: Contemporary Concepts and Methods, 2nd edn. Pearson/Prentice Hall, Upper Saddle River (2007)
Down, M., Czubak, F., Gruska, G., Stahley, S., Benham, D.: Measurement Systems Analysis Reference Manual. Automotive Industry Action Group, Detroit (2010)

Elster, C. et al.: A Guide to Bayesian Inference for Regression Problems. Deliverable of EMRP project NEW04. Novel Mathematical and Statistical Approaches to Uncertainty Evaluation (2015)

Hall, B.D., Willink, R.: Does Welch-Satterthwaite make a good uncertainty estimate? Metrologia. **38**, 9–15 (2001)

JCGM 100: Evaluation of Measurement Data - Guide to the Expression of Uncertainty in Measurement. (2008)

JCGM 101: Evaluation of measurement data - Supplement 1 to the Guide to the Expression of Uncertainty in Measurement - Propagation of distributions using a Monte Carlo method. (2008)

JCGM 102: Evaluation of Measurement Data - Supplement 2 to the -Guide to the Expression of Uncertainty in Measurement - Extension to any number of output quantities. (2011)

Kackar, R., Jones, A.: On use of bayesian statistics to make the guide to the expression of uncertainty in measurement consistent. Metrologia. **40**, 235–248 (2003)

Kuhl, D., Everhart, J., Hallissy, J.: Measurement and control of the variability of scanning pressure transducer measurements. In: Proceedings of the 21st AIAA Applied Aerodynamics Conference, Orlando, FL (2003)

Lira, I., Grientschig, D.: Bayesian assessment of uncertainty in metrology: a tutorial. Metrologia. **47**, R1–R14 (2010)

Lucas, J.M.: The design and use of V-mask control schemes. J. Qual. Technol. **8**(1), 1–12 (1982)

MARLAP: Multi-Agency Radiological Laboratory Analytical Protocols Manual. U.S. Nuclear Regulatory Commission, Washington, DC (2004)

Marsden, J.E., Tromba, A.: Vector Calculus. Freeman and Company, New York (2012)

Montgomery, D.: Introduction to Statistical Quality Control. Wiley, New York (2013)

NIST Handbook 143: State Weights and Measures Laboratories, 5th edn. Program Handbook (2007)

NIST/SEMATECH: e-Handbook of Statistical Methods. https://www.itl.nist.gov/div898/handbook/ (2013). Accessed 29 Apr 2020

Oakland, J., Oakland, R.: Statistical Process Control. Routledge, New York (2019)

Pereira da Cunha, C., Marcolino, M.: Implementing ASTM D-6299/99 in laboratory methods. Researchgate (2001)

Pruckler, R.E.: An assessment of pipette calibration stability using statistical process control charts. Master's Thesis, Boston University (2016)

Severn, D.E., Steiner, S.H., Mackay, R.J.: Assessing binary measurement systems: a cost-effective alternative to complete verification. J. Qual. Technol. **48**(2), 128–138 (2016)

Song, J., Vorburger, T.: Verifying measurement uncertainty using a control chart with dynamic control limits. NCSLI Measure. **2**(3), 76–80 (2007)

Ubersax, J.: Kappa Coefficients: Statistical methods for diagnostic agreement. http://john-uebersax.com/stat/kappa2.htm (2010). Accessed 29 Apr 2020

Van Wieringen, W.N., De Mast, J.: Measurement system analysis for binary data. Technometrics. **50**(4), 468–478 (2008)

Van Wieringen, W.N., Van den Heuvl, E.R.: A comparison of methods for the evaluation of binary measurement systems. Qual. Eng. **17**, 495–507 (2005)

Zhao, Y., et al.: Determining sample size in binary measurement system. Symposium on Precision Mechanical Measurements, vol. 7130 (2008)

Appendix A: Acronyms and Abbreviations

A2LA	American Association for Laboratory Accreditation
AC	Alternating Current
ANOVA	Analysis of Variance
ANSI	American National Standards Institute
AR-XPS	Angle-Resolved X-Ray Photoelectron Spectroscopy
ASME	American Society of Mechanical Engineers
BIPM	International Bureau of Weights and Measures
BMS	Binary Measurement System
CDF	Cumulative Distribution Function
CGPM	General Conference on Weights and Measures
CMM	Coordinate Measuring Machine
CoM	Center of Mass
DC	Direct Current
DMM	Digital Multimeter
DOD	Department of Defense
DOEX	Design of Experiments
EDA	Exploratory Data Analysis
EDS	Energy Dispersive X-ray Spectroscopy
EMF	Electromagnetic Field
FA	False Accept
FR	False Reject
GBF	Guard-banding Factor
GBI	Gage Block Interferometer
GUI	Graphical User Interface
GUM	Evaluation of Measurement Data—Guide to the Expression of Uncertainty in Measurement (JCGM 100 2008)
IEC	International Electrotechnical Commission
ILAC	International Laboratory Accreditation Cooperation
IPK	International Prototype of the Kilogram
ISO	International Organization for Standardization

(continued)

ITS	International Temperature Scale
KDE	Kernel Density Estimation
LCG	Linear Congruential Generator
M&TE	Measuring and Test Equipment
MLE	Maximum Likelihood Estimation
MPE	mise en pratique
MRA	Mutual Recognition Arrangement
NCSLI	National Conference of Standards Laboratories International
NIST	National Institute of Standards and Technology
NMI	National Metrology Institute
NPL	National Physical Laboratory
NSE	Nuclear Security Enterprise
NVLAP	National Voluntary Laboratory Accreditation Program
ODR	Orthogonal Distance Regression
OIML	International Organization for Legal Metrology
OJT	On-the-Job Training
OOT	Out of Tolerance
PDF	Probability Density Function
PMF	Probability Mass Function
PRNG	Pseudo-Random Number Generators
PRT	Platinum Resistance Thermometers
PSL	Primary Standards Laboratory
R&D	Research and Development
R&R	Repeatability and Reproducibility
REML	Restricted Maximum Likelihood
RMS	Root Mean Square
ROC	Radius of Curvature
SI	International System of Units
SNL	Sandia National Laboratories
SPC	Statistical Process Control
SRM	Standard Reference Material
TC	Thermocouple
TUR	Test Uncertainty Ratio
UTC	Universal Time Scale
UUT	Unit Under Test
VIM	International Vocabulary of Metrology—Basic and General Concepts and Terms (JCGM 200 2012)
W-S	Welch-Satterthwaite
XPS	X-ray Photoelectron Spectroscopy

Appendix B: Guidelines for Valid Measurements

This appendix provides related reading by discipline offering best practices for obtaining valid measurements.

Related Reading: Electrical Measurements

It is important to avoid common pitfalls when using electrical measuring and test equipment (M&TE), such as multimeters, oscilloscopes, power supplies, LCR meters, and spectrum analyzers. For information on electrical measurement best practices, see Tumanski (2006), Malaric (2011), and Callegaro (2013). For supporting information on microwave measurements, see Basu (2014). For characterization and measurement specific to the semiconductor industry, see Schroder (2006). The Tektronix Low Level Measurements Handbook (2020) is a good resource for low current, voltage, and resistance measurement practices. For general measurement and experimentation guidelines, see NIST Special Publication 672 by Youden (1997).

Related Reading: Time and Frequency Measurements

There are three different types of time and frequency measurements: (1) date and time-of-day, (2) time interval, and (3) frequency. Date and time-of-day are used to describe when an event occurred. Time interval is used to determine the duration between two events. Frequency is the rate of a repetitive event. Time and frequency are typically measured using oscilloscopes, digitizers, or counters depending on the required accuracy of the measurement. For further reading in the area of time and frequency measurement, see Maichen (2006) and Kajita (2018). For details on GPS receivers and their role in time and frequency measurements, see Doberstein (2012).

Related Reading: Physical Measurements

Dimensional, mass, force, pressure, and flow measurements constitute important areas of physical metrology. For additional information on gauge blocks, see the Gauge Block Handbook by Doiron and Beers (1995). For further reading on contact probing considerations, see Puttock and Thwaite (1969). For detailed information on mass measurement and calibration guidelines, see Harris and Torres (2019) and Fraley and Harris (2019). For force measurement and calibration infromation, refer to ASTM standard E74—18e1 (2018). Lastly, for guidance on pressure and flow measurement, refer to Benedict (2009).

Related Reading: Temperature Measurement

Many types of temperature sensors are available for use in different applications. Common sensors include thermocouples, platinum resistance thermometers, and thermistors. Most temperature sensors generate a voltage proportional to the temperature that must be converted into a temperature reading. Therefore the uncertainty in the temperature reading is a combination of uncertainty in the sensor and uncertainty in the voltmeter or readout. For related reading providing details on temperature measurement and best practices, refer to Michalski et al. (2001), Brewer et al. (2013), and Benedict (2009).

Related Reading: Radiation

Radiation metrology includes measurement of ionizing and non-ionizing radiation. Ionizing radiation measurements encompass measurement of alpha particles, beta particles, neutrons, gamma rays, and X-rays. Non-ionizing radiation includes optical, infrared, and microwave radiation. For additional information on measurement of ionizing radiation, see Knoll (2010), Tsoulfanidis and Landsberger (2010), and Ahmed (2007). For information on radioisotope gauges for industrial process measurement, see Johansen and Jackson (2004). For further reading on non-ionizing radiation measurement (solar and infrared), see Vignola et al. (2012).

Related Reading: General Measurement and Instrumentation Techniques

While not an exhaustive list, additional references on best practices in selecting instrumentation, performing measurements, and analyzing data include books by Webster and Eren (2014), Anderson (1998), and Dunn (2015).

Appendix C: Uncertainty Budget Case Study: CMM Length Measurements

This appendix introduces the steps involved in performing a detailed uncertainty budget analysis for CMM length measurements. Related reading is provided for formulating uncertainty budgets in other measurement areas including gauge block interferometry.

An uncertainty budget is a list of the components, usually associated with an indirect measurement, that contribute to the uncertainty in a measurement result. It identifies, quantifies, and characterizes each independent variable in the measurement equation. It also includes an overall uncertainty analysis incorporating Type A and Type B uncertainties for each variable and the effect of each variable on the measurand. The result is a formal record of an uncertainty analysis that can be used to both characterize and improve the measurement process.

Coordinate Measuring Machine (CMM) Measurements

Product Acceptance Uncertainty: Dimensional Part Inspection with a CMM

This section outlines the uncertainty budget for a spherical mirror measured with a CMM. It describes three types of uncertainty: fitting uncertainty (Type A), process uncertainty (Type A), and standard uncertainty (Type B).

The multivariate statistical methods used to calculate the fitting and process uncertainties of the spherical mirror measurements extend beyond the central methodologies outlined in this book. However, it is included for practitioners that may encounter real multidimensional measurement uncertainty problems requiring both multivariate measurements and analyses. The discussion includes an uncertainty budget (described above) for the spherical mirror measurement performed with a CMM. The measurement is the radius of curvature for a 100 mm diameter mirror.

Radius of Curvature of a Spherical Mirror

The Radius of Curvature (ROC) of a spherical mirror was measured with a CMM at a manufacturing plant. The nominal dimensions of the mirror are as follows:

1. Radius: 50 mm.
2. Thickness: 12.7 mm.
3. ROC 20 m.

The measurand for this calibration is the mirror's ROC. The uncertainty requirements were ± 0.125 m ($k = 2$).

An optical image of the mirror mounted for application is shown in Fig. C.1a. Schematics with the nominal dimensions of the mirror are shown in Fig. C.1b and Fig. C.1c.

The Measurement Model

The equation for the radius (squared) of a sphere is

$$R^2 = (x - H)^2 + (y - K)^2 + (z - L)^2, \quad (C.1)$$

where (x, y, z) are the coordinates at the edge of the sphere, R is the radius, and (H, K, L) are the origin of the sphere. A schematic of a sphere described by Eq. (C.1) is shown in Fig. C.2.

From Eq. (C.1), the maximum change in the z-coordinate (sag) from the center of the mirror to the edge of the mirror is calculated with a nominal ROC of 20 m and an origin of (0, 0, 0):

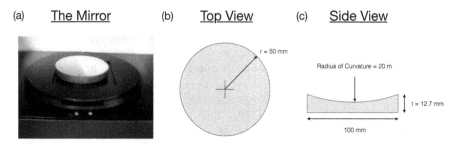

Fig. C.1 (a) Optical image, (b) top view schematic, and (c) side view schematic of the spherical mirror under test

Appendix C: Uncertainty Budget Case Study: CMM Length Measurements

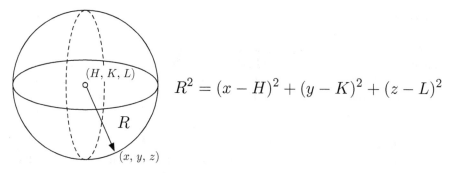

$$R^2 = (x - H)^2 + (y - K)^2 + (z - L)^2$$

Fig. C.2 Equation of a sphere in Cartesian coordinates (x, y, z), with radius R, and origin of (H, K, L)

$$z_1 = (20 \text{ m})^2 - (0 \text{ m})^2 - (0 \text{ m})^2$$
$$z_2 = (20 \text{ m})^2 - (0.05 \text{ m})^2 - (0 \text{ m})^2. \quad \text{(C.2)}$$

The estimation of total sag, Δz:

$$\Delta z = z_1 - z_2$$
$$\Delta z = (20 \text{ m}) - (19.9999375 \text{ m})$$
$$\Delta z = 0.0000625 \text{ m} \quad \text{(C.3)}$$
$$\Delta z = 62.5 \text{ μm} (2460 \text{ μin}).$$

With the estimated sag for the spherical mirror of 62.5 μm (2460 μin), measurement considerations should be analyzed to minimize the variations of the z-axis coordinate.

Measurement Considerations

The accuracy in the measurement of the mirror's ROC is dependent on:

1. Surface form of the mirror.
2. Sampling density.
3. CMM probing force.
4. CMM positioning error.

Items 1–3 are discussed below and the CMM positioning error is considered in the uncertainty analysis.

Surface Form of the Mirror

According to the ASME Y14.5 standard (ASME 2018), form tolerances control straightness, flatness circularity, and cylindricity. The form of the mirror is intrinsic to the part itself; however, the mirror mount at the time of measurement could introduce stress and deform the mirror. Line standards measured with a CMM should be mounted on the Airy points to minimize sag at the ends. On the other hand, the Bessel points should be used to minimize 2D sag. For this measurement, the mirror was mounted in its in-use position, i.e., on the mounting chuck for the mirror during use. Based on a degrees of freedom argument (Chap. 11), it was decided that 50 measurement points were adequate for this experiment.

Moreover, measurements near the edge of the mirror are the most sensitive to determining R because the sag, Δz, near the edge of the mirror is the largest. Figure C.3 shows the measurement points used in the calibration of the mirror.

CMM Probing Force

The force used to probe the mirror surface can introduce error in the measurement. Specifically, if the probing force is large, it is possible for probe skidding or probe deflection to occur. Therefore, it is important to make sure the data is collected normal to the surface using a surface normal probing direction. Surface normal probing will minimize the possibility of skidding, leaving only the worry about exceeding the elastic limit of the mirror.

For most glass surfaces, the elastic limit of glass in compression is approximately 50 MPa and is not a major concern. However, care is still taken when probing the surface and an overtravel force of about 0.3 Newtons is used with a backway and trigger of the probe at 0.1 Newtons. Coatings may require some investigation as to

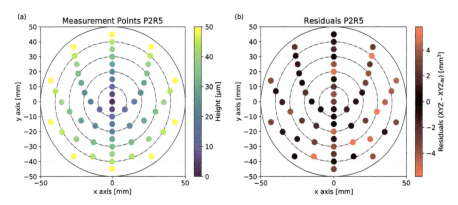

Fig. C.3 (a) Location of the 55 measurement points for each run. The measurements of height are taken from Run 5 in Position 2. (b) The difference between the measurement data and the measurement model

Appendix C: Uncertainty Budget Case Study: CMM Length Measurements 313

how they will hold up when touched in this fashion. The probe diameter size can be varied as well. A large probe diameter on a high-quality surface will provide good performance and minimize any surface damage.

The mirror must be stable and not deformed by the fixture method. It must not deflect, move, or bend under triggering forces. Any movement or stresses in fixturing will result in big errors on radius of curvature realized in a free state. Varying the triggering force and extrapolating the evaluated data to zero trigger force usually uncovers some of these problems and gives confidence in the process.

ROC Measurement

The mirror was measured five times at two different locations on the CMM bed, for a total of 10 runs. The temperature and relative humidity during the measurement ranged from 60.08 °F to 68.34 °F and 40.2% RH to 40.6% RH, respectively.

At each position, the operator identified the zero position and then took 5 measurements at the center of the mirror to verify z repeatability and to make sure the probe was set. The operator then took 19 measurements across the surface of the mirror along x-axis, which is the axis that runs perpendicular to the Koba bar, a calibrated step gage (the purpose of the Koba bar is explained in the uncertainty analysis). The operator then took 55 measurements approximately evenly spaced across the surface of the mirror. The operator started approximately 5 mm from the edge of the mirror for a total of 74 measurements per run. Figure C.3a shows the location of the 74 data points taken from Run 5 in mirror Position 2. The color indicates the measured height, which is synonymous to the sag, Δz.

At each measurement point, x, y, and z-coordinates were obtained. The measurement data at each point was transformed to create a new variable so that a linear least-square fit of the data and the measurement model given in Eq. (C.1) could be performed:

$$\left[x^2 + y^2 + z^2\right] = 2xH + 2yK + 2zL - D, \tag{C.4}$$

where H, K, and L are the origin coordinates of the x, y, and z-axis, respectively and D is

$$D = R^2 - H^2 - K^2 - L^2 \tag{C.5}$$

and R is the ROC. A least-squares fit of the data to Eq. (C.4) was performed to extract the parameters H, K, L, and D. As given by Eq. (C.5), the parameter D includes the value for the mirror's ROC. The difference between the measured data and the fit of Eq. (C.2) for Run 5 in mirror Position 2 is shown in Fig. C.3b. The units are in mm^2 because the variable used for the least-squares fit is $x^2 + y^2 + z^2$.

For each of the 10 runs, a mean value for R was extracted, as seen in Table C.1.

Table C.1 Mean value of the ROC for 10 measurements at two locations on the CMM bed

Position	Run	Mean ROC (m)
1	1	20.281
1	2	20.264
1	3	20.294
1	4	20.283
1	5	20.289
2	1	20.272
2	2	20.295
2	3	20.274
2	4	20.274
2	5	20.267

Table C.2 Uncertainty budget table of the ROC measurements of a spherical mirror

Uncertainty component description	Source	Standard uncertainty		Sensitivity coefficients	Type (A or B)	Distribution	Degrees of freedom
CMM process: uncertainty due to process	Process	$s(R_p)$	0.011 µm	1	A	Normal	10
Standards: z-axis repeatability	Process	$s(z)$	0.072 µm	$\dfrac{\partial R}{\partial z}$	A	Normal	86
Standards: x- and y-axis error	Koba bar	$u(x)$	0.39 µm	$\dfrac{\partial R}{\partial x}$	B	Normal	59
		$u(y)$	0.39 µm	$\dfrac{\partial R}{\partial y}$			
Fit: error in H, K, and L	Residuals of fit	$s(H)$	8.1 µm	$\dfrac{\partial R}{\partial H}$	A	Normal	70
		$s(K)$	6.6 µm	$\dfrac{\partial R}{\partial K}$			
		$s(L)$	9.5 mm	$\dfrac{\partial R}{\partial L}$			

Uncertainty Analysis

The uncertainty analysis is organized with an uncertainty budget table (Table C.2). There are four primary contributions to uncertainty: process uncertainty, CMM z-axis positional uncertainty, CMM x-y positional uncertainty, and the uncertainty in fitted values of H, K, L, and D.

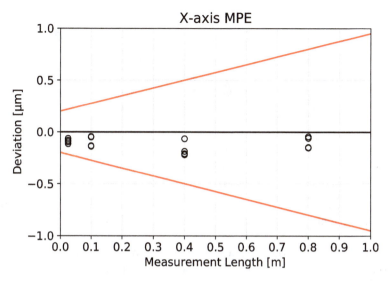

Fig. C.4 Example MPE of a CMM. The manufacturer MPE is plotted with red lines and the measurements in black

CMM Measurement Process Uncertainty

The mirror was placed in two locations on the CMM and the same procedure was run five times at each location. A total of 740 measurements were taken, where each run measured the mirror 74 times. The standard deviation of the mean value for each of the 10 procedure runs was taken to be the standard deviation of the process:

$$s(R_P) = 0.011 \text{ m}. \tag{C.6}$$

CMM Positioning Error (Standards)

The maximum permissible error (MPE), as per ISO 10360-2:2 (ISO 2009), specifies how well a CMM can measure the distance between any two points within the entire CMM measuring volume. The MPE must be valid over the entire CMM volume; therefore, it can overestimate the CMM error within a small measurement volume. For more strict calibrations with lower uncertainty requirements, map the error of the CMM within a small measurement volume. An example MPE plot is shown in Fig. C.4. The manufacturer MPE is plotted in red, while measurements are shown in black.

Due to the strict uncertainty requirements on this measurement, the MPE of the CMM could not be used. Instead, the positioning error of the CMM in the x, y and z-axis were determined at the time of measurement within the measurement volume.

Fig. C.5 A Koba Bar

The x and y-axis error of the CMM were determined at the time of measurement through comparative measurements of a Koba bar, which is a step gage with known distances between each pad. A Koba bar is shown in Fig. C.5.

To account for the CMM x and y-axis positional errors, the mirror was measured at two positions on the CMM bed and a Koba bar was mounted along the x-axis.

The Koba bar used in this measurement had an expanded uncertainty ($k = 2$) for the length between any two steps of $\pm(0.2\ \mu m + 0.4\ L/1000)$, where L is given in meters. There were 12 measurements performed five times at each position.

Figure C.6 shows the Koba Bar measurements in both Position 1 and Position 2. The blue squares are the measurement deviations from the nominal value. There is a clear up-down pattern due to the CMM probe approaching from the left and right side of the Koba Bar. The teal line across the origin represents the mean of the deviations. The red solid line is the uncertainty in the Koba Bar nominal values. The black dashed line is the uncertainty of the mean measured values. The blue dashed line represents the region where any future measurements may be located and is used as the uncertainty for the x and y-axis measurements. For these measurements $k = 2$ was used, which represents 95.45% confidence. The standard uncertainty for the x and y-axis CMM measurements was calculated at both mirror positions and the larger calculated error was used:

$$u(x, y) = 0.39\ \mu m. \tag{C.7}$$

The absolute value of the spherical mirror's z-axis measurements is not needed. Instead, only the change of the z-coordinate relative to the center of the mirror is needed. Therefore, assume the z-axis error to be equal to the repeatability of the z-axis measurements. To estimate the CMM z-axis error, a 100 mm diameter flat mirror was mounted in a similar way as the UUT. The standard deviation of the z-axis coordinates of the 87 measurements taken across the mirror was used to estimate the z-axis error.

The value of the standard deviation was

Appendix C: Uncertainty Budget Case Study: CMM Length Measurements

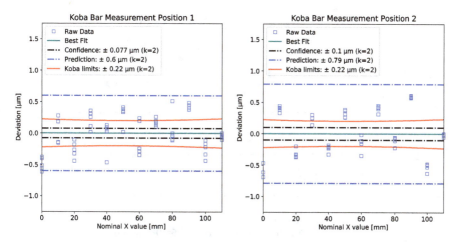

Fig. C.6 Koba bar measurements and best fit, confidence interval, prediction interval, and certified uncertainty

$$s(z) = 0.072 \text{ μm}. \tag{C.8}$$

The mirror was placed in two locations on the CMM and the same procedure was run five times at each location. A total of 740 measurements were taken, where each run measured the mirror 74 times. The standard deviation of the mean value for each of the 10 procedure runs was taken to be the standard deviation of the process, $s(R_P)$.

Fit Uncertainty

Type A uncertainty of the fit was determined from the residuals of the fit. Type B uncertainty for numerical error due to the disparity of numbers (20 m radius curvature on a 0.05 m diameter object) was not calculated because the disparity between inputs is within 2 orders of magnitude when the data is organized into the linearized equation. The linearization of the measurement model results in the low process uncertainty and enables CMM measurements on ROC comparable to optical techniques.

Combined Standard Uncertainty: $u(R)$

The final combined standard uncertainty is calculated through an RSS combination of the individual uncertainty components listed in Table C.1 using both the variance and covariance of the parameters and the final computed ROC. The computed standard uncertainty on the ROC due to the z-axis repeatability, x and y-axis error, and the error in H, K, and L is given:

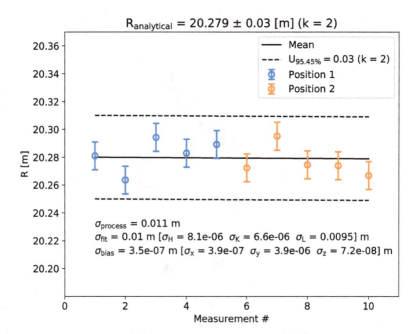

Fig. C.7 Final result on the mirror ROC

$$u(R_o) = f(s(z), u(x), u(y), s(H), s(K), s(L)) = 0.010 \text{ m}. \tag{C.9}$$

The final combined standard uncertainty for the UUT:

$$u^2(R) = \sqrt{u^2(R_o) + u^2(R_p)} = 0.015 \text{ m}. \tag{C.10}$$

With an expansion coefficient ($k = 2$), the final uncertainty: $U_R = 0.030$ m.

Final Results

The final mirror was calibrated with a ROC, $R = 20.279$ m ± 0.030 m ($k = 2$), which corresponds to a 95.45% confidence interval. The results of all 10 runs are shown in Fig. C.7. A Monte Carlo simulation with $N = 7{,}400{,}000$ was then run to verify the analytical uncertainty results. The data for each run was fit with a random distribution of noise in the x, y, and z-coordinates equal to the uncertainty in each coordinate. The simulation fits the 74 points of a run 10,000 times. There were 10 runs for a total of 7,400,000 points. The results are shown in Fig. C.8 and are consistent with the analytical determination of uncertainty.

Appendix C: Uncertainty Budget Case Study: CMM Length Measurements

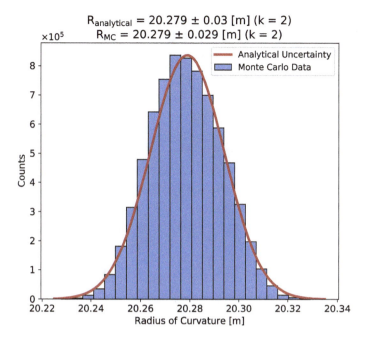

Fig. C.8 Results of Monte Carlo simulations with $N = 7{,}400{,}000$

Related Reading

This Appendix has explored in-depth approaches to uncertainty analysis using a case study in dimensional metrology. The case study itself is not intended to make the reader an expert in these measurements. Uncertainties associated with coordinated measuring machine (CMM) measurements can be especially complex given the dimensional degrees of freedom. ISO 10360-2:2 (ISO 2009) specifies acceptance testing to verify CMM performance and provides guidance on determining maximum permissible error (MPE). An understanding of geometric dimensioning and tolerancing (GD&T) is a requisite for interpreting CMM results, so ASME Y14.5 (ASME 2018) is recommended further reading for anyone using or applying CMM results. Although not discussed here, gage blocks and gage block interferometry form the basis of dimensional measurement traceability. For more detail on gage block interferometry, including development history and the principle of measurement, refer to Ikonen and Riski (1993) and the NBS/NIST publications by Beers (1975) and Doiron and Beers (1995).

Appendix D: Uncertainty Quick Reference

GUM Method for Measurement Uncertainty

Step			Equations
1.		Determine the measurement model equation	$y = f(X_1, X_2, \ldots, X_n)$
2.		Determine mean $\bar{x}_i$ and uncertainty $u(x_i)$ for each input variable	
	a.	Find mean of n repeated measurements of variable x_i ($x_{i,k}$ is the kth measurement on variable X_i)	$\bar{x}_i = \frac{1}{n}\sum_{k=1}^{n} x_{i,k}$
	b.	Compute Type A standard uncertainty $u_A(x_i)$ for each variable	$\sigma_i^2 = \frac{1}{n-1}\sum_{k=1}^{n}(x_{i,k} - \bar{x}_i)^2$ $u_A(x_i) = \frac{\sigma_i}{\sqrt{n}}$
	c.	Compute Type B standard uncertainty $u_B(x_i)$ given knowledge of measurement equipment. Common distributions and use cases:	
		Normal: (Calibration certificate provides uncertainty as $\pm U$ with k-value and/or confidence)	$u_B(x_i) = U/k$
		Uniform: (Equipment specification of $\pm U$ (without confidence given); Digital resolution of equipment, with least significant digit 2 U)	$u_B(x_i) = \frac{U}{\sqrt{3}}$
		Triangular: (Analog gauge resolution of equipment with smallest discernable interval U)	$u_B(x_i) = \frac{U}{\sqrt{6}}$
	d.	Combine Type A and Type B components for input variable x_i	$u(x_i) = \sqrt{u_A(x_i)^2 + u_B(x_i)^2}$
	e.	Determine degrees of freedom for variable x_i using Welch-Satterthwaite	$\nu_A = n - 1$ (Type A) $\nu_B = \infty$ (typical, Type B) $\nu_{\text{eff}} = \frac{u_c^4(y)}{\frac{u_A^4(y)}{\nu_A} + \frac{u_B^4(y)}{\nu_B}}$
3.		Compute combined standard uncertainty	
	a.	Compute sensitivity coefficients c_i, evaluate at the mean $(X_i = \bar{x}_i)$	$c_i = \left(\frac{\partial f}{\partial X_i}\right)$

(continued)

Step			Equations
	b.	Compute standard uncertainty, $u_c(y)$, of y	$u_c(y) = \sqrt{\sum_{i=1}^{N} c_i^2 u^2(x_i)}$
4.		Compute expected value of measurement model	$\bar{y} = f(\bar{x}_1, \bar{x}_2, \ldots)$
5.		Compute effective degrees of freedom of y using Welch–Satterthwaite (with sensitivity coefficients)	$\nu_{\text{eff}} = \dfrac{u_c^4(y)}{\sum_{i=1}^{N} \left(\dfrac{(c_i u(x_i))^4}{\nu_i} \right)}$
6.		Determine coverage factor for confidence level p with ν_{eff} degrees of freedom (see T-table below)	$k = t_p(\nu_{\text{eff}})$
7.		Report result as value and expanded uncertainty. Typically use 2 significant digits of uncertainty and report expected value to same number of decimal places. Always provide k-value and confidence level	$U = \bar{y} \pm k\, u_c(y)$

Percentage Points of the t Distribution

Confidence level for two-sided confidence interval

Degrees of freedom ν	68.27%	90%	95%	95.45%	99%	99.73%
1	1.84	6.31	12.71	13.97	63.66	238.80
2	1.32	2.92	4.30	4.53	9.92	19.21
3	1.20	2.35	3.18	3.31	5.84	9.22
4	1.14	2.13	2.78	2.87	4.60	6.62
5	1.11	2.02	2.57	2.65	4.03	5.51
6	1.09	1.94	2.45	2.52	3.71	4.90
7	1.08	1.89	2.36	2.43	3.50	4.53
8	1.07	1.86	2.31	2.37	3.36	4.28
9	1.06	1.83	2.26	2.32	3.25	4.09
10	1.05	1.81	2.23	2.28	3.17	3.96
11	1.04	1.80	2.20	2.25	3.11	3.85
12	1.04	1.78	2.18	2.23	3.05	3.76
13	1.04	1.77	2.16	2.21	3.01	3.69
14	1.03	1.76	2.14	2.20	2.98	3.64
15	1.03	1.75	2.13	2.18	2.95	3.59
16	1.03	1.75	2.12	2.17	2.92	3.54
17	1.03	1.74	2.11	2.16	2.90	3.51
18	1.03	1.73	2.10	2.15	2.88	3.48
19	1.03	1.73	2.09	2.14	2.86	3.45
20	1.03	1.72	2.09	2.13	2.85	3.42
25	1.02	1.71	2.06	2.11	2.79	3.33
30	1.01	1.70	2.04	2.09	2.75	3.27
35	1.01	1.70	2.03	2.07	2.72	3.23
40	1.01	1.68	2.02	2.06	2.70	3.20

(continued)

Appendix D: Uncertainty Quick Reference

Confidence level for two-sided confidence interval						
Degrees of freedom ν	68.27%	90%	95%	95.45%	99%	99.73%
45	1.01	1.68	2.01	2.06	2.69	3.18
50	1.01	1.68	2.01	2.05	2.68	3.16
100	1.005	1.660	1.984	2.025	2.626	3.077
∞	1.000	1.645	1.960	2.000	2.576	3.000

Guardbanding

Symmetric Specification Limits

For specification limits $\pm$ SL and measurement uncertainty $\pm$ U, calculate TUR = $\frac{SL}{U}$. Ensure SL and U are both reported to the same confidence level. See Chap. 5 for discussion on when TUR-based guardbanding is appropriate.

TUR	Description	Guardband
>4	The measurement is adequate	N/A
1.5–4	Must be guard banded to reduce risk	$SL_{GB} = SL \sqrt{1 - \frac{1}{TUR^2}}$
<1.5	Measurement is not adequate	N/A

Asymmetric Specification Limits

In the case of asymmetric limits, given as Value + X/ − Y, convert to absolute limits SL_U = Value + X and SL_L = Value − Y, and the equivalent symmetric specification limits SL = $\frac{SL_U - SL_L}{2}$. Calculate TUR and guardband, if needed, as described above.

One-Sided Specification Limits

For a one-sided limit, the measurement must be >SL with no lower limit, or the measurement must be <SL with no upper limit. In this case, always guard band by the measurement uncertainty U.

$$SL_{GB} = SL + U \text{ (for a minimum limit requirement)}$$
$$SL_{GB} = SL - U \text{ (for a maximum limit requirement)}$$

Metrology Reference Table

This table provides sources for special topics that may be of interest to practitioners of metrology. Chapter numbers refer to chapters of this book.

Metrology topic	Sources
Accuracy of measurement methods	ISO 5725
Bayesian methods in metrology	Chapter 11, JCGM 108 (in progress)
Binary measurement systems	Chapter 11, AIAG measurement systems analysis reference manual
Conformity assessment	ISO/IEC 17065:2012
Curve fitting	Chapter 10, ASTM E3080-19, ISO/TS 28037:2010
Establishing traceability	Chapter 3, ISO/IEC 17025
Experimental design in metrology	Chapter 9, ISO/WD TS 23471 (in progress)
Generating random variates	Chapter 8, ISO 28640:2010
Interlaboratory comparisons	NCSLI RP-15, ISO/IEC 17043
Legal metrology	OIML V1 (VIML), OIML recommendations
Managing calibration laboratories	ISO 17025, ANSI Z540.3
Modelling and equation building	Chapters 9, 10, JCGM 103 (in progress)
Proficiency testing	ISO/IEC 17025, ISO/IEC 17043, ISO 13528
Reference materials	ISO guide 35, ISO guide 33
Risk in measurement decision making	Chapter 5, JCGM 106, NCSLI RP-18, ISO 10576-1
Setting optimal calibration intervals	Chapter 10, NCSLI RP-1, ILAC-G24/OIML D-10
The SI, reporting measurement units	Chapter 3, ISO 80000-1:2009, NIST SP-811
Software in metrology	NPL reports DEM-ES-010, DEM-ES-011, NCSLI RP-13
Verification of geometrical product specifications	ISO/TC 213, ASME Y14.43
Vocabulary of metrology	Chapter 2, ISO/IEC 99 (JCGM 200:2012, the VIM)

Appendix E: R for Metrology

Introduction

R (R Core Team 2019) is an open-source programming language and environment that is used for data analysis and statistical computing. R was originally developed in the mid-90s as a free statistical analysis tool influenced by two other programming languages—S and Scheme. It has since grown in popularity, primarily due to the number of R packages that are available for a wide range of applications, including linear and nonlinear regression, machine learning, graphical analysis, and more. There are many advantages to using R, including that it is:

- Open-source and free to download.
- Specifically designed for statistical analysis.
- Easily extended through R packages.
- Supported by a large community of users.
- Excellent for data visualization.

Although there is no official dedicated support for R, there is a group of users known as the Core R Team that maintains and edits the base R source code. There are also many available resources for learning and using R; for more information, see Hornik (2018).

Installation of R

R can be downloaded from the Comprehensive R Archive Network (CRAN) at www.r-project.org (R Core Team 2019). It is available for Windows, Unix-Like, and Mac operating systems with installation instructions provided on the CRAN site. R Studio is a recommended and commonly used (but optional) user interface that can be downloaded from www.rstudio.com (R Studio Team 2015).

R Packages

An R package is a group of code containing functions, data, etc. that can be used for a specific purpose (e.g., time series analysis). These packages can be developed by individual R users and submitted to CRAN for public release. As mentioned above, R packages extend the functionality of base R by providing code for a spectrum of purposes. Packages can be installed from CRAN via the console in R by using the install.packages() function.

R for Metrology

There are several existing R packages that perform many of the analyses that are used in metrology. Several packages are used throughout this book, such as Deriv for symbolic differentiation, and onls for orthogonal distance regression. NIST has developed a package specifically for metrology called metRology (Guthrie et al. 2015) which performs many of the analyses outlined in the GUM. Pertinent functions in the metRology package include:

- GUM()—Evaluation of uncertainty using the GUM method.
- uncertMC()—Evaluation of uncertainty using the Monte Carlo method.
- welch.satterthwaite()—Calculation of the Welch–Satterthwaite effective degrees of freedom.

The package also includes built-in example data sets and methods for plotting the uncertainty results. However, since the goal of this book was to show the step-by-step calculations used in the GUM, the metRology package was not used.

R can be used in numerous areas of metrology; in this handbook, it was used for the following:

- Exploratory data analysis (Chap. 4).
- Fitting and assessing probability distributions (Chap. 4).
- Numerical integration (Chap. 5).
- Symbolic partial differentiation for the GUM method (Chap. 7).
- Monte Carlo sampling (Chap. 8).
- Performing Analysis of Variance (ANOVA) with both fixed and random effects (Chap. 9).
- Regression (linear, nonlinear, etc.) (Chap. 10).
- Statistical process control (Chap. 11).

Summary

R is an open-source statistical software that can be used for many metrology applications, and its functionality continues to increase with the large number of packages that are developed by R users. While there is a learning curve to programming in R, it provides a powerful tool for creating automated, reproducible analyses using a variety of statistical methodologies. Additional resources for learning R can be found in Wickham and Grolemund (2017) and Teetor (2011).

References

Abernathy, R.B., Benedict, R.P., Dowdell, R.B.: ASME measurement uncertainty. J. Fluids Eng. **107**, 161–164 (1985)

Addelman, S., Kempthorne, O.: Orthogonal Main Effects Plans. Aeronautical Research Laboratory Office of Aerospace Research. Wright-Patterson Air Force Base, OH (1961)

Agamloh, E.A.: Comparison of direct and indirect measurement of induction motor efficiency. https://doi.org/10.1109/IEMDC.2009.5075180 (2009)

Ahmed, S.N.: Physics and Engineering of Radiation DetectionAcademic Press Inc., San Diego (2007)

Ahmed, Z., et al.: Towards Photonics Enabled Quantum Metrology of Temperature, Pressure, and Vacuum. https://arxiv.org/abs/1603.07690v1 (2016). Accessed 29 Apr 2020

Akdogan, A.: MetrologyIntechOpen Limited, London (2018)

Alain Rüfenacht, N.E.-J.: Impact of the latest generation of Josephson voltage standards in ac and dc electric metrology. Metrologia. **55**, S152–S173 (2018)

Althin, T.K.: C.E. Johansson, 1864–1943: The Master of MeasurementNordsk Rotogravyr, Stockholm (1948)

Anderson, N.A.: Instrumentation for Process Measurement and Control3rd edn, CRC Press, Boca Raton (1998)

ANSI/ASME PTC 19.1: Measurement Uncertainty. (1983)

ANSI/NCSLI Z540.3: Requirements for the Calibration of Measuring and Test Equipment. (2006)

Arendacka, B.: Linear mixed models: GUM and beyond. Measure. Sci. Rev. **4**(2) (2014). https://doi.org/10.2478/msr-2014-0009

ASME Y14.43: Dimensioning and Tolerancing Principles for Gages and Fixtures. (2011)

ASME Y14.5: Dimensioning and Tolerancing. (2018)

ASTM Committee E20: Temperature Measurement Manual on the Use of Thermocouples in Temperature Measurement, 4th edn. Philadelphia, PA (1993)

ASTM D6299: Standard Practice for Applying Statistical Quality Assurance and Control Charting Techniques to Evaluate Analytical Measurement System Performance. (2013)

ASTM E74 - 18e1: Standard Practices for Calibration and Verification for Force-Measuring Instruments. (2018)

ASTM E882: Standard Guide for Accountability and Quality Control in the Chemical Analysis Laboratory. (2016)

ASTM MNL7: Manual on Presentation of Data and Control Chart Analysis. (2018)

Baldo, V., et al.: The new pandemic influenza a/(H1N1)pdm09 virus: is it really "new"? J. Prev. Med. Hyg. **57**(1), E19–E22 (2016)

Balhorn, R., Lebowsky, F., Ullrich, D.: Beat frequency between two axial modes of a He–Ne laser with internal mirrors and its dependence on cavity Q. Appl. Opt. **14**(12), 2955–2959 (1975)

Ball, P.: Precise atomic clock may redefine time. Nature. https://www.nature.com/news/precise-atomic-clock-may-redefine-time-1.13363 (2013). Accessed 29 Apr 2020

Ballico, M.: Limitations of the Welch-Satterthwaite approximation for measurement uncertainty calculations. Metrologia. **37**, 61–64 (2000)

Barari, F., Morgan, R., Barnard, P.: A design of experiments approach to optimise temperature measurement accuracy in solid oxide fuel cell (SOFC). J. Phys. Conf. Ser. **547**(1), 012004 (2014)

Barini, E., Tosello, G., De Chiffre, L.: Uncertainty analysis of point-by-point sampling complex surfaces using touch probe CMMs. Precis. Eng. **34**, 16–21 (2010)

Barreto, H., Howland, F.M.: Introductory Econometrics Using Monte Carlo Simulation with Microsoft ExcelCambridge University Press, Cambridge (2006)

Barry, N. Taylor, P.J.: NIST Reference on Constants, Units, and Uncertainty. https://physics.nist.gov/cuu/Constants/ (2015). Accessed Apr 28, 2020

Basnet, C., Case, K.E.: The effect of measurement error on accept/reject probabilities for homogeneous products. Qual. Eng. **4**(3), 383–397 (1992)

Basu, A.: An Introduction to Microwave MeasurementsTaylor & Francis Group, Boca Raton (2014)

Becerra, L., Hernandez, I.: Evaluation of the air density uncertainty: the effect of the correlation of input quantities and higher order terms in the Taylor series expansion. Meas. Sci. Technol. **17**, 2545–2550 (2006)

Beers, J.S.: A Gauge Block Measurement Process Using Single Wavelength Interferometry. NBS Monograph 152. National Bureau of Standards, Washington, DC. (1975)

Bendavid, E., et al.: COVID-19 Antibody Seroprevalence in Santa Clara County, CAL. https://doi.org/10.1101/2020.04.14.20062463 (2020). Accessed 29 Apr 2020

Benedict, R.P.: Fundamentals of Temperature, Pressure, and Flow Measurements 3rd edn, Wiley, Hoboken (2009)

Benham, D.: Non-Replicable Gauge R&R Study. Automotive Industry Action Group (2002)

Berger, J.: Statistical Decision Theory and Bayesian Analysis. Springer Series in StatisticsSpringer, New York (1985)

Berger, J.: The case for objective Bayesian analysis. Bayesian Anal. **1**, 385–402 (2006)

Bhise, V., Rajan, S., Sittig, D., Morgan, R., Chaudhary, P., Singh, H.: Defining and measuring diagnostic uncertainty in medicine: a systematic review. J. Gen. Intern. Med. **33**(1), 103–115 (2018)

Birch, K.P., Downs, M.J.: Correction to the updated Edlén equation for the refractive index of air. Metrologia. **31**(4), 315–316 (1994)

Blackman, N.J.-M., Koval, J.J.: Interval estimation for Cohen's kappa as a measure of agreement. Stat. Med. **19**(5), 723–741 (2000)

Boggs, P.T., Rogers, J.E.: Orthogonal Distance Regression. National Institute of Standards and Technology, Applied Computational Mathematics Division, Gaithersburg, MD: U.S. Department of Commerce (1990)

Box, G.E., Muller, M.E.: A note on the generation of random normal deviates. Ann. Math. Stat. **29**(2), 610–611 (1958)

Box, G., Hunter, J., Hunter, W.: Statistics for Experimenters2nd edn, Wiley, New York (2005)

Boyles, R.A.: Gauge capability for pass-fail inspection. Technometrics. **43**(2), 223–229 (2001)

Brewer, E.J., Pawlik, R.J., Krause, D.L.: Contact Thermocouple Methodology and Evaluation for Temperature Measurement in the Laboratory. NASA/TM—2013-216580. National Aeronautics and Space Administration, Cleveland, OH (2013)

Brooks, S., Gelman, A., Jones, G.L., Meng, X.: Handbook of Markov Chain Monte CarloChapman & Hall/CRC, Boca Raton (2011)

Brown, W.E.: Random number generation in C++11 (WG21 N3551). Standard C++ Foundation. https://isocpp.org/files/papers/n3551.pdf (2018). Accessed 28 Apr 2020

Bucher, J.L.: The Metrology Handbook2nd edn, ASQ Quality Press, Milwaukee (2012)

References

Burdick, R., Borror, C., Montgomery, D.: Design and analysis of Gauge R&R studies: Making decisions with confidence intervals in random and mixed ANOVA models. ASA-SIAM Series on Statistics and Applied Probability (2005)

Bureau International des Poids et Measures (BIPM): The International System of Units (SI), 9th edn. https://www.bipm.org/en/publications/si-brochure (2019). Accessed 29 Apr 2020

Bureau International des Poids et Mesures: Guide to Expression of Uncertainty in Measurement. International Organization for Standardization (1995)

Cable, M.: Calibration: A Technician's GuideISA, Research Triangle Park (2005)

Callegaro, L.: Electrical Impedance – Principles, Measurement, and ApplicationsTaylor & Francis Group, Boca Raton (2013)

Carroll, R.J., Ruppert, D.: The use and misuse of orthogonal regression in linear errors-in-variables models. Am. Stat. **50**(1), 1–6 (1996)

Casella, G., Berger, R.L.: Statistical Inference2nd edn, Duxbury, Pacific Grove (2002)

Castrup, H.: Estimating bias uncertainty. NCSL Workshop and Symposium, Washington DC (2001)

Castrup, H.: Calibration Intervals from variables data. NCSLI Workshop and Symposium, Washington DC (2005)

Castrup, H.: A Welch-Satterthwaite relation for correlated errors. Measurement Science Conference, Pasadena, CA (2010)

Cividino, S., Egidi, G., Zambon, I., Colantoni, A.: Evaluating the degree of uncertainty of research activities in industry 4.0. Fut. Inter. **11**, 196 (2019)

Clarke, S., Engelbach, R.: Ancient Egyptian Construction and ArchitectureDover Publications, Inc., New York (1990)

Cohen, J.: A coefficient of agreement for nominal scales. Educ. Psychol. Meas. **20**, 37–46 (1960)

Comte de Buffon, G.: Essai d'arithmetique morale. Supplement a l'histoire naturelle (1774)

Cox, M.G., Harris, P.M.: Software Specifications for Uncertainty Evaluation, DEM-ES-010Middlesex, National Physical Laboratory (2006)

Cox, C., Ma, G.: Asymptotic confidence bands for generalized nonlinear regression models. Biometrics. **51**(1), 142–150 (1995)

Cox, M.G., Siebert, B.R.L.: The use of a Monte Carlo method for evaluating uncertainty and expanded uncertainty. Metrologia. **43**(4), S178–S188 (2006)

CRAN Task View: Probability Distributions. https://cran.r-project.org/web/views/Distributions.html. Accessed 10 June 2018

Crowder, S.V.: Computation of ARL for combined individual measurement and moving range charts. J. Qual. Technol. **19**(2), 98–102 (1987a)

Crowder, S.V.: A simple method for studying run length distributions of exponentially weighted moving average charts. Technometrics. **29**(4), 401–407 (1987b)

Crowder, S.V.: Design of exponentially weighted moving average schemes. J. Qual. Technol. **21**(2), 155–162 (1989)

Crowder, S.: An introduction to the Bernoulli CUSUM. Proceedings of the IEEE Annual Reliability and Maintainability Symposium, Orlando (2017)

Crowder, S.V., Moyer, R.D.: A two-stage Monte Carlo approach to the expression of uncertainty with non-linear measurement equation and small sample size. Metrologia. **43**(1), 34–41 (2006)

Daniel, C.: Use of half-normal plots in interpreting factorial two-level experiments. Technometrics. **1**, 311–340 (1959)

Danila, O., Steiner, S.H., McKay, R.J.: Assessing a binary measurement system. J. Qual. Technol. **40**, 310–318 (2008)

Danila, O., Steiner, S.H., McKay, R.J.: Assessment of a binary measurement system in current use. J. Qual. Technol. **42**, 152–164 (2010)

Danila, O., Steiner, S.H., McKay, R.J.: Assessing a binary measurement system with varying misclassification rates using a latent class random effects model. J. Qual. Technol. **44**, 179–191 (2012)

Danila, O., Steiner, S.H., McKay, R.J.: Assessing a binary measurement system with varying misclassification rates when a gold standard is available. Technometrics. **55**, 335–345 (2013)

De Groot, M.: Optimal Statistical DecisionsMcGraw-Hill, New York (1970)

De Mast, J., Trip, A.: Gauge R&R studies for destructive measurement. J. Qual. Technol. **37**(1), 40–49 (2005)

De Silva, G.M.S.: Basic Metrology for ISO 9000 Certification. Routledge, London (2002)

De Vor, R., Chang, T., Sutherland, J.: Statistical Quality Design and Control: Contemporary Concepts and Methods2nd edn, Pearson/Prentice Hall, Upper Saddle River (2007)

Deaver, D.: How to maintain your confidence. Proceedings of the NCSL Workshop and Symposium, Washington DC (1993)

Deaver, D.: Guardbanding with confidence. Proceedings of the NCSL Workshop and Symposium, Albuquerque (1994)

Deaver, D.: Managing calibration confidence in the real world. Proceedings of the NCSL Workshop and Symposium, Dallas, TX (1995)

Decker, J.E., Schödel, R., Bönsch, G.: Considerations for the evaluation of measurement uncertainty in interferometric gauge block calibration applying methods of phase step interferometry. Metrologia. **41**(3), L11–L17 (2004)

Delker, C.J.: Evaluating risk in an abnormal world: how arbitrary probability distributions affect false accept and false reject. Proceedings of the NCSLI Workshop and Symposium, Aurora, CO (2020)

Delker, C.J., Auden, E., Solomon, O. Calculating interval uncertainties for calibration standards that drift with time. Proceedings of the NCSLI Workshop and Symposium, Portland, OR (2018)

Delker, C.J., Auden, E., Solomon, O.: Calculating interval uncertainties for calibration standards that drift with time. NCSLI Meas. [accepted]. (2020)

Dicker, R.C., Coronado, F., Koo, D., Parrish, R.G.: Principles of Epidemiology in Public Health Practice: An Introduction to Applied Epidemiology and Biostatistics, 3rd edn. U.S. Center for Disease Control and Prevention (CDC). https://www.cdc.gov/csels/dsepd/ss1978/ (2012). Accessed 29 Apr 2020

Dobbert, M.: Understanding measurement risk. Proceedings of the NCSL International Workshop and Symposium, St. Paul, MN (2007)

Dobbert, M.: A guard-band strategy for managing false-accept risk. Proceedings of the NCSL International Workshop and Symposium, Orlando, FL (2008)

Doberstein, D.: Fundamentals of GPS Receivers: A Hardware Approach Springer, New York (2012)

DOE-STD-1054-96: Guideline to Good Practices for Control and Calibration of Measuring and Test Equipment (M&TE) at DOE Nuclear Facilities (1995)

Doiron, T., Beers, J.: The Gauge Block Handbook. NIST Monograph 180. National Institute of Standards and Technology, Gaithersburg, MD (1995)

Down, M., Czubak, F., Gruska, G., Stahley, S., Benham, D.: Measurement Systems Analysis Reference ManualAutomotive Industry Action Group (2010)

Dunn, P.F.: Measurement and Data Analysis for Engineering and Science3rd edn, CRC Press, Boca Raton (2015)

Easterling, R., Johnson, M., Bement, T., Nachtsheim, J.: Statistical tolerancing based on consumer's risk considerations. J. Qual. Technol. **23**(1), 1–11 (1991)

Elster, C., Toman, B.: Bayesian uncertainty analysis for a regression model versus application of GUM supplement 1 to the least-squares estimate. Metrologia. **48**, 233–240 (2011)

Elster, C., Wübbeler, G.: Bayesian regression versus application of least squares – an example. Metrologia. **53**, S10–S16 (2016)

Elster, C. et al.: A Guide to Bayesian Inference for Regression Problems. Deliverable of EMRP project NEW04, Novel Mathematical and Statistical Approaches to Uncertainty Evaluation (2015)

Emil, B., Tamar, G.: Some statistical aspects of binary measurement systems. Measurement. **46**, 1922–1927 (2013)

Exponent: The Effect of Various Environmental and Physical Factors on the Measured Internal Pressure of NFL Footballs. Exponent Engineering (2015)

Fauci, A.S., Lane, H.C., Redfield, R.R.: Covid-19—navigating the uncharted. N. Engl. J. Med. **382**(13), 1268–1269 (2020)

Feng, C., Saal, A., Salsbury, J., Ness, A., Lin, G.: Design and analysis of experiments in CMM measurement study. Precis. Eng. **31**, 94–101 (2007)

Ferguson, N., et al.: Impact of Non-pharmaceutical Interventions (NPIs) to Reduce COVID-19 Mortality and Healthcare Demand (2020). https://doi.org/10.25561/77482

Fewell, T.R.: An evaluation of the alpha counting technique for determining 14-MeV neutron yields. Nucl. Inst. Methods. **61**(1), 61–71 (1968)

Filipe, E.: Evaluation of standard uncertainties in nested structures. Adv. Math. Comput. Tools Metrol. **VII**, 151–160 (2006)

Flicker, C., Tran, H.: Calculating measurement uncertainty of the "conventional value of the result of weighing air". NCSLI Meas. **11**(2), 22–37 (2016)

Fong, J., de Wit, R., Marcal, P., Filliben, J., Heckert, N.: A design-of-experiments plug-in for estimating uncertainties in finite element simulations. SIMULIA Customer Conference (2009)

Fong, J., Heckert, N., Filliben, J., Ma, L., Stupic, K., Keenan, K., Russek, S.: A design-of-experiments approach to FEM uncertainty analysis for optimizing magnetic resonance imaging RF coil design. Proceedings of the 2014 COMSOL Conference, Boston (2014)

Forrest, E., Schulze, R., Liu, C., Dombrowski, D.: Influence of surface contamination on the wettability of heat transfer surfaces. Int. J. Heat Mass Transf. **91**, 311–317 (2015)

Fraley, K.L., Harris, G.L.: Advanced Mass Calibration and Measurement Assurance Program for State Calibration Laboratories. NISTIR 5672. National Institute of Standards and Technology, Gaithersburg, MD (2019)

Fraser, C., et al.: Pandemic potential of a strain of influenza A (H1N1): early findings. Science. **324**(5934), 1557–1561 (2009)

Fuller, W.A.: Measurement Error ModelsWiley, New York (1987)

Gauge Block Interferometer. http://www.npl.co.uk/instruments/products/dimensional-mass-force/gauge-block-interferometer/ (n.d.). Accessed 28 Apr 2020

Gauss, C.F.: Theoria combinationis observationum erroribus minimis obnoxiae. In: Carl Friedrich Gauss Werke, pp. 95–100. Dieterichsche universitäts-Druckerei, Göttingen (1821)

Gavin, H.P.: The Levenberg-Marquardt Method for Nonlinear Least Squares Curve-Fitting Problems. Duke University: Department of Civil and Environmental Engineering (2017)

Gentle, J.E.: Computational StatisticsSpringer, New York (2009)

Ghani, A.C., et al.: Methods for estimating the case fatality ratio for a novel. Emerg. Infect. Dis. Am. J. Epidemiol. **162**(5), 479–486 (2005)

Glantz, S., Slinker, B.: Primer of Applied Regression & Analysis of Variance2nd edn, McGraw-Hill, New York (2001)

Goodell, R.: 2014 Official Playing Rules of the National Football LeagueNational Football League (2014)

Gopinath, G.: The Great Lockdown: Worst Economic Downturn Since the Great Depression. International Monetary Fund. https://blogs.imf.org/2020/04/14/the-great-lockdown-worst-economic-downturn-since-the-great-depression/ (2020). Accessed 29 Apr 2020

Graham, C., Talay, D.: Strong Law of Large Numbers and Monte Carlo MethodsSpringer, Berlin (2013)

Gust, J.C., Harris, G.L.: Weights and Measures Division Quality Manual for Proficiency Testing and Interlaboratory Comparisons. NISTIR 7214. National Institute of Standards and Technology, Gaithersburg, MD (2005)

Guthrie, W.F., et al.: NIST: metRology Software Project. https://www.nist.gov/programs-projects/metrology-software-project (2015)

Gyllenbok, J.: Encyclopaedia of Historical Metrology, Weights, and Measures, vols. 1–3. Birkhäuser, Switzerland (2018)

Hall, B.D., Willink, R.: Does Welch-Satterthwaite make a good uncertainty estimate? Metrologia. **38**, 9–15 (2001)

Hamilton, C.A.: Josephson voltage standards. Rev. Sci. Instrum. **71**(10), 3611–3623 (2000)

Hamilton, C.A., Burroughs, C.J., Benz, S.P.: Josephson voltage standard- a review. IEEE Trans. Appl. Supercond. **7**(2), 3756–3761 (1997)

Harben, J., Reese, P.: Risk mitigation strategies for compliance testing. NCSLI Meas. **7**(1), 38–49 (2012)

Harrell, F.E.: Regression Modeling Strategies 2nd edn, Springer, New York (2015)

Harris, G.L., Torres, J.A.: Selected Laboratory and Measurement Practices and Procedures to Support Basic Mass Calibrations (2019 Ed). NISTIR 6969. National Institute of Standards and Technology, Gaithersburg, MD (2019)

Hartley, H.O.: The modified Gauss-Newton method for the fitting of non-linear regression functions by least squares. Technometrics. **3**, 269–280 (1961)

Hays, W., Petersen, A.: Eli Whitney and the Machine AgeF. Watts, New York (1959)

Hellekalek, P.: Don't trust parallel Monte Carlo! ACM SIGSIM Simul. Dig. **28**(1), 82–89 (1998)

Hendricks, J.H., Ricker, J.E., Egan, P.F., Strouse, G.F.: In search of better pressure standards. Phys. World. **67**(8), 13–14 (2014)

Heumann, C., Schomaker, M., Shalabh: Introduction to Statistics and Data AnalysisSpringer, New York (2016)

Historical Context of the SI. https://physics.nist.gov/cuu/Units/history.html (1998). Accessed 29 Apr 2020

Ho, C.J., Chen, M.W., Li, Z.W.: Numerical simulation of natural convection of nanofluid in a square enclosure: effects due to uncertainties of viscosity and thermal conductivity. Int. J. Heat Mass Transf. **51**, 4506–4516 (2008)

Holman, J.: Experimental Methods for Engineers 8th edn, McGraw-Hill Companies, Inc., New York (2012)

Hornik, K.: The R FAQ. https://CRAN.R-project.org/doc/FAQ/R-FAQ.html (2018)

Hund, L.B., Campbell, D.L., Newcomer, J.T.: Statistical guidance for setting product specification limits. Proceedings of the IEEE Annual Reliability and Maintainability Symposium (2017)

Iamsasri, T., et al.: A Bayesian approach to modeling diffraction profiles and application to ferroelectric materials. J. Appl. Crystallogr. **50**, 211–220 (2017)

IEC 31010: Risk Management – Risk Assessment Techniques. (2009)

Ikonen, E., Riski, K.: Gauge-block interferometer based on one stabilized laser and a white-light source. Metrologia. **30**(2), 95–104 (1993)

ISO 1: Geometrical Product Specifications (GPS)—Standard Reference Temperature for the Specification of Geometrical and Dimensional Properties. (2016)

ISO 10360-2: Geometrical Product Specifications (GPS)—Acceptance and Reverification Tests for Coordinate Measuring Machines (CMM)—Part 2: CMMs Used for Measuring Linear Dimensions. (2009)

ISO 10576-1: Statistical Methods—Guidelines for the Evaluation of Conformity with Specified Requirements. (2013)

ISO 13528: Statistical Methods for Use in Proficiency Testing by Interlaboratory Comparison. (2015)

ISO 14253-1: Geometrical Product Specifications (GPS)—Inspection by Measurement of Workpieces and Measuring Equipment. (2017)

ISO 17043: Conformity Assessment—General Requirements for Proficiency Testing. (2010)

ISO 21748: Guidance for Use of Repeatability, Reproducibility, and Trueness Estimates in Measurement Uncertainty and Evaluation. (2017)

ISO 3534-1: Statsitcs Vocabulary and Symbols—Part 1: Probability and General Statistical Terms. (1993)

ISO 3534-3: Statistics Vocabulary and Symbols—Part 3: Design of Experiments. (2013)

ISO 5725-2: Accuracy (trueness and precision) of Measurement Methods and RESULTS—Part 2: Basic Method for the DETERMINATION of Repeatability and Reproducibility of a Standard Measurement Method. (1994)

ISO/IEC 17025: General Requirements for the Competence of Testing and Calibration Laboratories. (2017)

ISO/IEC Guide 98: Guide to the Expression of Uncertainty in Measurement. (1993)

ISO/TR 10017: Guidance on Statistical Techniques for ISO 9001:2000. (2003)

Izenman, A.J.: Modern Multivariate Statistical Techniques – Regression, Classification, and Manifold LearningSpringer, New York (2013)

Jackson, D.: A surprising link between measurement decision risk and calibration interval analysis. Proceedings of the NCSLI Workshop and Symposium, Cleveland, OH (2019)

References

Javamex: How does java.util.Random work and how good is it? https://www.javamex.com/tutorials/random_numbers/java_util_random_algorithm.shtml (2018). Accessed 29 Apr 2020

JCGM 100: Evaluation of Measurement Data—Guide to the Expression of Uncertainty in Measurement. (2008)

JCGM 101: Evaluation of Measurement data—Supplement 1 to the Guide to the Expression of Uncertainty in Measurement—Propagation of Distributions Using a Monte Carlo Method. (2008)

JCGM 102: Evaluation of Measurement Data—Supplement 2 to the—Guide to the Expression of Uncertainty in Measurement—Extension to Any Number of Output Quantities. (2011)

JCGM 103: Guide to the Expression of Uncertainty in Measurement—Developing and Using Measurement Models (Committee Draft). (2018)

JCGM 106: Evaluation of Measurement Data - The role of Measurement Uncertainty in Conformity Assessment. (2012)

JCGM 200: International Vocabulary of Metrology – Basic and General Concepts and Terms (VIM), 3rd edn. (2012)

Johansen, G.A., Jackson, P.: Radioisotope Gauges for Industrial Process MeasurementsWiley, Chichester (2004)

Johansson, I.: Metrological thinking needs the notions of parametric quantities, units, and dimensions. Metrologia. **47**(3), 219–230 (2010)

Josephson, B.D.: Possible new effects in superconductive tunneling. Phys. Lett. **1**(7), 251–253 (1962)

Kackar, R., Jones, A.: On use of Bayesian statistics to make the guide to the expression of uncertainty in measurement consistent. Metrologia. **40**, 235–248 (2003)

Kajita, M.: Measuring Time – Frequency Measurements and Related Developments in PhysicsIOP Publishing Ltd, Bristol (2018)

Kamigaki, T., Oshitani, H.: Epidemiological characteristics and low case fatality rate of pandemic (H1N1) 2009 in Japan. PLoS Curr. **1**, RRN1139 (2009)

Kelvin, L.: Ninteenth century clouds over the dynamical theory of heat and light. Phil. Mag. **6**(2), 1–40 (1901)

Kilogram: The Kibble balance. https://www.nist.gov/si-redefinition/kilogram/kilogram-kibble-balance (2018). Accessed 29 Apr 2020

Kim, J.Y., Byung, R.C., Kim, N.: Economic design of inspection procedures using guard band when measurement errors are present. Appl. Math. Model. **31**(5), 805–816 (2007)

Kimothi, S.K.: The Uncertainty of Measurements: Physical and Chemical Metrology and AnalysisASQ Quality Press, Milwaukee (2002)

Kindratenko, V.: Numerical Computations with GPUsSpringer, New York (2014)

King, M.L., Giles, D.E.: Specification Analysis in the Linear ModelRoutledge, New York (2018)

Klauenberg, K., Wübbeler, G., Mickan, B., Harris, P., Elster, C.: A tutorial on Bayesian normal linear regression. Metrologia. **52**, 878–892 (2015)

Klein, H.A.: The Science of Measurement: A Historical SurveyDover Publications, Inc., New York (1988)

Kline, S.J.: The purposes of uncertainty analysis. J. Fluids Eng. **107**, 153–160 (1985)

Kline, S.J., McClintock, F.A.: Describing uncertainties in single sample experiments. Mech. Eng. **75**, 3–8 (1953)

Knoll, G.F.: Radiation Detection and Measurement4th edn, Wiley, Hoboken (2010)

Knuth, D.: The Art of Computer Programming, Volume 2: Semi-Numerical Algorithms3rd edn, Addison Wesley Longman (1998)

Ku, H. NBS Special Publication 747. National Bureau of Standards (1988)

Kuhl, D., Everhart, J., Hallissy, J.: Measurement and control of the variability of scanning pressure transducer measurements. Proceedings of the 21st AIAA Applied Aerodynamics Conference, Orlando, FL (2003)

Kuster, M.: Applying the Welch-Satterthwaite formula to correlated errors. NCSLI Meas. **8**(1), 42–55 (2013)

Kutner, M.H., Nachtsheim, C.J., Neter, J., Li, W.: Applied Linear Statistical Models5th edn, McGraw-Hill, New York (2005)

Lee, J.L. FLOC Takes Flight: First Portable Prototype of Photonic Pressure Sensor. National Institute of Standards and Technology. https://www.nist.gov/news-events/news/2019/02/floc-takes-flight-first-portable-prototype-photonic-pressure-sensor (2019). Accessed 29 Apr 2020

Lenth, R.: Quick and easy analysis of unreplicated factorials. Technometrics. **31**, 469–473 (1989)

Lipe, T.E.: A Re-evaluation of the NIST low-frequency standards for AC-DC difference in the voltage range 0.6-100 V. IEEE Trans. Instrum. Meas. **45**(6), 913–917 (1996)

Lira, I., Grientschig, D.: Bayesian assessment of uncertainty in metrology: a tutorial. Metrologia. **47**, R1–R14 (2010)

Lira, I., Woger, W.: Comparison between the conventional and Bayesian approaches to evaluate measurement data. Metrologia. **43**, 249–259 (2006)

Lombardi, M.A.: Frequency measurement and analysis service (FMAS). https://www.nist.gov/programs-projects/frequency-measurement-and-analysis-service-fmas (2009). Accessed 29 Apr 2020

Lucas, J.M.: The design and use of V-mask control schemes. J. Qual. Technol. **8**(1), 1–12 (1982)

Lugli, E.: The Making of Measure and the Promise of SamenessUniversity of Chicago Press, Chicago (2019)

Machin, G.: The Kelvin redefined. Meas. Sci. Technol. **29**(2), 022001 (2018)

Maichen, W.: Digital Timing Measurements: from Scopes and Probes to Timing and JitterSpringer, Dordrecht (2006)

Malaric, R.: Instrumentation and Measurement in Electrical EngineeringBrown Walker Press, Boca Raton (2011)

Mallapaty, S.: Antibody tests suggest that coronavirus infections vastly exceed official counts. Nature. https://doi.org/10.1038/d41586-020-01095-0 (2020). Accessed 29 Apr 2020

Mandel, J.: The Statistical Analysis of Experimental DataDover Publications, Inc., New York (1984)

MARLAP: Multi-Agency Radiological Laboratory Analytical Protocols Manual. U.S. Nuclear Regulatory Commission, Washington, DC (2004)

Marsaglia, G.: Random numbers fall mainly in the planes. Proc. Natl. Acad. Sci. **61**(1), 25–28 (1968)

Marsden, J.E., Tromba, A.: Vector CalculusFreeman and Company, New York (2012)

Marshall, K.: AC Voltage Standards with Quantum Traceability. NPL Electromagnetics (2007)

Mathworks Inc.: RandStream—random number stream. https://www.mathworks.com/help/matlab/ref/randstream.html (2018). Accessed 29 Apr 2020

Matsumoto, M., Nishimura, T.: Mersenne Twister: A 623-dimensionally equidistributed uniform pseudo-random number generator. ACM Trans. Model. Comput. Simul. **8**(1), 3–30 (1998)

May, W., et al.: Definitions of Terms and Modes Used at NIST for Value-Assignment of Reference Materials for Chemical Measurements. NIST Special Publication 260-136 (2000)

McCullough, B.D.: Microsoft excel's 'Not the Wichmann-Hill' random number generator. Comput. Statis. Data Anal. **52**, 4587–4593 (2008)

Measurement Precision Task Group: Measurement Precision Estimation for Binary DataIPC, Northbrook (2003)

Measurement Systems Analysis Working Group: MSA Reference Manual, 4th edn, 104 (2010)

Melard, G.: On the accuracy of statistical procedures in Microsoft excel 2010. Comput. Stat. **29**, 1095–1128 (2014)

Merkel, W.R., White, V.R.: National Voluntary Laboratory Accreditation Program: Procedures and General Requirements. NIST Handbook 150, 2016 edn. National Institute of Standards and Technology, Gaithersburg, MD (2016)

Metrological Traceability. https://www.bipm.org/en/bipm-services/calibrations/traceability.html (n.d.). Accessed 29 Apr 2020

Metropolis, N.: The beginning of the Monte Carlo method. Los Alamos Sci. **15**, 125–130 (1987)

Michalski, L., Eckersdorf, K., Kucharski, J., McGhee, J.: Temperature Measurement2nd edn, Wiley, Chichester (2001)

Mimbs, S.M.: Measurement decision risk - The importance of definitions. Proceedings of the NCSLI Workshop and Symposium, St. Paul, MN (2007)

References

Mimbs, S.M.: Using reliability to meet Z540.3's 2% rule. Proceedings of the NCSL Workshop and Symposium, National Harbor, MD (2011)

Montgomery, D.: Introduction to Statistical Quality ControlWiley, New York (2013)

Montgomery, D.: Design and Analysis of Experiments9th edn, Wiley, New York (2017)

Moré, J.: The Levenberg-Marquardt algorithm. Numerical Analysis: Lecture Notes in Mathematics, vol. 630. Springer, Berlin (1978)

Mottonen, M., Belt, P., Harkonen, J., Haapasalo, H., Kess, P.: Manufacturing process capability and specification limits. Open Indust. Manuf. Eng. J. **1**, 29–36 (2008)

Mueller, F.H., Astin, A.V.: Research highlights of the National Bureau of Standards Annual Report, Fiscal Year 1959. Miscellaneous Publication 229. National Bureau of Standards, Washington, DC (1959)

National Conference of Calibration and Standards Laboratories International: Recommended Practice 12 - Determining and Reporting Measurement Uncertainties. (1995a)

National Conference of Calibration and Standards Laboratories International: Determining and Reporting Measurement Uncertainties. (1995b)

National Conference of Calibration and Standards Laboratories International: Handbook for Application of ANSI/NCSL Z540.3–2006. (2009)

National Conference of Calibration and Standards Laboratories International: Recommended Practice 1 - Establishment and Adjustment of Calibration Intervals. (2010)

National Conference of Calibration and Standards Laboratories International: Recommended Practice 18 - Estimation and Evaluation of Measurement Risk. (2014)

National Conference of Calibration and Standards Laboratories International: Recommended Practice 5 - Measuring and Test Equipment Specifications. (2016)

National Institute of Standards and Technology: FLOC takes flight: First portable prototype of photonic pressure sensor. https://www.nist.gov/news-events/news/2019/02/floc-takes-flight-first-portable-prototype-photonic-pressure-sensor (2019). Accessed 29 Apr 2020

National Institute of Standards and Technology: Standard reference materials. https://www.nist.gov/srm (n.d.). Accessed 29 Apr 2020

National Instruments: What is the algorithm used by the LabVIEW Random Number (0–1) Function? http://digital.ni.com/public.nsf/allkb/9D0878A2A596A3DE86256C29007A6B4A (2018). Accessed 29 Apr 2020

NIST Handbook 143: State Weights and Measures Laboratories. Program Handbook, 5th edn (2007)

NIST/SEMATECH: e-Handbook of Statistical Methods. https://www.itl.nist.gov/div898/handbook/ (2013). Accessed 29 Apr 2020

Oakland, J., Oakland, R.: Statistical Process ControlRoutledge, New York (2019)

Papananias, M., Fletcher, S., Longstaff, A., Forbes, A.: Uncertainty evaluation associated with versatile automated gauging influenced by process variations through design of experiments approach. Precis. Eng. **49**, 440–455 (2017)

Pereira da Cunha, C., Marcolino, M.: Implementing ASTM D-6299/99 in laboratory methods. Researchgate (2001)

Phillips, S.D., Eberhardt, K.R.: Guidelines for expressing the uncertainty of measurement results containing uncorrected bias. J. Res. NIST. **102**, 577–585 (1997)

Piratelli-Filho, A., Di Giacomo, B.: CMM uncertainty analysis with factorial design. Precis. Eng. **27**, 283–288 (2003)

Possolo, A.: Simple guide for evaluating and expressing the uncertainty of NIST measurement results. https://doi.org/10.6028/NIST.TN.1900 (2015). Accessed 29 Apr 2020

Press, W.H., Teukolsky, S.A., Vetterling, W.T., Flannery, B.P.: Numerical Recipes in C—The Art of Scientific Computing2nd edn, Cambridge University Press, Cambridge (2002)

Preston-Thomas, H.: The international temperature scale of 1990 (ITS-90). Metrologia. **27**(1), 3–10 (1990)

Pruckler, R.E.: An assessment of pipette calibration stability using statistical process control charts. Master's Thesis, Boston University (2016)

Puttock, M.J., Thwaite, E.G.: Elastic Compression of Spheres and Cylinders at Point and Line Contact. National Standards Laboratory Technical Paper No. 25. Commonwealth Scientific and Industrial Research Organization, Melbourne, Australia (1969)

Python Software Foundation: Python documentation. https://docs.python.org/3/library/random.html (2018). Accessed 29 Apr 2020

Quinn, T.J.: Practical realization of the definition of the metre, including recommended radiations of other optical frequency standards. Metrologia. **40**(2), 103–133 (2001)

R Core Team: R: A Language and Environment for Statistical. R Foundation for Statistical Computing. https://www.R-project.org (2019)

R Studio Team: RStudio: Integrated Development for R. RStudio, Inc. http://www.rstudio.com (2015)

Rabinovich, S.G.: Measurement Errors and Uncertainties3rd edn, Springer, New York (2005)

Reed, C., et al.: Estimates of the prevalence of pandemic (H1N1) 2009, United States, April–July 2009. Emerg. Infect. Dis. **15**(12), 2004–2007 (2009)

Refraction Redefines the Pascal. https://ws680.nist.gov/publication/get_pdf.cfm?pub_id=916352 (n.d.). Accessed 29 Apr 2020

Regalado, A.: Blood tests show 14% of people are now immune to covid-19 in one town in Germany. MIT Technol. Rev. https://www.technologyreview.com/2020/04/09/999015/blood-tests-show-15-of-people-are-now-immune-to-covid-19-in-one-town-in-germany/ (2020). Accessed 29 Apr 2020

Research Highlights of the National Bureau of Standards. National Bureau of Standards (1959)

Richtmyer, R.D., von Neumann, J.: Statistical Methods in Neutron DiffusionLAMS-551, Los Alamos National Laboratory, Los Alamos (1947)

Robert, C.P., Casella, G.: Monte Carlo Statistical Methods2nd edn, Springer, New York (2004)

Robinson, I.A., Schlamminger, S.: The watt or kibble balance: a technique for implementing the new SI definition of the unit of mass. Metrologia. **53**(5), A46–A74 (2016)

Rossi, G.B. (ed.): Probability in metrology. Data Modeling for Metrology and Testing in Measurement Science. Springer, New York (2008)

Rothman, K.J., Greenland, S., Lash, T.L. (eds.): Modern Epidemiology3rd edn, Lippincott Williams & Wilkins, Philadelphia (2008)

Ruby, L., Rechen, J.B.: A fast-neutron activation detector for 14-MeV pulsed neutron sources. Nucl. Inst. Methods. **15**(1), 74–76 (1962)

Rüfenacht, A., Flowers-Jacobs, N.E., Benz, S.P.: Impact of the latest generation of Josephson voltage standards in ac and dc electric metrology. Metrologia. **55**(5), S152–S173 (2018)

Saltelli, A., Tarantola, S., Campolongo, F., Ratto, M.: Sensitivity Analysis in Practice: A Guide to Assessing Scientific ModelsWiley, Chichester (2004)

Sankle, R., Singh, J.R.: Single sampling plans for variables indexed by AQL and AOQL with measuremetn error. J. Mod. Appl. Stat. Methods. **11**(2), 12 (2012)

Satterthwaite, F.: An approximate distribution of estimates of variance components. Biom. Bull. **2**(6), 110–114 (1946)

Schroder, D.K.: Semiconductor Material and Device Characterization3rd edn, Wiley, Hoboken (2006)

Schweitzer, W., Kessler, G., et al.: Description, performance, and wavelengths of iodine stabilized lasers. Appl. Opt. **12**(12), 2927–2938 (1973)

Searle, S.R., Gruber, M.H.J.: Linear ModelsWiley, New York (2016)

Severn, D.E., Steiner, S.H., Mackay, R.J.: Assessing binary measurement systems: a cost-effective alternative to complete verification. J. Qual. Technol. **48**(2), 128–138 (2016)

Sharp, D.B.: Measurement standards. In: Webster, J.G. (ed.) The Measurement, Instrumentation, and Sensors HandbookCRC Press, Boca Raton (1999)

Shrestha, S.S., et al.: Estimating the burden of 2009 pandemic influenza A (H1N1) in the United States (April 2009–April 2010). Clin. Infect. Dis. **52**(suppl_1), S75–S82 (2011)

Silverman, B.: Density Estimation for Statistics and Data AnalysisChapman & Hall, Boca Raton (1986)

Simonsen, L., et al.: Global mortality estimates for the 2009 influenza pandemic from the GLaMOR project: a modeling study. PLoS Med. **10**(11), 1–17 (2013)

References

Smith, W.: A New Classical Dictionary of Greek and Roman Biography, Mythology and GeographyHarper & Brothers, New York (1851)

Smith, J.R.: Statistical aspects of measurement and calibration. Comput. Inustr. Eng. **18**(3), 365–371 (1990)

Song, J., Vorburger, T.: Verifying measurement uncertainty using a control chart with dynamic control limits. NCSLI Meas. **2**(3), 76–80 (2007)

Spang, S.: Calibration: Philosophy in PracticeFluke Corporation, Everett (1994)

Spencer, C.E., Jacobs, E.L.: The lead activation technique for high energy neutron measurement. IEEE Trans. Nucl. Sci. **12**(1), 407–414 (1965)

Srinivasan, V., Sharkarji, C.M., Morse, E.P.: On the enduring appeal of least-squares fitting in computational coordinate metrology. J. Comput. Inform. Sci. Eng. **12**(1), 1–15 (2012)

Stewart, A.: Basic Statistics and Epidemiology: A Practical Guide4th edn, CRC Press, Boca Raton (2016)

Stone, J.A., et al.: Advice from the CCL on the use of unstabilized lasers as standards of wavelength: the helium–neon laser at 633 nm. Metrologia. **46**(1), 11–18 (2008)

Suga, N.: The Metrology Handbook: The Science of Measurment2nd edn, Mitutoyo America Corporation, Aurora (2016)

Suzuki, T., Takeshita, J., Ono, J., Lu, X.: Designing a measurement precision experiment considering distribution of estimated precision measures. J. Phys. Conf. Series. **1065**(21) (2018)

Tang, Y., Belecki, N., Mayo-Wells, J.: A practical Josephson voltage standard at one volt. https://nvlpubs.nist.gov/nistpubs/sp958-lide/315-318.pdf (n.d.). Accessed 28 Apr 2020

Taubes, G.: Cold fusion conundrum at Texas A&M. Science. **248**(4961), 1299–1304 (1990)

Taylor, B.N.: NIST Special Publication 330—The International System of Units (SI). National Institute of Standards and Technology (1991)

Taylor, J.R.: An Introduction to Error Analysis University Science Books, Sausalito (1997)

Taylor, B.N., Kuyatt, C.E.: Guidelines for Evaluating and Expressing the Uncertainty of NIST Measurement Results. NIST Technical Note 1297. National Institute of Standards and Technology, Gaithersburg, MD (1994)

Teetor, P.: R CookbookO'Reilly Media, Inc (2011)

Tektronix: Low Level Measurements Handbook. 7th edn. http://download.tek.com/document/LowLevelHandbook_7Ed.pdf (2018). Accessed 29 Apr 2020

Thwaite, E.G., Leslie, R.T.: Variation in contact error between repeatedly wrung surfaces. Br. J. Appl. Phys. **14**(10), 711–713 (1963)

Time and Frequency from A to Z. https://www.nist.gov/pml/time-and-frequency-division/popular-links/time-frequency-z (2010). Accessed 29 Apr 2020

Tsoulfanidis, N., Landsberger, S.: Measurement and Detection of Radiation4th edn, CRC Press, Boca Raton (2010)

Tumanski, S.: Principles of Electrical MeasurementsTaylor and Francis, New York (2006)

Ubersax, J.: Kappa Coefficients: Statistical methods for diagnostic agreement. http://john-uebersax.com/stat/kappa2.htm (2010). Accessed 29 Apr 2020

Vaillant, L., La Ruche, G., Tarantola, A., Barboza, P.: Epidemiology of fatal cases associated with pandemic H1N1 influenza 2009. Eur. Secur. **14**(33), 1–6 (2009)

Van Wieringen, W.N., De Mast, J.: Measurement system analysis for binary data. Technometrics. **50**(4), 468–478 (2008)

Van Wieringen, W.N., Van den Heuvl, E.R.: A comparison of methods for the evaluation of binary measurement systems. Qual. Eng. **17**, 495–507 (2005)

Vardeman, S., Jobe, J.: Statistical Methods for Quality AssuranceSpringer, New York (2016)

Verity, R., et al.: Estimates of the severity of coronavirus disease 2019: a model-based analysis. Lancet Infect. Dis. (2020) (in press)

Vignola, F., Michalsky, J., Stoffel, T.L.: Solar and Infrared Radiation Measurements2nd edn, CRC Press, Boca Raton (2012)

Vrba, I., et al.: Different approaches in uncertainty evaluation for measurement of complex surfaces using coordinate measuring machine. Meas. Sci. Rev. **15**(3), 111–118 (2015)

Walkup, N.: Eratosthenes and the Mystery of the Stades. Mathematical Association of America. https://www.maa.org/press/periodicals/convergence/eratosthenes-and-the-mystery-of-the-stades-introduction (2005). Accessed 29 Apr 2020

Walsh, D., Crowder, S., Burns, E., Thacher, P.: Estimating Uncertainty in Laboratory neutron Measurements Made with a Lead Probe Scaler Detector. SAND2017-6863J, Sandia National Laboratories (2017)

Wang, C.M., Iyer, H.K.: Propagation of uncertainties in measurements using generalized inferences. Metrologia. **42**, 145–153 (2005a)

Wang, C.M., Iyer, H.K.: On higher-order corrections for propagating uncertainties. Metrologia. **42**, 406–410 (2005b)

Webster, J.G., Eren, H.: Measurement, Instrumentation, and Sensors Handbook2nd edn, CRC Press, Boca Raton (2014)

Wehr, R., Saleska, S.R.: The long-solved problem of the best-fit straight line: application to isotopic mixing lines. Biogeosciences. **14**, 17–29 (2017)

Welch, B.: The generalization of Student's problem when several population variances are involved. Biometrika. **34**, 28–35 (1947)

Wells Jr., T.V., Karp, B.S., Reisner, L.L.: Investigative Report Concerning Footballs Used During the AFC Championship Game on January 18, 2015Paul, Weiss, Rifkind, Wharton & Garrison LLP (2015)

What is a Check Standard?. https://www.itl.nist.gov/div898/handbook/mpc/section1/mpc12.htm (2003). Accessed 29 Apr 2020

Wichmann, B.A., Hill, I.D.: Algorithm AS 183: an efficient and portable pseudo-random number generator. J. R. Stat. Soc. Ser. C. **31**(2), 188–190 (1982)

Wichmann, B.A., Hill, I.D.: Generating good pseudo-random numbers. Comput. Statist. Data Anal. **51**, 1614–1622 (2006)

Wickham, H., Grolemund, G.: R for Data Science: Import, Tidy, Transform, Visualize, and Model DataO'Reilly Media, Inc, Sebastopol (2017)

Willink, R.: On using the Monte Carlo method to calculate uncertainty intervals. Metrologia. **43**(6), L39–L42 (2006)

Willink, R.: A generalization of the Welch-Satterthwaite formula for use with correlated uncertainty components. Metrologia. **44**, 340–349 (2007)

Wolfram Research: Wolfram Language & System Documentation Center. http://reference.wolfram.com/language/tutorial/RandomNumberGeneration.html (2018). Accessed 29 Apr 2020

World Health Organization: WHO Director-General's opening remarks at the media briefing on COVID-19 - 3 March 2020. https://www.who.int/dg/speeches/detail/who-director-general-s-opening-remarks-at-the-media-briefing-on-covid-19%2D%2D-3-march-2020 (2020). Accessed 29 Apr 2020

York, D., et al.: Unified equations for the slope, intercept, and standard errors of the best straight line. Am. J. Phys. **72**(3), 367–375 (2004)

Youden, W.J.: Experimentation and Measurement, vol. 672. National Bureau of Standards (1984)

Youden, W.J.: Experimentation and Measurement. NIST Special Publication 672. National Institute of Standards and Technology, Gaithersburg, MD (1997)

Zhang, S., et al.: Error analysis of data acquisition of reverse engineering process using design of experiment. Advances in Manufacturing Technology XVI, Professional Engineering Publishing Limited, London (2002)

Zhao, Y., et al.: Determining sample size in binary measurement system. Symposium on Precision Mechanical Measurements, vol. 7130(2008)

Zuckerwar, A.J.: Handbook of the Speed of Sound in Real GassesAcademic, San Diego (2002)

Index

A
Acceptance limits (AL), 88, 89, 91, 95, 97–100
Accreditation, 7, 50, 53–55, 95, 303, 304
Accuracies, 3, 6, 15, 21–22, 28, 35, 37, 46, 48, 51, 69, 83, 126, 157, 183, 221, 249, 262, 275, 281, 305, 311, 324
Adaptive sampling, 165–166, 176
Air pressure, 4–8, 38, 81
Alias, viii, 199, 200
Alpha radiation, 306
Analysis of variance (ANOVA), ix, 7, 22, 125, 181, 182, 184, 195, 203, 206–222, 224, 225, 292, 303, 326
ANSI/NCSL Z540, 53, 54, 82
Artifact, 42, 43, 47, 49, 50, 55
As-read, 95
Assignable cause, 60, 267–270, 300
Asymmetric, 91, 92, 323
Autocorrelation, 261, 270, 272
Avogadro's constant, 42, 44

B
Base, viii, 28, 31, 37, 41–45, 55, 56, 197, 325, 326
Basic measurement model, 107–114
Bayesian analysis, ix, 261, 267, 292–296, 298–300
Bernoulli CUSUM, 298
Beta radiation, 306
Between-operator agreement, 275, 278–280, 300
Bias, 9, 10, 21, 23, 24, 38, 46, 50, 81, 83, 92–94, 96, 97, 105, 107, 108, 110, 112–114, 126, 187, 191, 196, 206, 209, 218–220, 224, 268
Bias, selection, 10
Binary data, 60, 63
Binary Measurement System (BMS), ix, viii, 267, 274–284, 300, 303, 324
Binomial, 258, 282
Bivariate, 146
Boltzmann constant, 42, 44
Box-Muller Transformation, 156
Boxplots, 64, 65, 78, 217, 268
Bureau International des Poids et Mesures (BIPM), 41–43, 45–47, 55, 261, 303

C
Calibration, 2, 20, 28–29, 31, 46, 51–53, 70, 81, 93–96, 108, 150, 169, 227, 250, 258, 274
Calibration intervals, 51, 52, 74, 227, 250–253, 255, 258–261, 274, 324
Candela (cd), 41, 44
Case fatality rate (CFR), 9–13
Categorical data, 60, 63, 78
Cauchy-Schwartz inequality, 288
Central limit theorem, 119
Certification, 29, 51, 52, 253
Certified Reference Material (CRM), 49–50
Check standard, 50, 215, 269, 270, 273, 274, 298
Chi-square, 106, 214, 215, 286
CMM positioning error, 311, 316–317
Coded coefficients, 190, 191, 194, 195, 203
Cohen's kappa value, 280, 297, 300

Combined standard uncertainties, ix, 105–107, 111, 112, 116, 120, 121, 124, 127, 133, 136, 140, 141, 174, 208, 209, 269, 285, 287, 288, 291, 317–318, 321
Components of variation, 125, 128, 184
Conference Générale des Poids et Mesures (CGPM), 41, 46, 47
Confidence bands, 235, 236, 245, 251, 253
Confidence level, 51, 83, 105, 106, 111, 113, 117, 127, 135, 141, 258, 322, 323
Conformity assessment, 97, 324
Confounding, 197–199, 205, 222
Connectors, viii, 50, 91, 268
Continuous probability distribution, 68, 69, 72
Continuous variables, 60–62, 67, 74, 274
Convergence, 165–166, 176, 177
Coordinate measuring machine (CMM), x, 29, 30, 145, 184, 185, 187, 188, 190, 191, 193, 196, 210, 211, 222, 261, 303, 309–322
Correctness, 275, 281, 282, 284, 298, 300
Correlated inputs, ix, 116, 126, 137–143, 176
Correlation, 61–63, 65–67, 78, 126, 128, 138–141, 147, 149, 155, 179, 210, 233, 238, 261, 270, 272
Correlation coefficient, 61, 62, 65, 78, 138–141, 179, 238
Correlation plots, 65, 67
Count data, 60, 63, 67
Covariances, 105, 132, 133, 137–139, 141, 146, 149, 156, 245, 263, 317
Coverage factor, 5, 7, 26, 105, 106, 111, 113, 117, 123, 127, 134, 147, 236, 250, 270, 322
Coverage interval, 26–27, 163, 172, 175–177, 238
Coverage probability, 26
Credible interval, 293, 295
Crossed classification, 184
Cross-tabulation, 278–280, 300
Current, 28, 30, 31, 36, 37, 43–45, 47, 48, 50, 55, 94, 107, 115, 139, 145, 248, 249, 251, 254, 303, 305

D

Decay, 24, 33, 228, 229, 249, 264
Decision rules, 95–96
Deflategate, 3–8, 52
Degrees of freedom, 5, 7, 71–74, 106, 108, 109, 111, 113, 122, 124, 126, 127, 135, 137, 141, 146, 188, 190, 191, 193, 205–207, 212–218, 220, 225, 235, 255, 257, 260, 270, 285–287, 289–292, 295, 312, 321–323

Density, 45, 48, 64, 68, 72, 74, 77, 84, 85, 126, 128, 131, 135, 137, 144, 145, 157, 160–164, 166–168, 228, 292, 304, 311
Derived, viii, 9, 10, 12, 41, 42, 45, 48, 54, 56, 105, 114, 115, 143, 193, 244, 292, 298
Derived units, viii, 42, 44–45
Design of experiments (DOEx), vii, viii, 115, 181–225, 268, 303
Design patterns, 182–184, 199, 205
Destructive testing, 284–285
Device under test (DUT), 28, 51–53, 150
Digital multi-meters (DMMs), 35, 303
Digitizers, 109–114, 216, 294, 305
Dimensional measurements, 30, 32, 150, 309, 319
Direct measurements, 6, 27, 103, 107–109, 123, 125, 126
Discrete probability distribution, 63, 67–69, 72
Discrete variables, 60, 66, 78
Drift, 8, 29, 33, 42, 47, 50, 51, 57, 98, 104, 227, 228, 249–258, 268, 274

E

Effective degrees of freedom, 7, 111, 113, 117, 122, 127, 135, 141, 146, 213, 214, 257, 285–289, 291, 292, 322, 1069
Electrical measurements, 29, 30, 35, 305
End-of-period reliability (EOPR), 95, 97
Equipment specifications, 321
Eratosthenes, 2, 148
Error approach, 20, 23–24, 34, 35
Errors, 10, 20, 23–24, 50, 62, 107, 108, 134, 188, 207, 210, 228, 233
Estimates, 6, 10, 20, 54, 59, 62, 64, 70, 75, 85, 108, 115, 131, 133, 164, 181, 184, 215, 220, 228, 235, 272
Expanded uncertainties, 5, 6, 52, 74, 83, 105–107, 111, 113, 114, 117, 123, 124, 127, 128, 134, 146, 220, 252, 257, 270, 285, 298, 316, 322
Expectation operator, 132
Expected mean squares, 209, 212, 217, 218
Expected value, 133, 140, 160, 207–210, 214, 228, 286, 293, 322
Exploratory data analysis (EDA), 59–69, 75, 303, 326
Exponentially weighted moving average (EWMA), 274, 297
Exponential model, 229
Extrapolation, 11, 12

Index

F

Factorial designs, 182–185, 188, 197–199, 202, 204–206, 221, 222, 225
Factorial experiments, 182–206
Factorial regression table, 188, 191, 193, 195, 203
False negative, 281, 282, 300
False positive, 281, 282, 300
Fixed effects, 184, 206, 208, 215
Flow measurements, 306
Force measurements, 306
Four-to-one Ratio/Rule, 93, 95, 96
Fractional factorials, ix, 181, 183–185, 188, 196–205, 221, 222, 225
Frequency measurements, 305
Frequency plot, 66–68
Full factorials, ix, 183–196

G

Gamma function, 71
Gauge block interferometer (GBI), 46, 303, 321
Gauge blocks, 3, 46, 306, 309
Gauge R&R Studies, 7, 77, 206, 210–216, 222, 224, 225, 285
Gaussian distribution, 69, 77
Gauss-Newton Method, 241, 261
Gold standard system (GSS), 281, 297, 299, 300
Go/no-go gauge, 60, 267, 274, 298
Goodness-of-fit, 59, 115, 227, 230, 239
Gradient descent, 241
Gregorian date, 256
Grounding, 33
Guardband, viii, 53, 81, 83, 88–93, 95, 96, 98–100, 323
Guardband factor (GBF), 88, 89, 91, 100
Guarded acceptance, 95
Guide to the expression of uncertainty in Measurement (GUM), x, ix, vii, 19, 23, 24, 34, 51, 62, 77, 103, 104, 109, 113–116, 125–128, 131, 132, 134, 143, 145, 147–149, 159, 162, 164, 166–168, 172–175, 178, 183, 206, 221, 249, 251, 261, 263, 285, 290–292, 296, 298, 303, 321, 326
GUM validity test, 167–168, 178

H

Hellenic system, 2
Histograms, 64, 66, 67, 69–73, 78, 79, 157, 161–163, 167, 177–179, 238, 263, 268

Humidity, 33, 36, 37, 149, 313
Hybrid model, 115

I

Improper priors, 292
Independence, 23, 34, 35, 210
Indirect measurement, ix, viii, 9, 27, 34, 68, 103–105, 107, 114–128, 131, 159, 182, 206, 227, 309
Indirect measurement model, ix, viii, 34, 114–125, 131, 227
Individuals control chart, 273
Individuals-Moving Range chart, 268
Informative prior, 292, 298
Interactions, 183, 185, 188–191, 193–205, 207, 209, 221, 222
Interlaboratory comparisons (ILCs), viii, 55, 77, 324
International Prototype of the Kilogram (IPK), 42, 43, 47, 55, 56, 303
International system of units (SI), viii, 14, 28, 31, 41–57, 148, 304, 324
International Vocabulary of Metrology (VIM), 19–24, 26–28, 34, 304, 324
324Inter-quartile range, 61
Interval adjustment, 258
Interval analysis, 258–260
In-tolerance probability (ITP), 85–88, 90, 95, 96, 99, 100
Intrinsic, 33, 46, 48, 60, 312
Inverse normal distribution function, 95
ISO 17025, 53–54, 221, 253, 324

J

JCGM, ix, vii, 20, 23, 34, 49, 51, 77, 79, 82, 97, 103, 125, 126, 131, 145, 147, 159, 164, 167, 176, 178, 221, 267, 296, 298, 299, 303, 304, 324
Joint distribution, 163
Josephson junctions, 46
Josephson volt, 46, 115

K

Kappa value, 279, 280, 300
KDE Method, 157
Kelvin, 41, 44, 48, 55, 153
Kernel Density Estimation (KDE), 64, 70–73, 157, 304
k-factor, 142, 235
Kibble balance, 43, 46–48, 55

Kilograms, 41–43, 45–47, 55–57, 147, 304
Kline-McClintock Method, 131, 147, 148

L
Lack-of-fit, 115, 188, 191
Law of Large Numbers, 153, 165
Law of propagation of uncertainty, 116
Lead probe, 115, 117, 118, 122–124, 169, 284
Leaks, 32, 33, 264
Least squares fit, 231, 246, 261, 313
Legal metrology, 52, 297, 304, 324
Length, 2, 3, 27, 35, 36, 41, 43, 45–48, 56, 79, 82, 107, 126, 139–141, 145, 148–150, 155, 158, 159, 163, 177, 197, 198, 200, 204, 210–212, 224, 237, 243, 244, 247–249, 258, 309–322
Level of confidence, 5–7, 25–28, 50, 123, 236, 270
Levenberg-Marquardt Method, 241, 261
Likelihood, 75, 292, 293, 295, 296, 301, 304
Linear approximation, 133, 142
Linear combination, 133, 134, 213, 214
Linear congruential generator (LCG), 158, 159, 176, 177, 304
Linearity of the measurement system, 166, 173, 219
Linear model, 229, 232, 237, 242, 248, 249
Linear regression, 77, 244, 255, 256, 259
Los Alamos National Laboratory, 153
Luminous efficacy, 42
Luminous intensity, 44

M
Main effects, 187, 188, 190, 192–194, 198–200, 202, 203, 205, 223, 224
Main effects plot, 187, 192, 200, 223
Markov-Chain Monte Carlo, 246, 261
Mass, 3, 29, 37, 41, 43, 47, 48, 55–57, 68, 131, 135, 147, 160, 306
Mass measurements, 56, 135, 160, 306
Maximum Likelihood Estimation (MLE), 75
Mean, 60–63, 68–69, 84, 94, 105, 117, 119, 133, 146, 157, 160, 187, 235, 268, 293–295
Mean squares, 207–210, 212–214, 217, 218, 240
Measurand, 9, 20–24, 26, 27, 29, 30, 33, 59, 72, 81, 105–108, 111, 113–118, 124, 126–128, 131, 133, 142, 182–184, 197, 292, 293, 295, 296, 298, 309, 310

Measurement accuracy, 3, 21, 28, 35, 37, 126, 183, 249, 305, 311, 324
Measurement capability index, 97
Measurement, direct, 6, 27, 68, 103, 105, 107–114, 123, 125, 126
Measurement error, 23, 24, 213
Measurement, history of, viii, 2
Measurement, indirect, ix, viii, 9, 27, 34, 68, 103–105, 107, 114–128, 131, 159, 182, 206, 227, 309
Measurement model, ix, viii, 9–13, 15, 25, 27–28, 31, 33–35, 49, 103–128, 131–134, 144–146, 148, 149, 160, 166, 167, 173, 175, 178, 206, 221, 227, 310–313, 317, 321, 322
Measurement precision, 21, 22, 291, 296
Measurement repeatability, 6, 22, 35, 36, 193, 207, 208, 223, 284, 317
Measurement standards, viii, 3, 14, 28, 41, 48–51, 98
Measurement uncertainty, viii, 1, 3, 5–9, 14, 23–26, 28, 31, 33, 35, 36, 48, 49, 53, 54, 61, 81–100, 105, 107, 109, 137, 149, 182, 184, 196, 205, 208, 220, 222–224, 253–255, 267, 285, 291, 293, 294, 297, 309, 321, 323
Measuring and test equipment (M&TE), 6, 28, 31–33, 37, 50, 52–55, 93–95, 305
Median, 61, 62, 64, 65, 78
Mersenne twister, 158, 159
Meter, 30, 36, 41–46, 56, 99, 133, 140, 148, 179, 262, 305, 316
Metre convention, 3, 41
Metrologist, 14, 30, 35, 42, 55, 167
Metrology, 1, 19, 44, 46, 49, 52, 59, 60, 69–72, 81, 103, 181, 182, 206, 258, 267, 295
Microwave, 46, 50, 145, 305, 306
Mixed effects model, 184, 206, 208–210, 216–220, 222
Mixed-level factorials, 184
Mole, 41, 44
Monte Carlo method, ix, 77, 127, 144, 146, 153–179, 245, 246, 298, 326
Monte Carlo simulation, 77, 298, 318, 319
Monte Carlo uncertainty propagation, 159–167, 169, 175–178
Moving range, 268, 273, 274, 297, 300

N
Nanowire, 139, 140, 142, 247, 249

Index 345

National Conference of Standards Laboratories International (NCSLI), viii, 53, 54, 82, 96, 97, 147, 258, 261, 324
National Institute of Standards and Technology (NIST), 7, 34, 35, 45, 48, 50, 52, 55, 126, 147, 221, 254, 297, 299, 305, 319, 324, 326
National Metrology Institute (NMI), 43, 49, 51
Nested classification, 184
Neutron yield, ix, 103, 115, 117–124, 169–173
Non-binary decision rule, 96
Noninformative prior, 292, 298
Nonlinear, ix, 76, 117, 126, 132, 133, 142, 144, 164, 165, 227, 229, 230, 241–246, 261–263, 325, 326
Nonlinear least squares, ix, 241–244
Nonlinear model, 133, 164, 229, 239, 246
Nonlinear regression, ix, 241–245, 325
Nonlinear sensitivity coefficient, 164, 165
Normal distribution, 68–70, 72–76, 78, 79, 83, 85–87, 97, 100, 108, 146, 155–156, 167, 170, 171, 173, 177, 178, 263, 268, 292
Normal effects plot, 188
Normalized error, 1
Normal probability plot, 188, 190–195, 202, 204, 238
Nuclear security enterprise (NSE), vii, 83, 114, 117, 210, 284

O
Ohm's law, 27–28, 115
One-sided, 91, 323
Optical radiation, 41
Orthogonal distance regression (ODR), 243–244, 261, 263, 326
Orthogonal fractional factorial, 205
Oscilloscopes, 305
Outliers, 60, 62–65, 78, 238
Out-of-tolerance (OOT), 52, 85–87, 98, 99, 251, 258

P
Pass/Fail, 60, 94, 95, 258, 274
Pearson's correlation coefficient, 62, 65
Physical measurement, 29, 34, 49, 306
Planck constant, 42, 43, 46–48
Platinum resistance thermometers (PRT), 306
Poisson distribution, 63, 72, 74, 75, 119, 171
Pooled standard deviation, 254
Posterior distribution, 292, 293, 295, 296, 301
Posterior mean, 293, 295

Posterior variance, 293, 295
Precision, 21–23, 104, 199, 205, 210, 221, 278, 285–288, 291, 296
Prediction band, 227, 235–237, 244–245, 249–252, 255–259
Pressure gauge, 5–8, 37, 98, 261, 262
Pressure measurement, 4, 6, 7, 29, 38, 48
Pressure standard, 46, 48
Primary, vii, 14, 31, 48–51, 59, 66, 104, 117, 125, 209, 314, 325
Primary standard, x, ix, vii, 48–51, 104, 117
Primary standards laboratory (PSL), x, ix, vii, 50, 51, 104, 117, 118
Prior distribution, 292–296, 301
Probability density function (PDF), 68–71, 77, 85, 111, 120, 128, 146, 157, 167, 169, 178, 179, 292, 294
Probability distribution, viii, 59–61, 63, 64, 67–78, 94, 116, 132, 153, 160, 166, 175, 176, 298, 326
Probability mass function (PMF), 68
Probability of false accept and reject (PFA, PFR), 86–90, 92, 93, 96–100
Process distribution, 85, 87, 92, 98, 99
Process monitoring, 268, 297
Product distribution, 85, 86, 94, 98, 99
Proficiency testing, viii, 55, 221, 324
Propagation of uncertainties, x, ix, viii, 116, 126, 131–150, 153–179, 227
Pseudo random number generator (PRNG), 154, 155, 157–159, 176
Pure error, 188, 190, 191, 193

Q
Quadratic model, 243, 262
Quantile-quantile plot, 76
Quantity of substance, 44, 49
Quartiles, 61, 65

R
Radiation, 14, 30, 33, 41, 42, 44, 118, 119, 153, 228, 262, 263, 306
Radius of curvature (ROC), 29, 309–311, 315, 318–322
Random, 23, 24, 33, 68, 97, 107, 132, 153, 154, 176, 184, 229, 280
Random effects, 206–216, 220, 222, 224, 326
Random fluctuation, 23, 230, 252
Random numbers, ix, 97, 153, 155, 158, 167, 176, 177
Random sampling, 153–156

Random variables, 26, 68, 77, 115, 132, 133, 156, 207, 209, 217, 298
Random walk, 246
Range, 24, 26, 61, 62, 82, 155, 227, 251, 268
Realization, viii, 45–46, 48–50, 115, 133
Rectangular distribution, 70, 71, 109, 150
Reference materials, 49–50, 324
Regression, ix, 77, 188, 190, 191, 193–195, 203, 230, 232, 234–237, 240–245, 255, 256, 259, 261, 325, 326
Regression analysis, 190, 194, 232
Regression sum of squares, 240
Repeatability, 5, 6, 22, 24, 35, 36, 94, 126, 147, 155, 185, 187, 193, 199, 205–210, 215, 222, 223, 225, 274, 284, 297, 313, 314, 316, 317
Reproducibility, 5, 6, 22, 24, 32, 35, 126, 147, 155, 206, 208–210, 213–215, 222, 225, 274, 284, 297
Residual analysis, 238–241
Residuals, 77, 191, 192, 194, 195, 204, 210, 228, 231–235, 238–241, 244, 248, 249, 259, 262, 317
Resistance, 27, 30, 49, 91, 94, 97, 107, 139, 140, 142, 145, 148, 150, 155, 177, 178, 228–230, 232–237, 241, 245, 247–249, 252–254, 257, 262, 264, 305, 306
Resistor, 33, 49, 51, 97, 115, 150, 169, 173, 177, 178, 228, 229, 231–233, 235–237, 241, 242, 249, 251, 253–257
Resistor capacitor circuit, 173
Resolution, 6, 21, 22, 35–37, 109, 150, 197–199, 204, 205, 222, 225, 254, 321
Restricted maximum likelihood (REML), 304
Risk, ix, viii, 3, 7, 51, 53, 54, 81–98, 258, 260, 261, 323, 324
Risk, consumer's, 3, 86, 96
Risk, false accept, ix, 7, 81–83, 86–88, 96–98
Risk, false reject, 7, 81–83, 86–88, 96, 97
Risk, global, 86
Risk, producer's, 87, 96
Risk, specific, 98
Root sum of squares (RSS), 88, 90, 91, 98–100, 105, 106, 121, 235, 254, 255, 257, 317
R Packages, x, 325, 326
R-Squared, 191, 195, 203, 212, 218

S

Sample allocation, 285–291, 300
Sample size, ix, vii, viii, 5, 63, 71, 75, 76, 114, 126, 128, 176, 181, 188, 205, 206, 220, 260, 267, 274, 275, 282–292

Sampling distribution, 154, 178, 292–296
Scatterplot, 65, 66, 78, 156, 177, 179
Second, 41–43, 55, 148, 246, 250, 269
Secondary, 49, 50, 104, 112, 117, 123, 124
Semiconductor, 139, 247–249, 263, 305
Sensitivity coefficient, 116, 119–121, 123, 135, 139–141, 164, 165, 300, 321–322
Sensitivity matrix, 146
Sequence length, 155, 158, 159
Shielding, 33, 153
Signal to noise ratio, 126
Significant digits, 22, 113, 145, 168, 173, 322
Simple acceptance, 95
Slope test, 232, 238, 239, 252, 253
Software in metrology, 296, 324
Specification limits, 7, 82–86, 88, 91, 92, 94, 95, 98–100, 258, 259, 273, 323
Stade, 2, 148, 149
Standard, 2, 3, 24, 28–29, 31, 41, 45, 46, 48–50, 94, 104, 107, 117, 206, 228, 253
Standard deviation, 26, 27, 59–62, 69–71, 74, 78, 79, 84, 85, 94, 98, 99, 104, 105, 108, 109, 112, 132, 146, 160, 162, 164, 170, 171, 178, 185, 187, 192–196, 219, 220, 240, 245, 254, 255, 263, 268, 269, 272, 295, 297, 316, 317
Standard error, 12, 62, 78, 108, 235, 263
Standard error of the estimate, 263
Standard uncertainty, 10, 77, 92, 104–106, 108–109, 116, 131, 134, 160, 208, 242, 269
Statistical process control (SPC), ix, viii, 267–274, 296–298, 326
Student's t-distribution, 5, 68, 69, 71–74, 77, 106, 111, 113, 122–124, 135, 270, 285, 322
Sum of squares, 88, 231, 240, 241
Systematic, 24–25, 92, 105, 126

T

Taylor series, 116, 126, 128, 132–134, 137, 142, 146, 166, 182, 244
t-distribution, 5, 68, 69, 71–74, 77, 106, 111, 113, 122–124, 135, 270, 285, 322
Temperature, 24, 29, 32, 41, 44, 60, 115, 138, 178, 228
Temperature coefficient of resistance, 148, 178
Temperature correction, 24, 33
Temperature dependence, 115
Temperature measurement, 36, 306
Test accuracy ratio (TAR), 83, 97
Test distribution, 98

Index   347

Tester drift, 104, 268, 274
Tester qualification, 206
Test uncertainty ratio (TUR), viii, 7, 8, 81–83, 86, 87, 89–93, 95–100, 215, 323
Thermistors, 306
Thermocouple (TC), 127, 306
Time, 22, 42, 51, 65, 107, 139, 181, 215, 227, 249–250, 268
Time measurements, 269
Time-of-test uncertainty, 51, 114, 254, 255, 257
Tolerance interval, 82, 127
Tolerance limits, 82, 97, 98
Tolerance test, 29, 51, 52, 93, 99, 258
Total sum of squares, 240
Traceability, x, viii, 3, 5, 7, 8, 28, 41–57, 95, 123, 319, 324
Traceability chain, x, 28, 56, 95, 123
Triangular distribution, 70, 71, 74, 109, 178
True value, 6, 7, 10, 20–21, 23–26, 59, 84, 92, 98, 179
t-table, 5, 322
Two-way interaction, 188, 189, 193, 195, 198, 200–202, 222
Type A evaluation of uncertainty, 7, 106, 118
Type B evaluation of uncertainty, 6, 25, 104–106, 116, 118

U

Uncertainty, 1, 19, 24–25, 29, 31–33, 42, 51, 59, 81, 103–107, 116–117, 131, 133–134, 159, 215, 227, 231–235, 267, 290

Uncertainty approach, 19, 20, 24–28, 34, 227, 291
Uncertainty budget, x, 253, 309–319
Uncertainty calculator, ix, viii, 83
Uncertainty region, 145, 146, 176
Uniform distribution, 71, 78, 109, 127, 155, 159, 170, 173, 177, 178, 254
Units, 2, 28, 41–45, 48, 133, 252

V

Variance components, 206, 208–210, 215, 218, 220, 222, 224, 225
Variance operator, 132, 207

W

Weibull distribution, 76
Weighted least-squares, 233–234
Welch-Satterthwaite approximation, 126–128, 135, 147, 213, 220, 224, 295
Wichmann Hill Generator, 158
Within-operator agreement, 275, 277–278, 300
Working, 49, 50, 91, 250, 299

Z

Zeta score, 221
z-prime score, 221

Printed in the USA
CPSIA information can be obtained
at www.ICGtesting.com
CBHW061848131024
15794CB00003B/89